RELATING MATERIALS PROPERTIES TO STRUCTURE

RELATING MATERIALS PROPERTIES TO STRUCTURE

Handbook and Software for Polymer Calculations and Materials Properties

D. J. DAVID
ASHOK MISRA

TECHNOMIC
PUBLISHING CO., INC.
LANCASTER · BASEL

Relating Materials Properties to Structure
aTECHNOMIC publication

Technomic Publishing Company, Inc.
851 New Holland Avenue, Box 3535
Lancaster, Pennsylvania 17604 U.S.A.

Printed in the United States of America
10   9   8   7   6   5   4   3   2   1

Main entry under title:
    Relating Materials Properties to Structure: Handbook and Software for Polymer
    Calculations and Materials Properties

A Technomic Publishing Company book
Bibliography: p.
Includes index p. 683

Library of Congress Catalog Card No. 99-63892
ISBN No. 1-56676-797-0

*To our colleagues and all those scientists and engineers
who over the years have attempted to understand
the mysteries of nature, inspired us,
and made the fruits and knowledge of
their efforts available to us*

and

*To our wives Honoré and Rashmi
who somehow tolerate our idiosyncrasies*

To see a world in a grain of sand
And a heaven in a wild flower,
Hold infinity in the palm of your hand
And eternity in an hour

William Blake (1757–1827)

Table of Contents

Preface

Over the years, we observed that, for the most part, little attempt was made to convey the commonalties or explicit distinctions between different classes of materials. As a consequence, a study of one field leaves the impression that much of the body of knowledge is unique to that field or class of materials. In many instances, this is indeed the case, while in others useful parallels are ignored or forgotten.

The myriad of new materials we see around us enrich all of our lives and make possible the creature comforts of television, automobiles, communication devices, and health benefits of new medical devices. These materials are engineered into a given application taking into account their chemical, physical, and mechanical properties. In this way fundamental knowledge from first class research and engineering are joint partners. This has been, and will continue to be an evolutionary process. There are exceptions to this, as in the case of the transistor which was revolutionary, but even this device continues to evolve. Much of this evolution is dependent upon an understanding of how the properties of a material depend upon the structure. Without this understanding, the costs and time to arrive at new materials that meet the requirements of a given application would be daunting.

This book is arranged to provide a comparison between the structure of different classes of materials and their attendant properties where appropriate, but largely concentrates on the structural differences between individual polymers and the resultant properties, since this is the only class of materials where data and techniques allow properties to be estimated. It is with these thoughts in mind that the accompanying software for this book was developed. However, as an easy, comprehensive reference, and approach to appreciating the differences between classes of materials, we have included physical and mechanical properties databases for a wide range of metals, ceramics, and organic polymers, which are included with the software. We hope that this integrated approach will indeed make this book valuable and unique.

A number of methods and techniques have appeared in recent years for estimating the properties of polymers, two outstanding texts of which are those by Van Krevelen [1] and Bicerano [2]. There is no easy, straightforward way to estimate the properties of polymers. It is our intention to provide such a technique via simple software that permits one to see the effects of changing a structure, and also to estimate the properties of a polymer which might not otherwise be available, at the most, or very time consuming to find, at the least. We also consider the ability of the software to estimate the miscibility of various polymer blends to be a cornerstone of the software and therefore one of its more valuable aspects. Unfortunately, most methods that are extremely easy make simplifying assumptions that adversely affect accuracy. Although that is the case here, we believe that the inaccuracies introduced do not obviate the usefulness of the software or techniques.

In this way, the software for estimating the properties of polymers should prove to be valuable for obtaining properties not available, understanding the relationship of structure to properties, and asking any number of "What if we change the structure from this to that...." questions. We leave the follow-up on "So What" questions to the reader.

November 1998

References

1. Van Krevelen, D.W., *Properties of Polymers*, Elsevier, New York, N.Y., 1990.
2. Bicerano, Jozef, *Prediction of Polymer Properties*, Marcel Dekker, Inc., New York, N. Y.,1993.

CHAPTER 1

Structural Considerations/Why Materials Differ

1.1   INTRODUCTION

Materials are basic building blocks that all technologies use in one or more ways. The advent of newer materials has had a major impact on industries and has widely influenced the broad design and applications of consumer goods. It is therefore important to understand the structure of the various classes of materials and how they affect the properties and long term performance of the resulting products. In general, materials of importance in the modern world can be classified as metals, ceramics, and polymers. All three classes possess unique structures and properties. An appreciation of these fundamental differences in those classes, based on structural considerations, lead to new applications, improved processing, increased property control, and enhanced quality and value to the user.

The ultimate tensile strength of a material can be related simply to its structural bond strength and to the number of existing bonds per unit volume. Although this approach is an oversimplification, it is generally good to a first approximation. Other complications like surface imperfections, shear, dislocations, the presence of other atoms, molecules, or ions greatly complicate the determination of the ultimate strength of a material.

The structure and properties of materials are understood and are subject to control in ways that were unheard of a decade ago. Structures can now be prepared by laying down a single layer of molecules at a time which has broad applicability to the biological and electronic sciences. Plasma deposition is now used routinely to deposit layers of angstrom thickness that allow tailoring of the properties of solid state devices and many other applications such as the scratch resistant coatings placed on eyeglasses and automobile headlight lenses.

Applications of these new technologies to contemporary problems continue to reshape the concept of what materials science and engineering is. A common thread that links many contemporary applications in materials science is the controlled combination of atoms and molecules. Building on the knowledge developed in the classic fields of metals and ceramics, these materials can now be combined with polymers in controlled ways that endow resulting materials with properties that depend not only on the nature of the atomic and molecular constituents but also on their interactions with surfaces and bulk constituents.

A few examples of applications which utilize such sophisticated technologies include integrated circuit preparation, surface active agents, cermets (ceramics plus metals), composite materials, prosthetic devices ranging from artificial limbs to artificial organs, and coatings and adhesives whose applications cover the range from structural to medical and dental uses.

Since properties are determined by structure, it is obvious that a knowledge of the structure will allow an appreciation and understanding of the differences in behavior between various materials. The elements for understanding structure property relationships can be visualized as follows:

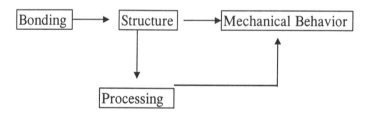

The bonds holding materials together can be metallic, covalent, ionic, or van der Waals. In the following sections bonding will be discussed in a general way only that presupposes an understanding of interatomic forces, chemical bonding, and the basis of statistical mechanics.

The following sections discuss basic differences in structures which are responsible for distinctive materials behavior of the major classes of materials, namely, metals, ceramics, and polymers. On the other hand these quite different materials exhibit and share some common characteristics which might not initially be anticipated.

In subsequent chapters, when specific properties are discussed and estimated, detailed treatment of significant variables such as bonding, composition, and phase effects, will be compared for the three major classes of materials.

Calculation of materials properties from first principles is a developing science and one that we will see used more and more by scientists and engineers to understand the origin of properties and to estimate how properties might be affected if specific atomic or molecular structural changes can be achieved. Our intention is to allow an appreciation of the structural differences between different classes of materials that account for their differences in behavior while providing a more detailed examination of synthetic polymers in which field the estimation of properties is more advanced than in metals or ceramics. Polymer science is in reality a study of the science of mixtures since in only special cases of preparation does a high polymer consist of only a single molecular weight. Thus, we also feel it is appropriate to include a chapter devoted solely to the miscibility of polymers.

To achieve these goals, we have included a software program which includes databases that allow diverse materials properties to be compared, and polymer properties and polymer miscibilities to be estimated. In this way, it will be possible to better appreciate the enormous differences in properties that slight structural changes in a polymer's composition can sometimes make. A key requirement in any software used for property estimation is that it function as a learning tool and should be simple to learn and use since the emphasis and concentration should be on the relationship of structure to properties and not on obtaining an exact number.

The property estimation software accompanying this book is based on principles of group additivity theory and is used only for estimating properties of organic based materials, and more specifically of polymers. Presently, the application of group additivity theory for estimating properties and miscibilities is only applicable to hydrocarbon based materials and, again, specifically polymers. Thus present computational techniques for property estimation are applicable mainly to polymers.

1.2   ATOMIC PACKING AND STRUCTURE

The short-range order of atoms or molecules in a solid are usually regular and predictable even though the long-range structure may be either regular (crystalline) or irregular (amorphous or glassy). The arrangements of patterns depend partly on whether or not the bonding is directional. For directional bonding, the local atomic arrangement is determined by the bond angles. For non-directional bonding, the packing arrangement depends on the relative sizes of the atoms. The long-range structure of a solid then depends upon repetition of the polyhedra formed by the bond angles (directional) or atom packing (non-directional). Covalent and permanent dipole bonds are directional while metallic, ionic, and van der Waals bonds are non-directional.

Before we consider the major classification of bond types, we first address the question of how many ways can points in space be arranged in three dimensions so that each point has identical surroundings. The unit cell of a space lattice or crystal structure is defined as the smallest repeating volume which has a lattice point in each of its four corners. The lattice parameters are the lengths of the edges of the unit cell along the reference axes. The space lattice for a crystal is an infinite array of these unit cells.

It turns out that there are 14 different ways in which points can be arranged in three dimensions so that each point has identical surroundings. These spatial arrangements are know as *Bravais* lattices. The fourteen types of lattices are listed in Table 1.1 and are illustrated in Figure 1.1. These lattices are applicable to both types of bonds, i.e., to directional and non-directional bonding types.

Table 1.1
The Fourteen Bravais Space Lattices

Lattice Type	Reference Axes
1. Simple Cubic 2. Face-Centered Cubic 3. Body-Centered Cubic	Three axes at right angles, $a = b = c$
4. Simple Tetragonal 5. Body-Centered Tetragonal	Three axes at right angles, $a = b \neq c$
6. Simple Orthorhombic 7. Base-Centered Orthorhombic 8. Face-Centered Orthorhombic 9. Body-Centered Orthorhombic	Three axes at right angles, $a \neq b \neq c$
10. Simple Rhombohedral	Three axes equally inclined but not at right angles, $a = b = c$
11. Simple Monoclinic 12. Base-Centered Monoclinic	One axis at right angles to the other two, which are not at right angles to each other, $a \neq b \neq c$
13. Simple Triclinic	Three axes not at right angles, $a \neq b \neq c$
14. Simple Hexagonal	Three coplanar axes at $120°$ and a fourth at $90°$ to these, $a_1 = a_2 = a_3 \neq c$

Directions in a space lattice are represented by vectors. The indices of direction are the vector components resolved along each of the coordinate axes and reduced to the smallest set of integers. Because of the regular spacing of points, it is possible to pass sets of equidistant parallel planes upon which the points will have the same orientation. A widely used labeling system for these crystallographic planes is the Miller indexing method.

The capital letters in Figure 1.1 refer to the type of cell: *P*, primitive cell; *C*, cell with a lattice point in the center of two parallel faces; *F*, cell with a lattice point in the center of each face; *I*, cell with a lattice point in the center of the interior; *R*, rhombohedral primitive cell. All points indicated are lattice points.

The Miller indices are particularly useful because in cubic systems all the planes of a given set will have the same set of numbers defining them. Curly brackets {} denote a single, specific,

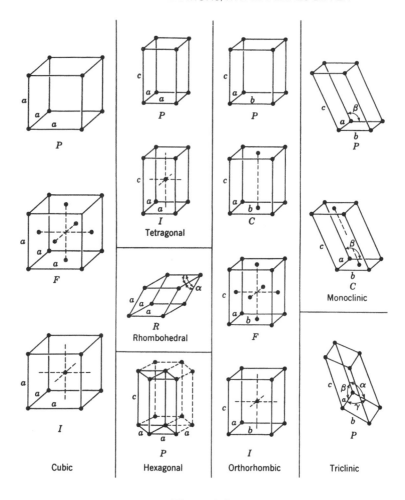

Figure 1.1
Conventional Unit Cells of the 14 Bravais Space Lattices

plane and parentheses () denote a family of symmetrically equivalent planes. A modified version of the Miller indices, the Miller Bravais indices, may be used to the same advantage with hexagonal systems.

A general rule for finding the Miller indices of a crystallographic direction in a Cartesian coordinate system is as follows: Draw a vector from the origin parallel to the direction whose indices are desired. Make the magnitude of the vector such that its components on the three axes have lengths that are simple

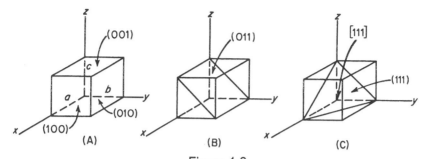

Figure 1.2
Cube Planes of a Cubic Crystal
(A) - a (100), b (010), c (001)
(B) - The (011) plane
(C) - The (111) plane

integers. These integers must be the smallest set of numbers or integers, placed in brackets, that will give the desired direction. Thus, the integers 1, 1, and 1 and 2, 2, and 2 represent the same direction in space, but by convention, the Miller indices are [111] and not [222]. Similarly, if the x, y, z, intercepts are ½, ¾, ½, the reciprocals of the intercepts are 2, 4/3, 2, and finally [323] are obtained by reducing to the smallest set of integers. These basic concepts are illustrated in Figures 1.2 and 1.3.

Planes and directions in hexagonal metals are defined almost universally in terms of Miller indices containing four digits instead of three. Three axes correspond to close-packed directions and lie in the basal plane of the crystal, making 120° angles with each other. The fourth axis is normal to the basal plane and is

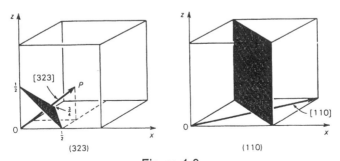

Figure 1.3
Crystallographic Planes in Cubic Lattices Identified by Miller Indices

called the c axis, where the three that lie in the basal plane are designated a_1, a_2, and, a_3 axes. Figure 1.4 should be consulted for illustration purposes.

The uppermost surface of the cell in Figure 1.4 corresponds to the basal plane of the crystal. Since it is parallel to the axes, a_1, a_2, and a_3, it intercepts the axes at infinity. Its c axis intercept, however, is equal to 1. The reciprocals are $1/\infty$, $1/\infty$, $1/\infty$, and $1/1$. This plane is therefore the (0001) plane. The six vertical surfaces of the unit cell are known as prism planes of Type I. The prism plane that forms the front face of the cell has intercepts of a_1 at 1, a_2 at ∞, a_3 at -1, and c at ∞. The Miller indices become $(10\overline{1}0)$. Another important plane in Figure 1.4 has intercepts a_1 at 1, a_2 at ∞, a_3 at -1, and c at ½ giving Miller indices of $(10\overline{1}2)$.

1.3   CONSEQUENCES OF PACKING

Next we consider some simple examples of the consequences of packing for a few structures.

- Simple Cubic (Cubic P) (referring to Figure 1.1):

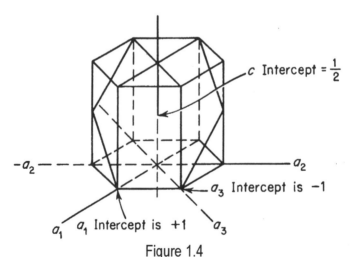

Figure 1.4
The $(10\overline{1}2)$ Plane of a Hexagonal Metal

Atoms/unit cell - There are eight atoms in the unit cell which are shared by adjacent cells.

∴ atoms/unit cell = $8 \times 1/8 = 1$

Coordination Number - There are six nearest neighbors of each atom.

∴ Coordination number = 6

Lattice Constant: (referring to Figure 1.5)
Lattice constant = $a = 2 \times r$ (radius of atom)
Close-packed direction is along the line of contact

Packing Factor (P.F)

$$P.F. = \frac{\text{Volume of spheres}}{\text{Volume of unit cell}}$$

$$= \frac{4\pi r^3}{3a^3} = \frac{4\pi r^3}{3(2r)^3} = 0.524$$

Densest Packed Plane (Figure 1.6)

$$P.F._{(100)} = \frac{\pi r^2}{a^2} = \frac{\pi r^2}{4r^2} = 0.785$$

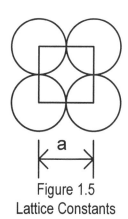

Figure 1.5
Lattice Constants

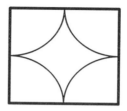

Figure 1.6
Densest Packed Plane

- Body-Centered Cubic (Cubic *I*)

Examples of body-centered cubic structures are Li, V, Cr, Fe, Na, Nb, Mo, K, Rb, Ta, W, Cs, and Fr. Referring to Figure 1.1 and Figure 1.7:

> *Atoms/unit cell* - There are eight atoms in the unit cell which are shared by adjacent cells and one atom in the center.
>
> Corners - atoms/unit cell = $8 \times 1/8$ = 1
>
> Center -   = 1
>
> ∴ 2 atoms/unit cell
>
> *Coordination Number*
> There are eight nearest neighbors of atom *a*.
> ∴ Coordination number = 8

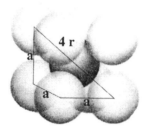

Figure 1.7
Lattice Constant
Lattice Constant = $(4r)^2 = a^2 + a^2 + a^2$

Lattice Constant (Figure 1.7)

$$a = \frac{4r}{\sqrt{3}}$$

Close Packed Direction
The close packed direction is along the body diagonals.
Packing Factor (P.F.)

$$P.F. = \frac{2 \times 4/3\pi r^3}{(4r/\sqrt{3})^3} = 0.68$$

Densest Packed Plane (Figure 1.8)

$$\rho_{(110)} = \frac{2\pi r^2}{\sqrt{2}a^2} = \frac{3\sqrt{2}\pi r^2}{16r^2} = 0.833$$

1.4　COMPARISON OF DENSITY OF METALS USING THE HARD BALL MODEL

In general, only the simpler structures provide densities that are in reasonable agreement with literature values. Nevertheless, this allows an appreciation of the importance of atomic packing/structure relationships and the influence on density. Consi-

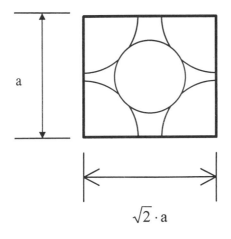

Figure 1.8
Densest Packed Plane for [110] Direction

der the example of iron which is body-centered cubic. In this case, the lattice constant $a = 2.87 \times 10^{-8}$ cm and the atomic weight = 55.84g/mol.

The density is calculated as:

$$\rho = \frac{M}{V} = \frac{nA}{N_A a^3} = \frac{2 \times 55.84}{6 \times 10^{23} \cdot (2.87 \times 10^{-8})^3}$$

$$= 7.87 \text{g/cm}^3 \text{ vs. } 7.86 \text{g/cm}^3 \text{ (literature value)}$$

- Face-Centered Cubic
    Examples of face-centered cubic structures are: Au, Ni, Ag, Cu, and Al. Referring to Figure 1.1 and 1.9:

    Atoms/unit cell
    There are eight atoms in the unit cell which are shared by adjacent cells and six atoms in the cube faces which are shared by adjacent cells.

Corners-	$8 \times 1/8 = 1$
Faces -	$6 \times 1/2 = \underline{3}$
	4

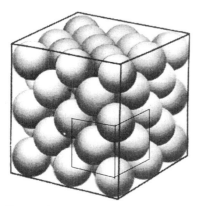

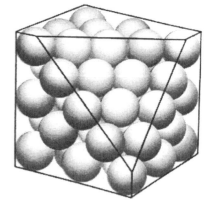

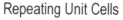
Repeating Unit Cells   Octahedral (close packed)
                                                        (111) Plane

Figure 1.9
Face-Centered Cubic Unit Cell (Hard Ball Model)

Coordination Number
There are twelve nearest neighbors.
$\therefore$ Coordination number $= 12$

Lattice Constant
$$(4r)^2 = 2a^2$$
$$a = \frac{4r}{\sqrt{2}}$$

Close-Packed Direction
Close-packed direction is along the face diagonals (Figure 1.9).

Packing Factor
$$\frac{4(4/3\pi r^3)}{(2\sqrt{2}r)^3} = 0.74$$
$$\rho_{(111)} = 0.907$$

- Hexagonal Close-Packed
  Examples are: Zn, Cd, Se, Ti, Sc, Co, Ru, Os, Zr, Hf.
  Referring to Figures 1.1, 1.4, and 1.10.

Atoms/unit cell - 2 or 6

Coordination Number - 12
From Figure 1.10, the three central atoms plus the one at the center of the top show this to be an equilateral tetrahedron with edges, $a = 2r$. From geometric considerations, $h = a\sqrt{2/3}$.

Lattice Constant
$$2r = a$$
$$c = 2h = 2a\sqrt{2/3} = 1.633a$$
$$c/a = 1.633$$

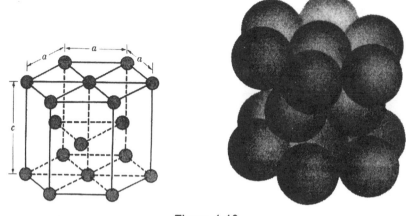

Figure 1.10
Hexagonal Close-Packed Structure Showing the Location of Atom
Centers and Hard Ball Model

Volume

$$V = \bar{c} \bullet (\bar{a} \times \bar{b})$$

$$\begin{vmatrix} 0 & 0 & c \\ 0.866a & 0.5c & 0 \\ 0 & a & 0 \end{vmatrix}$$

$$= 0.866a^2 c = 0.866a^2 (1.633a) = 1.414a^3$$

Packing Factor

$$P.F. = \frac{2(4/3\pi r^3)}{1.414a^3} = 0.74$$

1.5   BONDS AND BOND TYPES

1.5.1   Metals and Metallic Bonding

Metallic elements are good conductors of heat and electricity, in contrast to non-metallic elements, and they are opaque, lustrous, and can be deformed or shaped. These properties can be accounted for on the basis of the non-directionality of their bonding and their tendency to solidify in as closely a packed arrangement as possible. In this way the number of bonds per unit volume is maximized, and hence the bonding energy per unit volume is minimized. As a first approximation these atoms can be treated as hard spheres packed together.

The key feature of metal atoms which accounts for their different bonding characteristics is the looseness with which their valence electrons are held. The wave functions of these electrons in free atoms are spread out much more in space than those of nonmetal valence electrons with the same principle quantum number because the metal nuclei have fewer positive charges to attract electrons.

Typically the mean electron radius in a free atom is larger than the interatomic distance in a solid metal. Therefore, the valence electrons in a solid are always closer to one or another nucleus than they are in a free atom which lowers their potential energies in the solid. The kinetic energy of a metallic valence electron is also lowered in the solid due to the larger volume it occupies in space or spatial extension of the wave function. This lowers the probability ($\psi^2 dV$) of finding it in the same volume increment, dV, as in a free atom. This decrease in both potential and kinetic energies of valence electrons is responsible for metallic bonding.

Since each valence electron is not localized between only two ion cores, as in covalent bonding, metallic bonding is non-directional, and the electrons are more or less free to travel through the solid. This diffuse nature of metallic bonding is responsible for the easy deformability of metals and the looseness of these electrons is often described as an "electron gas" which holds the positive ion cores together. This "electron gas" is, of course, also responsible for the electrical conductivity of the metallic elements.

In general, the fewer valence electrons an atom has and the more loosely they are held, the greater the metallic bonding characteristics. Thus elements like sodium, potassium, copper, silver, etc., have high electrical and thermal conductivities because their valence electrons are so mobile. These mobile electrons easily absorb energy of light photons and are promoted to higher energy levels which accounts for their opaqueness. In a pure state, these elements have high reflectivity since the electrons re-emit this energy as they fall back to lower energy levels.

As the number of valence electrons and the tightness with which they are held to the nucleus increases, they become more localized in space which increases the covalent nature of the bonding. For example, the transition metals, i.e., those with incomplete d shells, such as iron, nickel, tungsten, etc., exhibit a significant fraction of covalent bonding which involves hybridized inner shell electron orbitals, which accounts in part for their high melting points. The competition between covalent and metallic bonding is evident in the fourth column of the periodic table (C, Si, Ge, Sn, and Pb). Diamond exhibits almost pure covalent bonding; silicon and germanium are more metallic; tin exists in two forms (gray and white): one is mostly metallic and the other is mostly covalent; and lead is mostly metallic.

We expect non-directionally bonded atoms like those of the metallic elements and the noble elements to solidify in as closely packed an arrangement as possible. This maximizes the number of bonds per unit volume. This allows us to treat these atoms as hard spheres packed together, as was mentioned previously.

1.5.2 Covalent Bonding

Before proceeding to ionic bonds, we consider covalent bonds in order to draw important distinctions between both metallic and ionic bonds.

Covalent and permanent dipole bonds produced by covalent bonds represent two types of directional bonds. Covalent bonds are the only ones that determine local packing arrangements to any significant degree since dipole bonds usually link groups of atoms rather than individual atoms. A prerequisite for strong covalent bonding is that each atom should have at least one half-filled orbital. Only then can the energy be lowered substantially by having each bonding electron in the orbitals of two atoms simultaneously. The greater the overlap, the more the overall energy is lowered which results in a stronger bond.

The amount of overlap is limited either by electrostatic repulsion or by the exclusion-principle, i.e., completely filled orbitals cannot overlap. The covalent bond formed by overlapping orbitals also tends to lie in the direction in which the orbital is concentrated, the maximum amount of overlap being obtained in this way. Similar arguments for the wave function (ψ^2) exist for covalent bonds as for metallic bonds. If ψ^2 is large and directional, the bonds will be stronger and concentrated in these directions. If ψ^2 is spherically symmetric, the bonds will be non-directional.

Carbon is the atomic building block for polymers and a multitude of compounds and precursors to polymers. Carbon, $(1s)^2(2s)^2(2p)^2$, sometimes forms four covalent bonds of equal strength as in methane or diamond which would be impossible with only two half-filled orbitals.

The explanation for carbon's four bonds is that one of the $2s$ electrons can be promoted to a $2p$ orbital if the expenditure of energy is more than compensated by the energy decrease accompanying bonding. This would result in one half-filled s orbital and three half-filled p orbitals. However, this would result in carbon having a weak spherically symmetric bond and three stronger, directional bonds. But carbon is observed to have four bonds of equal strength. This is a result of rearrangement of orbitals called hybridization of resulting orbitals which are called $(sp)^3$ hybrids. Each hybrid orbital has a larger maximum value of ψ^2 and, therefore, can overlap other orbitals to a greater extent, resulting in lower bond energies than for overlapping p orbitals.

These characteristics of the covalent bond which determine the arrangement of atoms are its discreteness and its spatial direction. Its spatial direction is, in turn, determined by the quantum state of the electrons taking part in the bond. These $(sp)^3$ hybrids result in bond angles close to 90° because this is the angle p orbitals make with each other.

Carbon bonds formed in diamond and many organic molecules represent extreme examples of hybridized s and p wave functions. The one $2s$ and three $2p$ orbitals form four bonds of equal strength directed to the corners of a tetrahedron (109°) as shown in Figure 1.11.

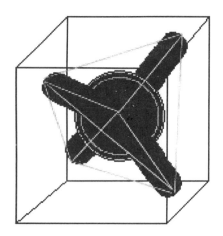

Figure 1.11
The Tetrahedral Directionality of the Four Bonds Resulting
from the Hybridization of One *s* and Three *p* Orbitals Inscribed
in a Reference Cube

The fact that a covalently bonded atom forms only a certain number of discrete bonds seems hardly worth mentioning, but it is this very aspect that accounts for the variety of organic structures and their unique properties. The discreteness and location account for nonequivalent structures depending upon which bond is used for which structure. The multiple of carbon in oligomers, (short chain) and long chain carbon compounds is tetrahedral (Figure 1.11) and a C-C bond may be pictured as two tetrahedra, joined corner to corner.

The structural variety of carbon compounds arises partly because these tetrahedra can be joined edge-to-edge (double bond) and face-to-face (triple bond). In the case of multiple bonds, the carbon-carbon distance is reduced from 1.54 Å for a single bond to 1.20 Å for a triple bond.

Many simple or small organic molecules are weakly bonded to each other via van der Waals forces. This is because the bonding electrons are very localized in space and therefore the bonding forces extend over very short distances. These facts account for the solidification of methane at very low temperatures. Longer chain hydrocarbons in this series are called paraffins, which means "little affinity".

1.5.3 Ionic Bonding

Ionic bonds result from electrostatic attraction between positive and negative ions due to the transfer (loss or gain) of electrons from one atom to another. Electronegative or non-metallic atoms tend to acquire electrons becoming negative ions. These elements have only a few half-empty p orbitals and attract an electron into one of the half-empty orbitals since the energy of the electrons is lowered with respect to the nucleus. Electropositive or metal atoms tend to loose one or more of the loosely held electrons in a higher energy level lying above a filled electron shell.

When an electropositive and an electronegative free atom are brought sufficiently close, they become two ions of opposite charge, the potential energy E of the ion pair will become more negative as the radial distance of separation decreases as shown by:

$$E_{attraction} = -\frac{e^2}{4\pi\varepsilon_o r}\text{joules} \qquad (1.1)$$

where e is the charge on the electron (1.6×10^{-19}) coulombs and ε_o is the permittivity of free space. The molecule becomes more stable as the ions approach but when the closed electron shells of the ions begin to overlap, a strong repulsive force arises. At this stage the Pauli exclusion principle requires that some electrons be promoted to higher energy levels. Work must then be done on the ions to force them closer together and the amount of work is inversely proportional to some power, n, of the distance between the ion centers. The total potential energy of the ion pair will be:

$$E_{total} = -\frac{e^2}{4\pi\varepsilon_o r} + \frac{B}{r^n} + \Delta E \qquad (1.2)$$

where B is an empirical constant and ΔE is the energy required to form the two ions from neutral atoms as represented in Figure 1.12.

The ionic bond, like the metallic bond, is non-directional and both differ from the covalent bond in this key feature. Each positive ion attracts all neighboring negative ions and vice versa, so that in a large aggregate each ion tends to be surrounded by as many ions of opposite charge as can touch it simultaneously.

For example, in the sodium chloride structure, each ion is surrounded by six of the opposite charge. The energy of the aggregate varies with interionic separation in much the same way as in Figure 1.12. The energy of one ion of charge $Z_i e$ may be obtained by summing its interaction with other j ions in the crystal, as given by Equation 1.3:

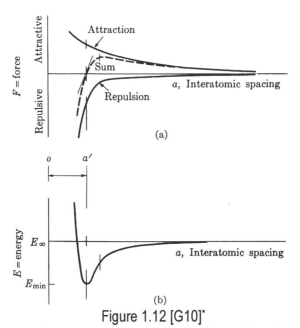

Figure 1.12 [G10]*
Overall Energy Changes Associated with the Formation of a Pair of Ions from Atoms with the Formation of an Ionic Bond
(a)   The equilibrium spacing o-a' is the distance at which the attractive forces equal the repulsive forces.
(b)   The lowest potential energy occurs at distance o-a'.

---

* The letters G and S accompanying reference numbers indicate whether the reference is part of the general or specific lists at the end of each chapter.

$$E_i = \sum_j \left(\frac{Z_i e \cdot Z_j e}{4\pi\varepsilon_o r_{ij}} + \frac{B_{ij}}{r_{ij}} \right) \tag{1.3}$$

where r_{ij} is the distance between the ion under consideration and its jth neighbor of charge $Z_j e$. The values of Z are explicitly the number of charges at each point, and may be either positive or negative. The ratio by which the potential energy of an ion within an ionic solid differs from a single pair of ions is known as the Madelung constant.

Equation 1.3 can be simplified to give Equation 1.4 when two unlike, monovalent ions which are separated by distance a are considered.

$$E_c = -\frac{e^2}{a} \tag{1.4}$$

For example, in the case of sodium chloride, the removal of a sodium ion from the interior of a NaCl crystal requires 1.747 times as much energy as it does to separate it from a single Cl⁻ ion.

The repulsive energy between electron shells of adjacent atoms may be expressed as:

$$E_r = \frac{b}{a^n} \tag{1.5}$$

Both equations conform to the representations in Figure 1.12. The quantity b is a constant and n has a value of approximately 9.

The sum (U) of the net coulombic attractive and electron repulsive energies is $E_c + E_r$ is :

$$U = -\frac{Ae^2}{a} + \frac{b}{a^n} \tag{1.6}$$

At the distance a in Figure 1.12,

$$\frac{dU}{dA} = 0 \tag{1.7}$$

$$\text{and} \quad \frac{b}{a^n} = \frac{Ae^2}{na} \tag{1.8}$$

Therefore, the energy per mole is:

$$U = -N_A Ae^2 \frac{1 - \dfrac{1}{n}}{a} \tag{1.9}$$

where N_A = Avogadro's number.

1.5.4   Van der Waals Bonds

In addition to the three primary bond types, covalent, metallic, and ionic, various secondary bonds exist in solids. They are secondary bonds in comparison to the primary bond types since they are relatively weak in comparison. These bonds result from the electrostatic attraction of dipoles. These bonds are of utmost importance in considering the interaction of high polymers and it will subsequently be seen in the chapter on miscibility that effective models that make use of these types of bonds are used to investigate and estimate miscibilities.

Covalently bonded atoms often produce molecules that behave as permanent dipoles. Permanent dipole bonds like the hydrogen bond are directional bonds, but other dipole bonds are not. For any atom or molecule there is a fluctuating dipole moment which varies with the instantaneous positions of electrons. These fluctuating dipole bonds are commonly called van der Waals bonds.

At any given moment there are a few more electrons on one side of a nucleus than on the other side and since the centers of positive and negative charges do not coincide at this moment and thus a weak dipole is produced. The bond produced by these fluctuating dipoles is non-directional and is weak (about 0.1 eV). These fluctuating dipoles, in turn, induce other dipoles in adjacent molecules.

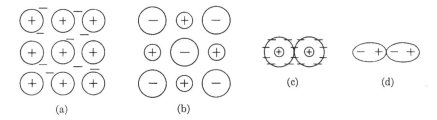

Figure 1.13 [G10]
Graphical Representation of Bond Types
(a) Metallic - Positive "ions" are held together by "free" electrons
(b) Ionic - Positive and negative ions are held together by coulombic attraction
(c) Covalent - Electrons are shared with electron pairs between two positive "ions" (centers)
(d) Van der Waals - Centers of positive and negative charges are not coincident resulting in induced dipoles and coulombic attraction

1.5.5   Summary of Bond Types

The bond types, discussed above, are graphically illustrated in Figure 1.13.

Although van der Waals bonds are weak compared to covalent bonds, the shear number of them are often strong enough to determine the final arrangements of groups of atoms or molecules in solids, especially in high polymers.

1.6   PACKING CONSEQUENCES OF NON-DIRECTIONALLY BONDED ATOMS OF UNEQUAL SIZE

While the lattices and packing arrangements discussed previously apply to ionic crystals, there are important consequences if the sizes of the atoms are appreciably different. Since cations and anions usually differ appreciably in size, the packing is greatly affected.

The cation-anion distance is determined by the sum of their radii. The "coordination number", i.e., the number of their radii. The "coordination number", i.e., the number of anions surrounding the cation is determined by the ratio of the radii of the two ions. A given coordination number is stable only when the ratio of cation

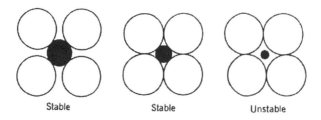

Stable   Stable   Unstable

Figure 1.14
Stable and Unstable Coordination Configurations

to anion radius is greater than some critical value. This stability can be visualized as shown in Figure 1.14.

The most stable structure always has the maximum permissible coordination number since the electrostatic energy of an array is obviously decreased as progressively larger numbers of oppositely charged ions are brought into contact. The limits for the critical ratios of cation to anion radii are listed along with their three dimensional representations in Figure 1.15. It should be noted that these critical ratios are not always followed since the packing considerations have treated the ions as being rigid spheres.

1.7   CERAMIC STRUCTURES

Ceramic materials consist of a multiplicity of structures with the stable array of ions in a crystal structure being the one of lowest energy. Generalizations which interpret the majority of ionic crystal structures are known as Pauling's rules.

Pauling's Rules

1. A coordination polyhedron of anions is formed about each cation in the structure. The cation-anion distance is determined by the sum of their radii. The coordination number (CN) is determined by the ratio of the radii of the two ions as illustrated in Figure 1.15 [G3].
2. The strength of an ionic bond donated from a cation to an anion is the formal charge on the cation divided by its coordination number. For example, Si with a valence of 4

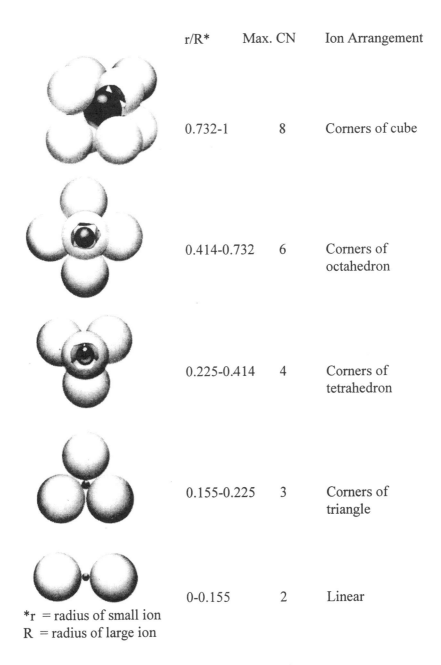

r/R*	Max. CN	Ion Arrangement
0.732-1	8	Corners of cube
0.414-0.732	6	Corners of octahedron
0.225-0.414	4	Corners of tetrahedron
0.155-0.225	3	Corners of triangle
0-0.155	2	Linear

*r = radius of small ion
R = radius of large ion

Figure 1.15 [G3]
Critical Radius Ratios for Various Coordination Numbers

and tetrahedral coordination has a bond strength 4/4=1.

3. In a stable structure, the corners rather than the edges and especially faces, of the coordination polyhedra tend to be shared. If an edge is shared, it tends to be shortened. The separation of the cations within the polyhedra decreases as the polyhedra successively share corners, edges, and faces and the repulsive interaction between cations accordingly increases.

4. The polyhedra formed about the cations of low coordination number and high charge tend to be linked by corner sharing.

5. The number of different constituents in a structure tends to be small. This follows from the difficulty encountered in efficiently packing into a single structure ions and coordination polyhedra of different sizes.

Examples of simple ceramic structures are listed in Table 1.2. Figure 1.16 illustrates combinations of simple cubic, face-centered cubic, and hexagonal close packed structures and their interstitial positions.

Table 1.2
Simple Ionic Structures Grouped According to Anion Packing

Structures	Anion Packing	Cation Interstitial Location	Locations Filled	Structure Name
NaCl, KCl, LiF, KBr, MgO, CaO, SrO BaO, CdO, VO, MnO, FeO, CoO, NiO	FCC	6-fold	All	Rocksalt
ZnS, BeO, SiC	FCC	4-fold	½	Zinc Blende
Li_2O, Na_2O, K_2O, Sulfides	FCC	4-8 fold	All	Anti-flourite
ZnS, ZnO, SiC	HCP	4-fold	½	Wurtzite
Al_2O_3, $Fe_2O_3, Cr_2O_3,, Ti_2O_3$	HCP	6-fold	2/3 oct.	Corrundum
CsCl	SC	8-fold	All	CsCl
CaF_2, ZrO_2, HfO_2	SC	8-fold	½	Fluorite
SiO_2, GeO_2	Connected Tetrahedra	2, 4-fold	½	Silica Types

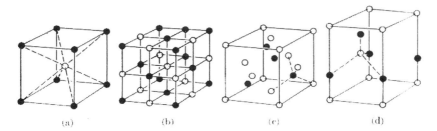

Figure 1.16 [G10]
Simple Ceramic Structures
(a) CsCl has simple cubic coordination
(b) NaCl has an octahedral unit with six nearest neighbors
(c) Sphalerite (ZnS structure tetrahedrally coordinated)
(d) Wurtzite (ZnS structure tetrahedrally coordinated)

1.7.1   Interstitial Positions
Each of the lattice types possess interstitial holes among the atoms. The largest interstitial hole in the simple cubic structure has eight neighboring atoms and a diameter equal to the minimum value given in Figure 1.15 for 8-fold coordination. The largest hole within a face-centered cubic or hexagonal close-packed structure has six neighboring atoms and a diameter equal to the minimum ratio for 6-fold coordination. There are also 4-fold holes for these structures for FCC or HCP. In a BCC structure, the largest interstices are not regular in shape and there is less opportunity for inclusion of small atoms.

1.8   POLYMER STRUCTURES

1.8.1   General
Polymers consist of a series of chain-like atomic arrangements of repeating units or monomers. The chain backbone usually consists of the tetragonal directionally bonded carbon atom discussed previously in section 1.5.2.
The simplest form of a polymer chain is one containing only one kind of a repeating unit arranged in a linear manner. In some cases there may be branches attached to the main backbone chain

making the polymer a branched one. In crystalline polymers the branches hinder the ordered arrangement of chains and hence result in a reduction of crystallinity. Lastly, the polymer chains may be inter-linked through the branches which connect two or more chains then a network structure is produced which will have an infinite molecular weight. Glassy polymers which form a network structure are referred to as thermosets and the rubbery polymers that are cross-linked are referred to as elastomers. All other linear or branched polymers fall under the category of thermoplastics. The repeat unit of the polymer is usually equivalent or nearly so to the monomer. Atoms other than carbon, such as oxygen, nitrogen, and sulfur constitute repeating sequences in the chain.

Example of groups that constitute a variety of polymer structures are shown in Table 1.3. A few examples of polymers are given also in Table 1.4.

Each amorphous polymer, due to its unique chemical structure, exhibits a glass transition temperature (T_g). Below the T_g, the Brownian motion of the chains stops and the polymer is in its glassy state in which the polymer is hard and brittle. Above the T_g, the polymer chains are mobile due to the onset of Brownian motion and the polymer is in its rubbery state in which the polymer is soft and elastic. Furthermore, polymer chains with a regular structure would allow the chains to arrange in an ordered fashion producing an ordered morphology making the polymer crystalline. In polymers with irregular chemical or geometric structures the polymer chains cannot be arranged in an ordered state and hence such polymers remain amorphous.

For a uniform structure of CH_2 repeating units the distance between carbon atoms is 1.54 Å while the linear distance between carbon atoms is 1.26 Å. These bond distances are illustrated in Figure 1.17.

However, this is an idealized picture and in actual practice, carbon chains are seldom extended to their full contour length but are present in many different shapes because of rotation of carbon-carbon bonds. Thus 10,000 repeat CH_2 units of an idealized completely extended polyethylene chain would have an end-to-end

Table 1.3
Important Structural Groups That Constitute Polymer Repeat Units[G8]

Groups	Monovalent	Bivalent	Trivalent	Tetravalent
Homogeneous Hydrocarbon Groups	$-CH_3$ $-CH=CH_2$	$-CH_2-$ $-CH=CH-$	$>CH-$ $-CH=C<$	$-C-$ $>C=C<$
Homogeneous Non-Hydrocarbon Groups	$-OH$ $-SH$ $-NH_2$ $-F$ $-Cl$ $-Br$ $-I$ $-C\equiv N$	$-O-$ $-S-$ $-NH-$ $-\overset{O}{\underset{O}{\overset{\|}{C}}}-$ $-\overset{O}{\underset{O}{\overset{\|}{\underset{\|}{S}}}}-$	$-N<$	$-Si-$
Heterogeneous Groups	$-COOH$ $-CONH_2$	$-O-\overset{O}{\overset{\|}{C}}-$ $-NH-\overset{O}{\overset{\|}{C}}-$ $-O-\overset{O}{\overset{\|}{C}}-O-$ $-O-\overset{O}{\overset{\|}{C}}-NH-$ $-NH-\overset{O}{\overset{\|}{C}}-NH-$ $-\overset{O}{\overset{\|}{C}}-O-\overset{O}{\overset{\|}{C}}-$ $-O-\overset{O}{\overset{\|}{C}}-O-\overset{O}{\overset{\|}{C}}-O-$		

Table 1.4(a)
Classification of Polymers
Homochain Polymers

Polymer Families	Basic Unit	Derivatives of Basic Unit
Polyolefins	Poly(ethylene) $-(CH_2CH_2)-$	$-(CHCH_2)-$ $\quad\quad\mid$ $\quad\quad R$
Polystyrenes	$-(CHCH_2)-$ (phenyl ring)	$-(CHCH_2)-$ (phenyl ring with R)
Polyvinyls	Poly(vinyl alcohol) $-(CHCH_2)-$ $\quad\mid$ $\quad OH$	Ethers $-(CHCH_2)-$ $\quad\mid$ $\quad OR$ Esters $-(CHCH_2)-$ $\quad\mid$ $\quad OOCR$
Polyacrylics	$-(CHCH_2)-$ $\quad\mid$ $\quad COOH$	$\quad R'$ $\quad\mid$ $-(CCH_2)-$ $\quad\mid$ $\quad COOR$
Polyhalo-olefins	$\quad R_1 \quad R_2$ $\quad\mid\quad\mid$ $-(C-C)-$ $\quad\mid\quad\mid$ $\quad R_3 \quad R_4$ R=H,F,Cl,Br	
· Polydienes	Polybutadiene $-(CH_2CH=CHCH_2)-$	$-(CH_2C=CHCH_2)$ $\quad\quad\mid$ $\quad\quad R$
Polynitriles	$-(CH_2CH)-$ $\quad\quad\mid$ $\quad\quad C\equiv N$	$\quad\quad R$ $\quad\quad\mid$ $-(CH_2C)-$ $\quad\quad\mid$ $\quad\quad C\equiv N$

Table 1.4(b)
Heterochain Polymers

Polymer Families	Basic Unit	Derivatives of Basic Unit
Polyoxides/ethers/ acetals	$—(OR)—$	$—(OR_1OR_2)—$
Polysulfides/ thioesters	$—(SR)—$	$—(SR_1SR_2)—$
Polyesters	$$—(O\overset{\displaystyle O}{\overset{\|}{C}}R)—$$	$$—(OR_1O\overset{\displaystyle O}{\overset{\|}{C}}R_2)—$$
Polyamides	$$—(NHC\overset{\displaystyle O}{\overset{\|}{\,}}OR)—$$	$$—(NHR_1NHC\overset{\displaystyle O}{\overset{\|}{\,}}R_2)—$$
Polyurethanes	$$—(NHC\overset{\displaystyle O}{\overset{\|}{\,}}OR)—$$	$$—(OR_1OCNHR_2NHC)—$$
Polyureas	$$—(NHC\overset{\displaystyle O}{\overset{\|}{\,}}NHR)—$$	$$—(NHR_1NHCNHR_2NHC)—$$
Polyimides	$$-(R\,\underset{\underset{\displaystyle O}{\|}}{\overset{\overset{\displaystyle O}{\|}}{\begin{matrix}C\\C\end{matrix}}}\,N)-$$	$$—(R_1N\,\underset{\underset{\displaystyle O}{\|}}{\overset{\overset{\displaystyle O}{\|}}{\begin{matrix}C\\C\end{matrix}}}R_2\underset{\underset{\displaystyle O}{\|}}{\overset{\overset{\displaystyle O}{\|}}{\begin{matrix}C\\C\end{matrix}}}\,N)—$$
Polycarbonates	$$-(O\overset{\displaystyle O}{\overset{\|}{C}}OR)—$$	$$—(OR_1O\overset{\displaystyle O}{\overset{\|}{C}}OR_2OC)—$$
Polyimines	$$—(\underset{R_1}{\overset{\|}{N}}R_2)—$$	$$—(\underset{R_1}{\overset{\|}{N}}R_2\underset{R_1}{\overset{\|}{N}}R_3)—$$
Polysiloxanes	$$—(\underset{\underset{R_2}{\|}}{\overset{\overset{R_1}{\|}}{Si}}O)—$$	$$—(\underset{\underset{R_2}{\|}}{\overset{\overset{R_1}{\|}}{Si}}O\underset{\underset{R_2}{\|}}{\overset{\overset{R_1}{\|}}{Si}}R_3)—$$

distance of 12,600 Å . However, a molecule's cloud size for a freely jointed chain model changes with molecular weight and the average square distance traveled increases linearly with the number of links, n, and their length, l, and is given by:

$$\overline{r_o^2} = nl^2 \tag{1.10}$$

In solution, e.g., this polymer would have a root mean end-to-end square distance $\left(\overline{r_o^2}\right)^{1/2}$ of 218 Å . Since tetrahedral carbon has a bond angle of 109.5°, $\overline{r_o^2}$ is approximated by $2nl^2$ (See Chapter 3 for a more complete description).

The rotation of bonds in a three-dimensional array produces a variety of geometric forms which are called conformations. The conformations permitted for a polymer chain are governed by the bond angles and the rotation of the chain around the bond angles. The conformations can be altered simply by rotation of the chain around the bond angles. For known conformations the actual end to end chain distance in polymers can be estimated.

1.8.2   Configurations of Polymer Chains

Polyethylene (PE) has the simplest chemical structure and hence it forms an ideal system to carry out conformational analysis. If, however, one of the hydrogen atoms on PE is substituted by another group such as CH_3, Cl, CN, OH or phenyl group the class of polymers that emerges is the vinyl class. The attachment of the substituent to alternating carbon atoms in vinyl polymers leads to a situation in which two structures are possible. Considering the molecule to be planar the substituent can be either above the plane or below the plane. Neither form can be converted from one to the other by possible internal rotation about the C-C bond. The two structures are distinctly different chemical structures called stereoisomers. Random arrangement of the substituents along the polymer chain gives the atactic isomer. When all the substituents are on one side of the plane, the resulting structure is the isotactic isomer form while regular alternation of substituents produce the syndiotactic form. The

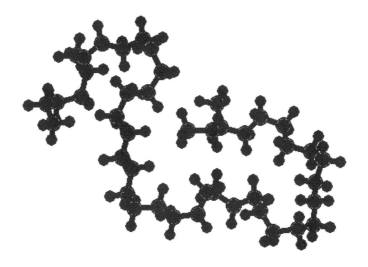

(a) Idealized CH_2 Chain, Distances in Å

(b)   Random CH_2 Polymer Chain

Figure 1.17
Hydrocarbon Chains

three isomers are illustrated for polypropylene in Figure 1.17.

If two different substituents are present on a carbon atom of the polymer chain, the carbon is asymmetric since the two parts of the chain with which it is connected are also different. As we have seen, carbon is tetrahedral and any asymmetric atoms can be in two different spatial configurations not interchangeable without bond breakage. The asymmetric carbons result in optical isomerism so familiar in organic chemistry and exhibited by sugars, amino acids, and many other compounds. An example illustrating the above occurs in the case of the polymer polypropylene since every other carbon is asymmetric.

$$\left(CH_2-\underset{\underset{H}{|}}{\overset{\overset{CH_3}{|}}{C}}\right)$$

Using appropriate catalysts and preparation techniques, polypropylene and other polymers can be prepared in stereoregular configurations. Examples of these structures, along with the terminology originally coined by Natta are shown in Figure 1.18a. A second type of stereoisomerism is of the cis-trans variety familiar in small ethenic molecules is also shown in Figure 1.18b. Because rotation is not free about a double bond, the substituents on either side can be in two configurations.

1.8.3   Classification of Polymers

A general classification of polymers can be done according to a) chain structure - linear, branched or network polymers, b) T_g, glassy or rubbery polymers, and c) ability to crystallize - amorphous or crystalline polymers. Another way to classify or group polymers would be according to their generic families.

While there are many different types and combinations of polymers that are commercially important, since the rapid expansion of polymer science over the past forty years, there are a number of polymer families or systems which are widely recognized and are commercially significant. These polymer families are used in a wide variety of applications.

Important homochain and heterochain polymers and their constituent groups are listed in Table 1.4(a) and 1.4(b). In subsequent chapters, beginning with Chapter 5, "Polymer Property Estimation and Estimated Properties of Selected Polymers", polymers from a number of different families will be used for property estimation. For all the properties that can be estimated from available data, comparisons of the estimated properties to literature values will be made in the appropriate chapters along with discussions of the advantages, deficiencies, and limitations of the estimates.

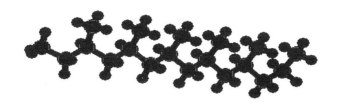

Isotactic

Syndiotactic

Figure 1.18a
Stereoisomers from the Asymmetric Carbon of Poly(propylene)

$$-C-C-C-C-C-C-C-C-C-$$

Atactic

**Stereoisomers from the Asymmetric Carbon of Poly(propylene)
(Figure1.18a continued)**

$$-(CH_2-CH=CH-CH_2)-$$

Trans

Cis

**Stereoisomers from Double Bonds in Chains of 1,4 Polybutadiene
Figure 1.18b**

1.8.4   Polymer Property Estimation

In subsequent chapters, proper group selection and the resultant estimated properties will be demonstrated and discussed. Group selection for the polymer families is of the utmost importance. A representative number of polymer systems will be used to demonstrate proper group selection and the resultant estimated properties so that the user will be prepared to estimate the properties (including polymer-polymer miscibility) of systems of interest. The software program included as an integral part of this book is MATPROP© and the specific program in this software suite used for property estimation is Molcalc.

Polymers selected from the following families will be used as examples in each chapter that addresses specific polymer topics.

1. Polyolefins
2. Poly-halo olefins
3. Polyvinyls
4. Polydienes
5. Polynitriles
6. Polyamides
7. Polyoxides/ethers/acetals
8. Polyacrylics

1.9   CERAMIC-GLASS STRUCTURE

1.9.1   Definition

It turns out that there is no single, precise way of defining a glass. Glassy materials are complex substances that are generally defined in terms of their properties or characteristics as opposed to their structures. Glasses are often described or referred to as viscous supercooled liquids, materials which possess a tetrahedral network with short-range structural order, or as a material which has cooled to a rigid condition without crystallizing.

All glasses have two characteristic structural features: 1) short-range or first neighbor order, and 2) a continuous (usually) three-dimensional networks of strong primary bonds. In the case of oxide glasses, these networks are composed of oxygen polyhedra.

Zachariasen studied the conditions for forming a random network such as shown in Figure 1.19 and suggested the following rules for the formation of an oxide glass:

1. Each oxygen ion should be linked to not more than two cations.
2. The coordination number of oxygen ions about the central cation must be small, 4 or less.
3. Oxygen polyhedra share corners, not edges or faces.
4. At least three corners of each polyhedron should be shared.

Silica glass is the most common of the glass structures consisting of a single oxide. The SiO_4 tetrahedral unit provides the short-range order in fused silica. Each tetrahedron is connected to another through the jointly shared apex oxygen, thus producing the three-dimensional network of strong Si-O-Si bonds (bridging oxygen). Silica glass produces a diffraction band which indicates an average interatomic spacing of 1.62 Å which corresponds to a value of 1.60 Å for the closest spacing (Si-O bonds) in crystalline silicas [Figure 1.19(a)]. These silicas begin to flow or change from a glassy to rubbery state at high temperatures ($\approx 1500°$ C - $1600°$ C).

Glass forming oxygen polyhedra are triangle and tetrahedra, and cations forming such coordination polyhedra have been termed network formers. In order to lower the temperature at which glass begins to flow and allow it to be worked and formed more easily the glass network is modified by the addition of oxides.

If alkali oxides, such as Na_2O, K_2O, CaO, etc. are added, the framework structure is depolymerized. The alkali ions, presumably occupying random positions distributed through the structure, are located to provide charge neutrality. In effect, the addition of Na_2O to a glass breaks into the framework to produce two O_v^- (non-bridging oxygens) branches and two Na^+ ions [Figure 1.19(c)]. Cations of higher valence and lower coordination number than the alkalis and alkaline earth metal elements may contribute in part to the network structure and are termed intermediates.

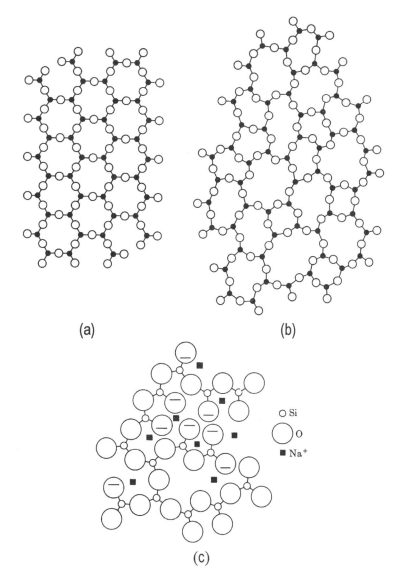

Figure 1.19 [G3, G10]
Schematic Representations of Ordered Crystalline, Random Network
Glassy Silicas, and Alkali Silicate Glass

(a) Ordered crystalline form
(b) Random network glassy form of the same composition
(c) Alkali silicate glass

Bond strength is the dissociation energy of the oxide divided by the coordination number, and this factor can be used as a qualitative index of the glass-forming tendencies of oxides. Network forming oxides have bond strengths of more than 80 kcal/mole while modifiers have bond strengths of less than 60 kcal/mole. A listing of network former, intermediates, and network modifiers, the dissociation energies, cation coordination number, and bond strengths are shown in Table 1.5.

Table 1.5
Bond Strengths of Oxide Glass Components

Oxide	Dissociation Energy, kcal/mole	Coordination Number of Cations	Bond Strength, kcal/mole
Network Formers			
B_2O_3	356	3	119
GeO_2	431	4	108
SiO_2	424	4	106
V_2O_5	449	4(5)	112-90
P_2O_5	442	4(5)	111-88
Al_2O_3	402-317	4	101-79
B_2O_3	356	4	89
As_2O_5	349	4(5)	87-70
Sb_2O_5	339	4(5)	85-68
ZrO_2	485	6	81
Intermediates			
TiO_2	435	6	73
PbO	145	2	73
ZnO	144	2	72
Al_2O_3	402-317	6	67-53
ThO_2	516	8	64
BeO	250	4	63
ZrO_2	485	8	61
CdO	119	2	60
Network Modifiers			
Sc_2O_3	362	6	60
Y_2O_3	399	8	50

continued

Table 1.15
(continued)

Oxide	Dissociation Energy, kcal/mole	Coordination Number of Cations	Bond Strength, kcal/mole
Network Modifiers			
SnO_2	278	6	46
Ga_2O_3	267	6	46
In_2O_3	259	6	43
PbO_2	232	6	39
MgO	222	6	37
Li_2O	144	4	36
PbO	145	4	36
ZnO	144	4	36
BaO	260	8	33
CaO	257	8	32
SrO	256	8	32
CdO	119	6	20
Na_2O	120	6	20
CdO	119	6	20
K_2O	115	9	13
Rb_2O	115	10	12
HgO	68	6	11
Cs_2O	114	12	10

1.10   EFFECTS OF STRUCTURE ON ELECTRICAL CONDUCTIVITY

1.10.1 Structural Imperfections

Many different types of imperfections can occur in materials. Electrons may be excited into higher energy levels, leaving vacant positions in the normally filled electronic energy-level bands. These vacant positions are called holes. There can also be atomic defects, including substitution of a different or a foreign atom for a normal one, the presence of interstitial atoms, vacant atom sites, and line imperfections called dislocations.

Displacement of an atom from its normal site to an interstitial position is called a Frenkel disorder. When cation and anion vacancies are simultaneously produced, the disorder is referred to as a Schottky disorder. Ceramic and metallic systems are rarely without impurities or purposeful additives for alloying as is the case for metals. These additional atoms may either substitute for host atoms on normal lattice sites and form substitutional solid solutions or be incorporated on interstitial sites that are normally unoccupied in the host lattice and thus form an interstitial solid solution. These concepts are demonstrated in Figure 1.20.

1.10.2 Electronic Structure

In an ideal crystal, in addition to all atoms occupying correct sites, with all sites filled, the electrons should be in the lowest-energy configuration. Because of the Pauli exclusion principle, the electron energy levels are limited to a number of energy bands up to some

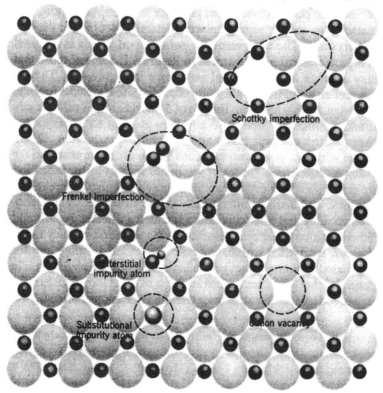

Figure 1.20 [G4]
Structural Imperfections That Can Occur in Materials

maximum cutoff energy at 0° K which is known as the Fermi energy $E_f(0)$. At higher temperatures thermal excitation gives an equilibrium distribution about the Fermi level $E_f(T)$ which is the energy for which the probability of finding an electron is equal to one-half.

The different temperature effects observed for metals, semi-conductors, and insulator are related to the electronic energy band levels as shown in Figure 1.21. In metals these bands overlap so that there is no barrier to excite electrons to higher energy states.

1.10.3 Metals

As we have seen, electrons in a metal are available to move freely through the matrix. The Sommerfield theory states that the electrons in a metal move with extremely high velocities and that they occupy definite energy levels with two electrons of opposite spins in each state. At absolute zero the electrons fill all of the lowest lying states up to the Fermi level.

The electron band energy at the Fermi level is given by:

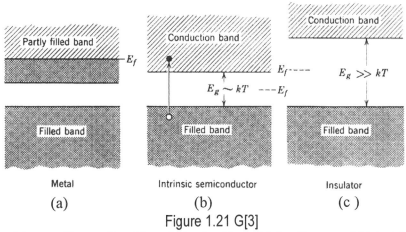

| Metal | Intrinsic semiconductor | Insulator |
| (a) | (b) | (c) |

Figure 1.21 G[3]
Electron Energy Band Levels for the Three Major Classes of Materials
(a) Metals with a partially filled conduction band
(b) Intrinsic semiconductors with a narrow band gap
(c) Insulators with a high $E_g \left(E_g - E_f \right)$

$$E_f = \frac{\hbar^2 \pi^2}{2m} \left(\frac{3N_A}{\pi L^3} \right)^{2/3} = \frac{\hbar^2 \pi^2}{2m} \left(\frac{3N_A}{\pi V} \right)^{2/3} \tag{1.11}$$

where:

$\hbar$ = Planck's constant/2 π

      = 1.06×10^{-27} erg sec

m = electron mass

      = 9.11×10^{-28}

V = L^3 (assumed volume)

N_A = Avogadro's number (6.02×10^{23} molecules / mole)

For example in the case of silver which has a gram molecular volume of 10.28 cm^3, we have:

$$E_f = \frac{\left(1.06 \times 10^{-27}\right)^2 \cdot \pi^2}{2 \cdot \left(9.11 \times 10^{-28}\right)} \cdot \left(\frac{3 \cdot \left(6.02 \times 10^{23}\right)}{\pi \cdot 10.28} \right)^{2/3} \tag{1.12}$$

$$= 8.9 \times 10^{-12} \, \text{ergs} \tag{1.13}$$

$$= 130 \, \text{kcal/mole} \tag{1.14}$$

At absolute zero, the average electron possesses an energy which is ¾ of the Fermi value,

$$\therefore E_{avg}^{Ag} = 78 \, \text{kcal / mole} \tag{1.15}$$

1.10.4 Semiconductors and Insulators

In semiconductors and insulators a completely filled energy band is separated from a completely empty conduction band of higher electron energy states by a band gap of forbidden energy levels as shown in Figure 1.21. In intrinsic semiconductors the energy difference between the filled and empty bands is not large compared with the thermal energy, so that a few electrons are thermally excited into the conduction band, leaving empty electron position (electron holes) in the normally filled band. The energy gap can also be bridged by introducing small levels of impurities with electron states in the gap.

In perfect insulators, the gap between bands is so large that thermal excitation cannot change the electron energy states, and at all temperatures the conduction band is completely devoid of electrons and the next lower band of energy is completely full, with no vacant sites. For materials that are intrinsic semiconductors, each electron whose energy is increased so that it goes into the conduction band leaves an electron hole, so that the number of holes equals the number of electrons, $p = n$. The positive electron hole concentration, p, is denoted by, $p = [h^{\bullet}]$, and the negative excess electron concentration is given by n, i.e., $n = [e']$.

Semiconductors are the building blocks of transistors, solid electronics, computers, and ceramic devices that operate at high temperatures $(\rangle 600°\ C)$ whose electrical conductivity depends on temperature and the gaseous atmosphere. Polymers are normally considered to be insulators but they can be doped to become semiconductors. Special classes of polymers, such as the polyacetylenes, are intrinsically conducting but generally to be useful, these materials must also be doped. Ceramics are also normally considered to be insulators but at high temperatures some become semiconductors because of their structures.

Electrical charges are transferred from one point to another by 1) anions, 2) cations, 3) electrons, and 4) electron holes. The transfer of charges by electrons and holes have been discussed under metals and above under semiconductors.

Ionic conduction, σ_i, is equal to the product of charge density nq and ionic mobility, μ_i:

$$\sigma_i = nq\mu_i \tag{1.16}$$

Charge density includes the number n of charge carries per unit volume, while q is the charge per carrier and may be expressed as $(1.6 \times 10^{-19})Z$ coulomb, where Z is the valence of the ion. The ionic mobility μ_i is the net velocity per voltage gradient (cm/sec/V/cm) and is given by the Einstein relationship:

$$\mu_i = qD_i / kT \qquad (1.17)$$

where:

D_i	= the ionic diffusion coefficient
k	= Boltzmann's constant
T	= temperature in Kelvin

Three factors are important: 1) the higher the temperature the greater the diffusion for vacancy (Schottky) defects and interstitial defects (Frenkel); 2) diffusion occurs more readily in structures containing defects; and 3) the smaller the ion the more readily diffusion occurs. These factors help to explain ionic conductivity in many ceramic materials.

Ion movements in defect containing structures such as that illustrated in Figure 1.20 are shown in Figure 1.22. An example of a ceramic which functions as a high temperature semiconductor is calcia doped zirconia. When zirconia $\left(ZrO_2\right)$ is treated with CaO, the charge balance is maintained by anion vacancies. The Ca^{2+} substitutes directly for Zr^{4+}, and therefore one less O^{2-} is required. An electrical gradient will thus move the anions toward the positive electrode, and the anion hole serves as a positive carrier as it moves toward the negative electrode (Figure 1.22). Ionic conductivity in glass depends on the numbers and kinds of modifier oxides as discussed above.

For semiconductors, the structure of the material accommodates the movement of electrons rather than ions. However, the principles are the same, i.e., mobility varies with the charge site and it is proportional to the gradient. The higher the temperature, the more charges are freed to move as a result of activation energy $e^{-E/kT}$.

In semiconductors, when the energy gap is small, it is possible for an electron to achieve sufficient energy by thermal activation. An example of this occurs in the case of the NiO lattice which has the NaCl structure. At low temperatures there is a regular

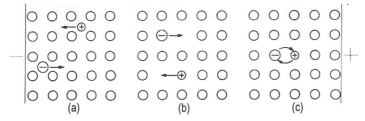

Figure 1.22
Schematic of Ion Movements
(a) Frenkel, (b) Schottky, (c) interchange

array of Ni^{2+} ions among O^{2-} ions. With increasing temperature, 3d-electrons of some atoms are transferred, leaving Ni^{3+} ions and forming Ni^+ ions. The end result is that the Ni^+ ion has an electron vacancy, thus producing a semiconductor. This is illustrated in Figure 1.23.

$$Ni^{2+} \Leftrightarrow Ni^{3+} + e \qquad (1.18)$$

$$e^= + Ni^{2+} \Leftrightarrow Ni^+ \qquad (1.19)$$

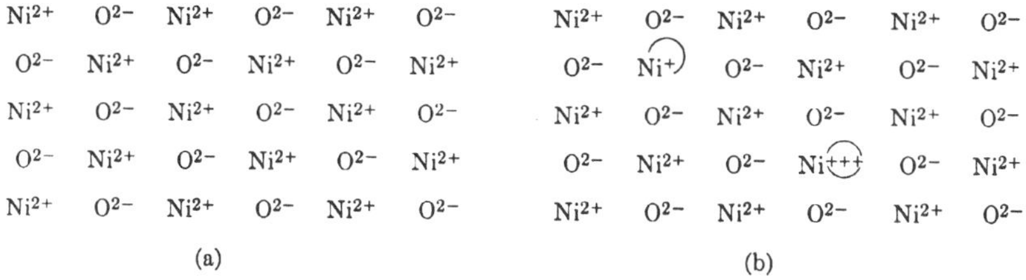

Ni^{2+}	O^{2-}	Ni^{2+}	O^{2-}	Ni^{2+}	O^{2-}
O^{2-}	Ni^{2+}	O^{2-}	Ni^{2+}	O^{2-}	Ni^{2+}
Ni^{2+}	O^{2-}	Ni^{2+}	O^{2-}	Ni^{2+}	O^{2-}
O^{2-}	Ni^{2+}	O^{2-}	Ni^{2+}	O^{2-}	Ni^{2+}
Ni^{2+}	O^{2-}	Ni^{2+}	O^{2-}	Ni^{2+}	O^{2-}

(a)

Ni^{2+}	O^{2-}	Ni^{2+}	O^{2-}	Ni^{2+}	O^{2-}
O^{2-}	Ni^{+}	O^{2-}	Ni^{2+}	O^{2-}	Ni^{2+}
Ni^{2+}	O^{2-}	Ni^{2+}	O^{2-}	Ni^{2+}	O^{2-}
O^{2-}	Ni^{2+}	O^{2-}	Ni^{+++}	O^{2-}	Ni^{2+}
Ni^{2+}	O^{2-}	Ni^{2+}	O^{2-}	Ni^{2+}	O^{2-}

(b)

Figure 1.23 [G10]
Intrinsic Valence Semiconduction of NiO
(a) low temperatures (b) high temperatures

1.11   ABBREVIATIONS AND NOTATION

A	constant
A	atomic weight
Å	angstroms
a	interatomic spacing
a	lattice constant
b	constant (~9)
B	empirical constant
D_i	ionic diffusion coefficient
e	electron charge, coulombs
E	potential energy
E_r	energy of repulsion
E_f	energy at the Fermi level
ε_o	permittivity of free space (dielectric constant)
h	height
$\hbar$	Planck's constant, 1.06×10^{-34} J sec
k	Boltzmann's constant, 1.381×10^{-23} J K^{-1}
m	electron mass, 9.1×10^{-28} g
M	molecular weight (mass)
n	negative electron hole concentration
N_A	Avogadro's number, 6×10^{23} molecules / mole
p	positive electron hole concentration
q	charge per carrier, 1.6×10^{-19} coulombs
r	atomic radius
r_{ij}	distance between ions
V	volume
U	energy sum of attractive and repulsive forces
ψ	wave function
σ_i	ionic conduction
μ_i	ionic mobility
Z	number of charges

1.12   GENERAL REFERENCES

1. Coleman, M., Graf, J.F., and Painter, P.C., *Specific Interactions and the Miscibility of Polymer Blends*, Technomic Publishing Co., Lancaster, PA, 1991.
2. DiBenedetto, A.T., *The Structure and Properties of Materials*, McGraw-Hill, Inc., New York, N.Y., 1967.
3. Kingery, W. D., Bowen, H.K., Uhlmann, D.R., *Introduction to Ceramics*, 2nd Ed., John Wiley & Sons, New York, N.Y., 1976.
4. Moffatt, W.G., Pearsall, G.W., and Wulff, J., *Structure and Properties of Materials*, Vol. 1, Structure, John Wiley & Sons, New York, 1966.
5. Pauling, L., *Nature of the Chemical Bond*, 3d ed., Cornell University Press, Ithaca, N.Y., 1960.
6. Reed-Hill, R.E., *Physical Metallurgy Principles*, 2nd Ed., D. Van Nostrand Company, New York, N.Y., 1973.
7. Seymour, R.B., Carraher, C.E. Jr., *Polymer Chemistry*, Marcel Dekker, Inc., New York, N.Y., 1988.
8. Van Krevelen, D.W., *Properties of Polymers*, Elsevier, New York, N.Y., 1990.
9. Van Vlack, L.H., *Elements of Materials Science and Engineering*, 4th Edition, Addison-Wesley Publishing Co., Reading, Massachusetts, 1980.
10. Van Vlack, L.H., *Physical Ceramics for Engineers*, Addison-Wesley Publishing Co., Reading, Massachusetts, 1964.
11. Zachariasen, W.H., "The Atomic Arrangement in Glass", *J. Am. Chem. Soc.*, **54**, 3941(1932).

CHAPTER 2

Morphology of Materials, Thermodynamics, and Relationship to Properties

2.1   FOUNDATIONS OF MATERIALS SCIENCE

Technological advances in materials science in the past fifty years have been no less than spectacular. The advent of World War II was largely the nucleating agent and catalyst for many technological advances. Technology is only one aspect of our contemporary times whose direction has been forever changed.

The war itself changed peoples' expectations of their roles in life as women permanently entered the workplace and mainly men (and fewer women) had higher education opened up to them under the G. I. Bill. It also changed our social structures as black people moved north in large numbers in the U.S., to take advantage of the higher paying jobs resulting from the war effort. These remarkable social changes, the accompanying increased levels of income, and the shortage of consumer goods during the war, created a pent-up demand for consumer products of all types that could only be satisfied by a rapid introduction of new products of better design and functionality, manufactured using improved materials. Thus the pull of the market place has been a catalyst which has pushed materials and related technologies to new levels of understanding and performance.

Again, WWII was a major factor in the development of new technology. The need for alloys in the market place that were able to withstand the high temperatures and conditions within jet engines resulted in the development of the cobalt based super alloys. The need for automobiles that are more reliable, stronger and safer, and yet lighter in order to meet mileage requirements has resulted in high strength alloy steels which can be used for automobile exteriors such as the hood, fenders, and top, and aluminum alloys for engine blocks.

Accompanying more fundamental technical developments have been improvements in manufacturing processes with their companion social changes. These changes and adjustments continue to evolve even today.

The war placed many demands on technology in order to ensure an allied victory and the required effort was carried out at whatever costs were necessary. The subsequent cold war pitted the West against the East and this started the revolution in electronics, transistors and other solid state devices, metal alloys, and polymer science to provide us with the array of products and services that surround us today.

With regard to materials science, the history of the space program is no less a history of the development of new materials and the proper use of materials, separately, or in some optimized configuration, to fill a specific need. It was this program that imposed the requirements of lightweight, low power consumption and reliability of solid state electronics. Also, metal alloys were developed for rocket engines and other parts along with the development of ceramic tiles for heat shields and skin of space vehicles.

The three major components of materials science, i.e., metals, ceramics, and polymers will be discussed briefly in the following paragraphs with major emphasis on polymers.

In recent years, new applications have emanated from the understanding of the structure of a material, its relationship to properties, and how to engineer and control those properties.

A few examples are cited to illustrate the advances that have been made in the field of materials science.

- The strength-to-density ratio of structural materials has increased dramatically. To suspend a 25-ton weight vertically from the end of a cast iron rod would require a rod with a cross section of $1'' \times 1''$ weighing about 4 lb/ft. This same load can be handled with a high-strength polymer fiber with a cross section of $0.1'' \times 0.3''$ weighing about 1 oz./ft.

- Conventional airplane steel combustion engine efficiency is about 60 percent whereas engines using superalloys which operate at much higher temperatures ($2000°F$) have an efficiency of about 80 percent.

- Today, permanent magnets have magnetic strengths more than 100 times greater than those available at the turn of the century.

- The transparency of fused silica has been improved over the centuries about 100 orders of magnitude and improvements since 1966 have resulted in the use of a single optical glass fiber of 0.01 mm to transmit thousands of telephone conversations.

- Improvements in photolithography, vapor deposition, and plating techniques, have led to our present day printed circuits that power not only our computers but allow us to control our automobiles and household appliances. The number of components a solid state "chip" contains has grown from slightly over 1 in 1960 to 10^7 in 1992.

- Biomaterials are now routinely used to replace body parts no longer functioning or worn out. These can be hip or joint replacements, catheters, shunts to maintain artery openings, and even artificial hearts.

2.1.1 Metals

As in many areas of material science, increased materials uniformity, process repeatability, durability, and overall quality resulted in improved performance.

Durability itself has been a large issue in contemporary times. The understanding of corrosion and the development of new and quantitative techniques to measure and compare a material's ability to withstand specific corrosive environments have been major factors in the development of new corrosion resistant materials. The Alaska pipeline and oil well drilling platforms are just two examples of applications for which materials must have the required mechanical properties and the necessary corrosion resistance to harsh environments.

Probes in coal gasification and chemical processing are additional areas where alloys have been developed to meet harsh, specific, requirements.

Specific examples of new applications that include the above as well as additional categories are:

- Fine Grained, Single Phase Materials - These are achieved through special processing techniques to provide excellent combinations of strength, ductility, and corrosion resistance.
- Amorphous Alloys - These materials are used for corrosion resistant applications and photovoltaic converters and are obtained using advanced processing techniques.
- Stable High Temperature Alloys - Unique microstructural alloys are obtained from dispersion-strengthening techniques. Mechanical alloys are obtained through processing techniques that provide unique properties otherwise unobtainable.
- Composites - Innovative combinations of metal matrix, intermetallic matrix, and ceramic matrix composites are produced by controlling the processing.
- Surface Modified Materials - These materials can be prepared to withstand specific corrosive environments, and radiation damage. They provide self-lubricating hard facings for joints and valves, and can be tailored for other specific application via coating composition technology.

In the case of metals, much of the focus for new materials has been on processing by themselves and in combination with diverse materials.

2.1.2   Ceramics

Ceramic products range from microscopic single-crystal whiskers, tiny magnets, and encapsulants for transistors to multi-ton refractory furnace blocks, to glasses and ceramic foams such as those used on the space shuttle.

Historically, ceramic technology dates from clayware about 6500 BCE. The Romans combined burned lime with volcanic ash to make a natural hydraulic cement. However, the manufacture of Portland cement has been practiced only for about 100 years.

In the ceramic industry, emphasis has been largely on various glass products, cement products, and whiteware. Whiteware includes pottery and porcelains which constitute a wide variety of products for specific uses. These are generally viewed as traditional ceramic products.

The following are a few examples that illustrate new applications that have resulted from improved understanding of structure/property relationships.

- Pure oxide ceramics - These materials have been developed for special electrical and refractory requirements. Alumina (Al_2O_3), zirconia ($ZrO_{2)}$), thoria (ThO_2), and beryllia (BeO), are examples.

- Nuclear Fuels - Uranium dioxide is a widely used nuclear fuel.

- Electrooptic Ceramics - Lithium niobate ($LiNbO_3$) is an example of such a PLZT material that allows optical impulses of a specific frequency to be encoded, amplified and translated into electrical impulses which can be amplified and decoded.

- Piezolectric Materials (PZT) - Barium titanate ($BaTiO_3$) is an example of a ceramic crystal that lacks a center of symmetry. This structure allows the changing of mechanical energy into electrical energy and vice versa.

- Magnets - Magnetic ceramics form the basis of magnetic memory units in computers and are useful in other applications because of their efficiency.

- Ceramic Crystals - Single crystals can be grown from a melt and serve the basis for lasers and silicon transistors when properly doped.

- Ceramic Carbides - This class of materials has been developed as abrasive materials.

- Pure Silica - This material constitutes the silica fibers from which fiber optic cables are manufactured. Low transmission losses require a high degree of purity of the silica.

Probably on the order of more than one hundred new ceramic products have been developed which were unknown ten or twenty years ago. Materials for new applications requiring enhanced performance continue to emerge. The Japanese have been pioneering in the development of a ceramic engine that can run at high temperatures and thus increase efficiency. Ceramic coatings can now be applied (plasma sprayed) onto metal parts such as jet engine blades or automobile piston rings to provide greater reliability, better performance, and longer life.

2.1.3 Polymers

The field of polymer science encompasses both natural and synthetic polymers. Of these two areas, we are concerned here with synthetic polymers. This field was not known before the nineteenth century and appears to be "days old" in comparison to the relative time scale of metals and ceramics. All major developments have come within the last 170 years. A brief look at this short but impressive history allows us to appreciate those events that have brought us to the place where we are now.

In the case of polymers, it is easier to delineate a time scale of major events which influenced the growth of polymer science than for metals or ceramics. We are more concerned here with the relationship of structure to properties and use the following to highlight those events as cited by Morawetz [G5] (see general references).

<u>Highlights of Polymer Science Developments</u>
<u>of Structure/Property Relationships</u>

- 1832 - Berzelius coined the term polymer for any compound with a molecular weight that is a multiple of the MW of another compound with the same composition.
- 1844 - Vulcanization patent was granted to Charles Goodyear.
- 1860 Pasteur characterized tartaric and "paratartaric" acids dextrorotary and levorotary.
- 1860-1880 - Bouchardat demonstrated thermal depolymerization of natural rubber to isoprene and reverse polymerization of isoprene.
- 1863 - Berthelot first discussed polymerization in Paris. This was a first step in legitimizing research in this area.

- 1870 - van't Hoff clarified the relationship between optical activity and structure. This was an important step since there was a deep prejudice against the notion that chemical properties can be interpreted in terms of the relative spatial positioning of atoms in a molecule.
- 1872 - von Baeyer discovered the acid-catalyzed condensation of phenol with formaldehyde.
- 1873 - Van der Waals formulated that deviation in gas ideality was due to attractive forces between molecules.
- 1877 - Kekulé provided foresight in how molecules combine to form oligomers. He stated, "The hypothesis of chemical valence further leads us to the supposition that a considerable number of single molecules may, through polyvalent atoms, combine to net-like, and, if we like to say so, to sponge-like masses ...... which resist diffusion and which, according to Grahms' proposal are called colloidal." This was part of the profound controversy of whether or not these materials consisted of colloidal association or were actually giant molecules.
- 1889 - Brown demonstrated the very high molecular weight of starch (20,000-30,000) by means of cryoscopy.
- 1907 - Bakeland patented a heat and pressure process for making "Bakelite".
- 1920 - Staudinger started his epoch making work in polymers, he coins the term macromolecule.
- 1925 - Staudinger first demonstrated on crystallographic grounds that the synthetic polymer polyoxymethylene consists of long chain molecules.
- 1925 - Svedberg proved the existence of macromolecules by means of the ultracentrifuge and develops the first precise method for obtaining molecular weight distribution.
- 1925 - Katz used X-ray diffraction to show that natural rubber is amorphous in the relaxed state and crystalline upon stretching.
- 1928 - Meyer and Mark characterized cellulose, silk, and rubber by X-ray diffraction and showed that the

crystallographic and chemical evidence for the chain concept are in agreement.

- 1930 - Herman formulated the concept of fringed micelles (molecular chains passing through crystalline and amorphous regions).

- 1930-1937 - Carothers' work proved by means of organic synthesis that polymers are giant stable molecules. In Carothers' group, Flory elucidated the mechanisms of radical and condensation polymerization.

- 1932 - Mark obtained crystallographic data on cellulose and showed that cellulose occupies a cross section of 25 $\overset{\circ}{A}^2$ and covalent bond energies of 70-90 kcal/mol and a strength of 800 kg/mm^2 for an ideal fiber vs. 100 kg/mm^2 experimentally. Meyer and Lotmar estimate elastic moduli. These two papers are pioneering contributions to understanding the relationship of structure to properties.

- 1942-1947 - Flory, James and Guth et al. further developed the theory of rubber elasticity.

- 1953 - Watson and Crick discovered the double helix conformation of DNA.

- 1955 - Williams, Landel, and Ferry introduced their WLF equation for describing the temperature dependence of relaxation times as a universal function of T and T_g.

- 1959-1960 - Gel permeation or size exclusion chromatography, magnetic resonance techniques, thermogravimetric and differential thermal analysis were introduced.

- 1960-1965 - Investigations over many years lead to the Ziegler catalyst, titanium chloride and triethylaluminum to produce linear poly(ethylene) at ordinary temperatures and pressures and Ziegler-Natta catalysts for controlling the structure of polyolefins.

- 1970-1985 - Further sophisticated instrumental techniques for elucidating structure were developed. Examples are: laser light scattering, magic angle spinning, neutron scattering, etc.

- 1970-1980 - De Gennes proposed scaling concepts for polymer solutions and melts and the concept of reptation movement of polymer chains in melts.

2.2   MORPHOLOGY/PROPERTY RELATIONSHIPS

2.2.1   Metals Morphology/Microstructure

Metals and metal alloys are much simpler than polymer molecules since they consist of single atoms, the structures of which were discussed in Chapter 1. Pure elements may be mixed in an infinite number of different proportions. For each composition, in the equilibrium state they form phases which coexist and whose composition (relative amounts of each) is a function of temperature and pressure. More often than not the pressure is specified as one atmosphere. Intermetallic compounds can also form which give rise to more complicated phase diagrams.

Only substitutional solid solutions can be formed in all proportions for both components of a binary mixture. Certain conditions know as the Hume-Rothery rules must be satisfied if a solution of this type is to be formed. The two atoms, forming the solution, must exhibit:

1.   Less than 15% difference in size
2.   The same crystal structure
3.   No appreciable difference in electronegativity
4.   The same valence

For binary diagrams, temperature is plotted as the ordinate and composition on the abscissa. Since one degree of freedom has been used in specifying pressure, the phase rule has the form:

$$P + F = C + 1 \qquad (2.1)$$

where:

P = number of phases
F = degrees of freedom
C = number of components

Since $C = 2$, by definition, one phase equilibrium has two degrees of freedom, i.e., the temperature and composition of the phase. If temperature is specified, the composition of both phases in equilibrium is determined. A schematic solid solution diagram between

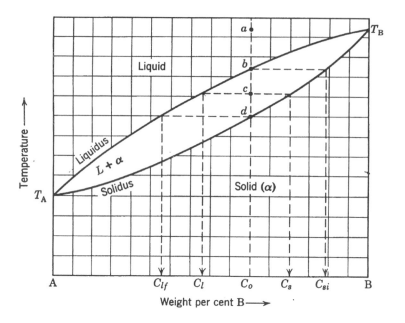

Figure 2.1
Binary Solid Solution Diagram

two components, A and B, which fit the Hume-Rothery rules are completely soluble in each other and form an alloy. A typical binary solid solution diagram is shown in Figure 2.1.

As the temperature of the liquid at point *a*, is lowered, the composition at point *b* consists of liquid of concentration C_o, and solid of concentration, C_{si}. At point *d*, the solid concentration becomes C_o and the liquid concentration is C_{lf}.

For a point of interest, at the temperature corresponding to the composition at *c*, not on the phase boundaries, the fraction of material in the solid phase, selected at the arbitrary concentration, C_o, is:

$$f_s = \frac{C_o - C_l}{C_s - C_l} \qquad (2.2)$$

and the fraction of material in the liquid phase is:

$$f_l = \frac{C_s - C_o}{C_s - C_l} \qquad (2.3)$$

A line at the temperature of interest extending through the solid-solution phase diagram is called a "tie" line and a two phase region may be considered as a lever with a fulcrum at C_o. These relationships (Equations 2.2 and 2.3), which are applicable in any two phase regions of a binary equilibrium diagram, are known as the lever rules.

The amounts of each element in a metal alloy along with its thermal treatment determines the properties of the material. Heat treatment, quenching, and impurities, control grain growth and therefore hardness of the final product.

The concentration of carbon in iron must be closely controlled along with other metallic elements to produce steels which are useful in manufacturing of automobiles and other useful and necessary commercial products. If the carbon content in iron is high, the resulting cast iron is very brittle. In lesser concentrations and with proper heat treatment a steel is obtained which is very tough and hard and can be used for objects such as hammers. In still lower concentrations of carbon $\approx$ (0.02-0.05), when additional alloying elements are added, high strength steel sheets and other products can be manufactured.

The various forms that carbon in iron can take along with the crystalline structures are illustrated in the carbon-iron phase diagram in Figure 2.2 which shows only a small portion of the carbon iron diagram. The α phase (BCC) is called ferrite and is less soluble than in the γ (FCC) called austenite because the interstitial holes between FCC packed atoms are larger than those between BCC packed atoms. It will be noted that carbon forms Fe_3C (cementite) which precipitates when the carbon content exceeds 0.02%.

As the temperature is lowered from melting for a steel containing 0.4% carbon, slow cooling produces microstructural changes with the formation of austinite at grain boundaries. The austinite phase grows and when the alloy is cooled below 723°C,

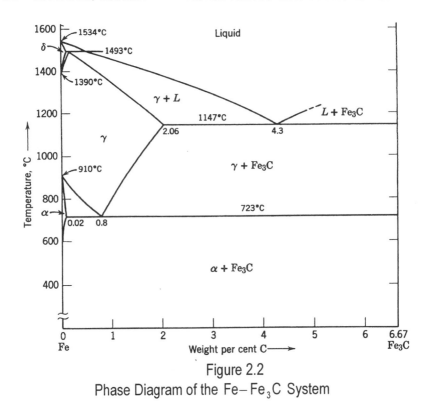

Figure 2.2

Phase Diagram of the Fe– Fe$_3$C System

the austinite transforms to a fine lamellar mixture of α and Fe$_3$C. This process is illustrated in Figure 2.3.

2.2.2 Ceramic Morphology/Microstructure

Ceramic materials, like metals, are capable of exhibiting complex microstructures and corresponding phase diagrams. The lever rule also applies to ceramics but since ceramic materials often consist of ionic bonds of two, three, or greater combinations of metallic and non-metallic elements, the phase diagrams for ternary systems are represented in the triangular coordinate system.

As we know, ceramic materials are brittle and introduce this additional consideration of great importance in comparison to metals and polymers. General properties can be found in the MATPROP© database of general material properties and polymers will be compared in Chapter 15. The high melting temperatures required, and the forming of parts from "green stock" introduces

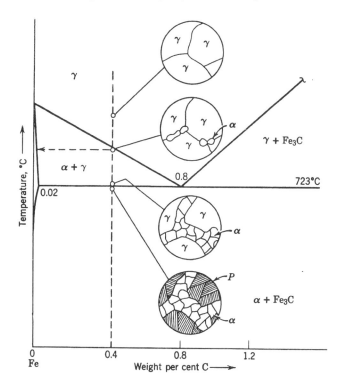

Figure 2.3 [G4]
Schematic Representation of the Microstructural Changes Which Occur
During Slow Cooling of a 0.4% C Steel

additional complications of the effects of porosity.

Since these brittle materials exhibit extremely little elongation or bending before failure, the presence of porosity decreases their strength to an even greater extent. Ceramic objects provide the greatest environmental durability and are used for parts requiring electrical insulation properties, high temperature applications such as furnace parts, and objects of all kinds requiring glazes for appearances or protective purposes.

An example of a simple binary system is the MgO/Al_2O_3 system. One phase of such a system is shown in Figure 2.4 which not only illustrates the richness and complexity of the microstructure but also shows porosity at the tips of the grain boundaries of the lighter appearing grain. The porosity occurs during sintering and vitrefication when the mixture is "fired".

The phase-equilibrium diagram illustrated in Figure 2.5 allows identification of a number of phases, phase composition, and amounts of phases present at any composition and temperature.

2.2.3 Polymer Morphology

2.2.3.1 Introduction

In Chapter 1, we saw how individual atoms and molecules are bound along with many of the resulting configurations. In addition to the conformations resulting from the presence of an asymmetric carbon, a number of important structural groups that constitute the repeat units of many important polymers were listed.

The polymer configurations (structures based on breaking and forming of primary bonds), polymer conformations (trans vs. gauche arrangements of consecutive carbon-carbon single bonds and helical structures), polymer processing conditions, and condi-

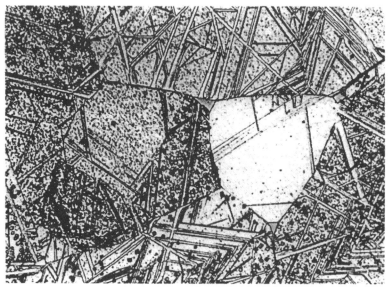

Figure 2.4[G3]
Precipitation of Al_2O_3 from Spinel Solid Solution on Cooling

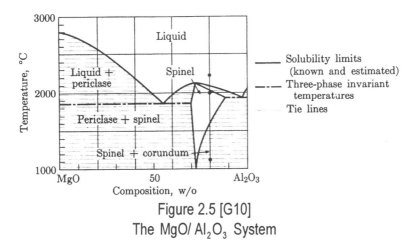

Figure 2.5 [G10]
The MgO/ Al$_2$O$_3$ System

tions used for manufacturing of the finished part(s), all play an important role in determining the morphology of the resulting polymer and the properties it will display in a particular application.

Thus polymers with the same chemical structure can exhibit different morphologies along with resulting different properties depending upon processing and final fabrication into a part. In many respects this is not too far different from what we see in the cases of ceramics and metals. Conditioning (heat treatment) is used extensively and routinely to change the grain structure of alloys which greatly influences hardness and other properties. Polymers, however, remain unique in the richness of the many different forms, appearance, and properties that they can display.

2.2.3.2 Crystal Structure of Polymers

Although we can separate polymers into crystalline and amorphous categories, in many instances this will prove to be an oversimplification since many polymers are only partially crystalline. The presence of crystallinity can be established from X-ray patterns. Crystalline polymers or regions show sharp features associated with regions of three dimensional order and more diffuse features characteristic of molecularly disordered substances.

Amorphous polymers are in the glassy state or rubbery state and may contain crosslinks to give a network. In both cases no three-dimensional order is possible. The usual model for representing macromolecular chains in the amorphous state is that of a random coil.

Many polymers are only partly crystalline which is borne out from property measurements such as density or enthalpy of melting. Typical crystalline polymers are chemically and geometrically regular in structure which allows the polymer chains to arrange in three-dimensional order and form crystalline entities. Occasional irregularities such as chain branching, copolymerization, or crosslinks interfere with the ability to form ordered three-dimensional arrays in a specific region and thus limit the extent of crystallization. The traditional model used to explain the properties of the (partly) crystalline polymer is the fringed micelle model of Herman et al. [S1] shown in Figure 2.6.

Although the fringed micelle model is the most useful model, there are other models which are known to be valid representations of polymer structures. There is no single structure which represents accurately a real life polymer. Other models re-

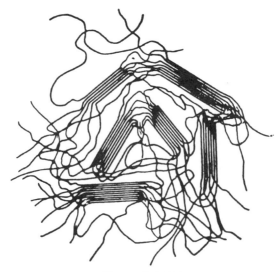

Figure 2.6 [S1]
The Fringed Micelle Model

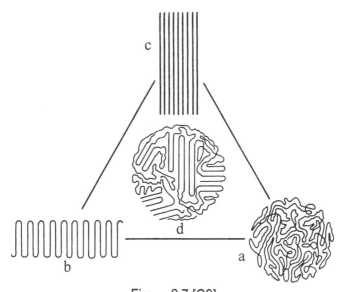

Figure 2.7 [G9]
Schematic Representations of the Different Polymer Conformations
Possible in Macromolecules

a) Random, glassy
b) Folded chain, lamellar
c) Extended chain equilibrium
d) Mixture of a, b, c

present conformations which exist in some polymers. One such structure is the folded chain or lamellar structure and another is an extended chain structure that occurs during drawing. These are shown in Figure 2.7. Bulk polymers are often mixtures of these conformations and are better represented by Figure 2.7(d).

·The fringed micelle model has been used for a long time since its simplicity allows us to visualize the amorphous structure co-existing with crystallites from the same or adjacent chains. This visualization provides an elementary understanding of experimental evidence from X-ray analysis and serves to explain other phenomena such as polymer elongation data.

The small angle X-ray analysis of typical crystal lamellae shows the thickness to be around 100 Å while the chain length is several times larger than this. Thus the chain must pass through several crystals or fold back and forth upon itself to form the crystal. Single crystals of polyethylene have been made from a dilute solution of polyethylene and the exposed surface etched. The analysis of the resultant material showed chains of length equivalent to twice the thickness of the crystal. This supports the folded chain model since etching would break the polymer chains at the folds of the surface of the crystal.

It should be recognized that only polymers with a regular structure can crystallize. Linear polyethylene is the simplest example of such a polymer and it crystallizes readily to give a high degree of crystallinity. However, branches along the main chain would hinder the arrangement of chains during the formation of crystals and thus would lead to a reduction of crystallinity when compared to that of the linear version. This has been illustrated for polyethylene in Table 2.1.

Vinyl polymers with a substituent group can crystallize if the group is small enough to be accommodated in the crystal repeat unit cell. Atactic poly(vinyl alcohol) is an example of such a polymer which is crystalline. However, if the substituent group is large and it cannot be accommodated in a unit cell, the polymer must be isotactic or syndiotactic for crystallization to occur. For polymers like poly(propylene), poly(styrene), and poly(vinyl chloride) their isotactic and syndiotactic versions crystallize while their atactic versions are amorphous. The absence of crystallinity in poly(vinyl acetate) and similar polymers is due to the combination of their atactic structure and the size of their substituent groups.

Polyacrylonitrile is an example of what can be viewed as an intermediate structure since it has a directionally ordered structure. In this polymer, steric and intramolecular dipole repulsions lead to a stiff, irregularly twisted, backbone chain conformation. The stiff chains pack like rigid rods in a lattice array where lateral order is present and longitudinal order is not. The effect of branching on crystallinity is shown in Table 2.1.

2.2.3.3 Polymer Crystals
Prior to the work of Katz and Mark and Meyer (see previous

Table 2.1
Types of Commercial Poly(ethylene)

Type	General Structure	Crystallinity, %	Density, g/cc
Low Density	Linear w/branching	≈50	0.4-0.92
Linear Low Density	Linear w/less branching	50	0.92-0.94
High Density	Linear w/little branching	90	0.95

section), it was believed that macromolecules did not exist in a crystalline form. The observation by Katz, that the stretching of a rubber band in an X-ray spectrometer produced an interference pattern typical of a crystalline substance was instrumental in changing this view.

It is now widely recognized that thick films of a semi-crystalline polymer such as LDPE are translucent because of the presence of crystallites along with the amorphous regions whereas amorphous polymer films are transparent. If the polymer could be made 100% crystalline, then also it would be transparent as in the

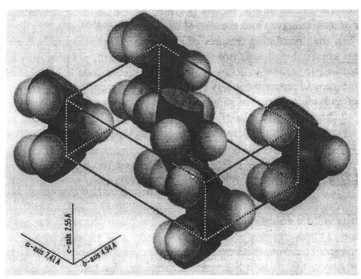

Figure 2.8 [S2]
The Unit Cell of Poly(ethylene)

case of highly oriented polyethylene. Crystals of polyethylene form unit cells similarly to metals and ceramics as discussed and illustrated in Chapter 1. The unit cell of polyethylene is rectangular with dimensions $a = 7.41$ Å, $b = 4.94$ Å, and $c = 2.55$ Å. The latter is the chain repeat distance, which is identical with the fully extended zigzag repeat distance. The arrangement of chains with respect to one another is shown in Figure 2.8 [S2].

Similarly, unit cell dimensions have been reported for many other polymers including gutta-percha [S3], poly(propylene) [S4], poly (oxy-methylene) [S5], polyamides [S6], and cellulose and its derivatives [S7-S9]. Thus ordered polymers form lamellar crystals with a thickness of 100 to 200 Å in which the polymer chains are folded back upon themselves to produce parallel chains perpendicular to the face of the crystals.

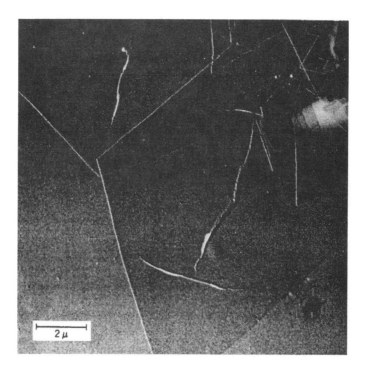

Figure 2.9 [S10]
Electron Micrograph Showing Pleats in a Single Crystal of Poly(ethylene)

All the structures described as polymer single crystals have the same general appearance, being composed of thin, flat platelets about 100 Å thick and often microns in lateral dimensions. They are usually thickened by the spiral growth of additional lamellae from screw dislocations.

Figure 2.9 [S10] shows secondary features such as corrugation and pleats of the single crystals of polyethylene. These pleats result from collapse of the pyramids as solvent is removed during preparation of the crystals as shown in Figure 2.10.

The crystallites in polymers are small and usually organized into larger, shallow pyramid like structures called spherulites which may be seen with the naked eye and viewed as Maltese cross-like structures with polarized light and crossed Nicol prisms in a microscope as shown in Figure 2.11 [S11]. Spherulites

Figure 2.10 [S10]
Electron Micrograph Showing Corrugation in a Single Crystal of Linear
Polyethylene Grown from a Solution in Perchloroethylene

apparently represent the crystalline portions of the sample and grow at the expense of the non-crystalline melt. The point of initiation of spherulitic growth, its nucleus, may be a foreign particle (heterogeneous nucleation) or may arise spontaneously in the melt (homogeneous nucleation) from temperature fluctuations and resulting undercooling in a particular region.

The rate of crystalline growth is given by the Avrami equation [12] (Equation 2.4) and the changes in specific volumes which are measured by dilatometry.

$$\frac{V_t - V_f}{V_o - V_f} = e^{-kt^n} \tag{2.4}$$

where: V_t = specific volume at time t

V_f = specific final volume

V_o = specific initial volume

k = kinetic constant related to the rate of nucleation and growth

n = integer from 1 to 4 for three dimensional growth

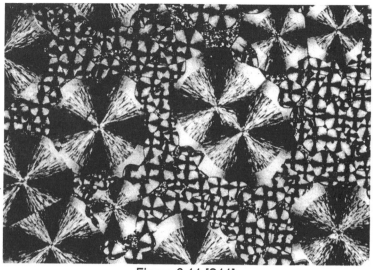

Figure 2.11 [S11]
Spherulites in a Silicone Polymer

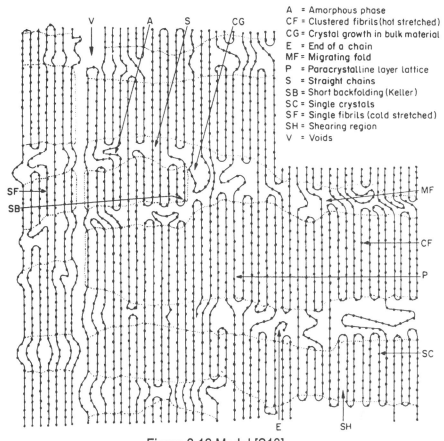

A = Amorphous phase
CF = Clustered fibrils (hot stretched)
CG = Crystal growth in bulk material
E = End of a chain
MF = Migrating fold
P = Paracrystalline layer lattice
S = Straight chains
SB = Short backfolding (Keller)
SC = Single crystals
SF = Single fibrils (cold stretched)
SH = Shearing region
V = Voids

Figure 2.12 Model [S13]
Diagrammatic Representation of the Paracrystallinity

Crystalline regions in "real" polymers are interrupted by disordered or amorphous regions which act as defect sites. This model is referred to as the paracrystallinity model as shown in Figure 2.12 [S13] which is generally applicable for oriented crystalline systems.

The crystalline regions are influenced greatly by processing techniques, such as thermoforming and extrusion of plastics and biaxial orientation during drawing of fibers (nylon) and film (PET).

The effects of orientation of a polymer during processing takes advantage of the inherent structure of the polymer to extend and align the chains to orient and thereby strengthen the polymer in a given direction. Another prime example of the effects of a polymer's structure occurs in the case of the use of acrylonitrile for beverage bottles. This application requires strength, clarity, a high heat distortion temperature, and resistance to gas permeation to maintain the quality of a carbonated beverage. Based on the PAN structure previously mentioned, the stiff polymer chains pack like rigid rods in a lattice array and provide lateral order and the required properties, particularly resistance to gas permeation, as a result of orientation during thermoforming. In practice nucleating agents such as benzoic acid are also used to control crystal growth.

Poly(vinyl chloride) is primarily syndiotactic but with considerable irregularity so that the crystallinity is poorly developed. The crystal structure is typical of syndiotactic polymers with a repeat distance of four chain carbon atoms as illustrated in Figure 2.13 [S14].

2.2.3.3.1 Helical Structures

Polymers with bulky substituents closely spaced along the chain often take on a helical conformation in the crystalline phase, since this allows the substituents to pack closely without appreciable distortion of chain bonds. Isotactic polymers often cry-

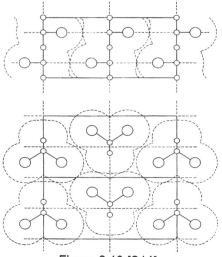

Figure 2.13 [S14]
Crystal Structure of Poly(vinyl chloride)

stallize with a helical conformation in which alternate chain bonds take trans and gauche positions as illustrated in Figure 2.14.

For the gauche position the rotation is always in that direction which relieves steric hindrance by juxtaposing *R* and *H* groups thus generating either a left hand or right hand helix. One example of this for vinyl polymers is illustrated in Figure 2.15 which shows helical conformations of isotactic vinyl polymers [S15].

2.2.3.4 Polymeric Liquid Crystals

Low molecular weight liquid crystals have been around for some time [S16,17]. Developments in the last twenty years have been responsible for their widespread use in electronic panel displays and commercial devices of all types.

These materials possess rigid molecular segments, called "mesogenic" groups and are able to exist in "mesophases". These mesophases are intermediate between the solid crystalline and isotropic liquid state. The rigid molecular segments allow these materials to align or be aligned in a field by imposing orientation of a substrate or using an electromagnetic field.

These alignment characteristics provide the properties required when coupled with an appropriate polarizer and reflector to transmit or reflect light when a voltage is applied which causes preferential orientation.

Polymers can also be prepared which exhibit these properties. The two major classifications are "main" chain and "side" chain liquid crystal polymers. This classification comes from whether or not the mesogenic group is present in the main chain or the side chain as Figure 2.16 shows [S18, S19].

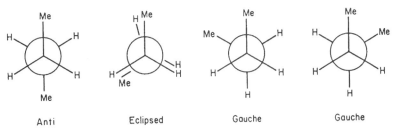

Figure 2.14
Newman Projections of Conformers of n-Butane

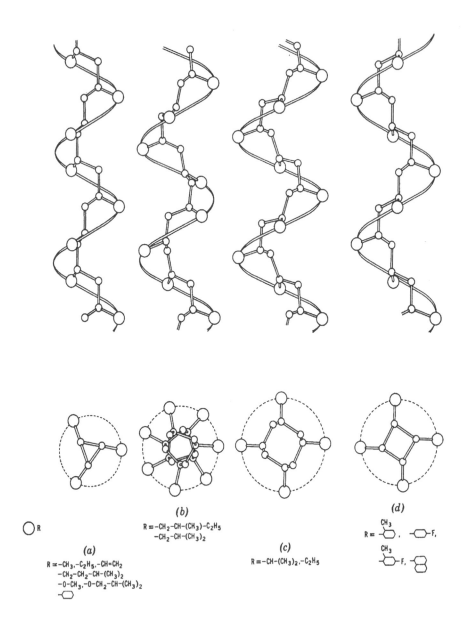

Figure 2.15 [S15]

Helical Conformations of Isotactic Vinyl Polymers

Classes of Polymeric Liquid Crystals	General chemical structure of the molecule	Physical behavior
Mesogenic groups in MAIN Chain		Thermotropic (e.g. Arylates) or Lyotropic (e.g. Aramids)
Mesogenic groups in SIDE Chains		Usually Thermotropic
Mesogenic groups both in MAIN and SIDE chains		Usually Lyotropic

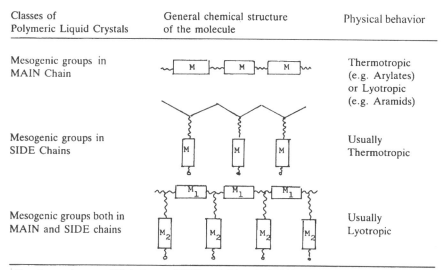

Based on schemes of Finkelmann (1982) and Ringsdorf (1981).

Figure 2.16
Classification of Polymeric Liquid Crystals

The mesogen or hard groups are generally aromatic rings or coupled aromatic rings. Polymeric liquid crystals which melt without decomposition are called thermotropic and those that decompose on heating before melting but can be dissolved in liquids are called lyotropic.

Thus even polymers with the correct structure display mesophases similar to those of the low molecular weight liquid crystals. The mesophases that have been identified are: nematic (from Greek nema or thread) with order in one direction; smectic (from the Greek smegma or soap) with a molecular arrangement in layers and can be ordered in two directions; and chiral or cholesteric that has a rotating order; and discotic, which are ordered disk like piles. The nematic types are generally used in displays. These configurations are shown in Figure 2.17 [S20].

2.3 POLYMER STRUCTURE/PROPERTY RELATIONSHIPS

The previous discussion of the morphology of polymers allows us to appreciate the various forms or configurations that a

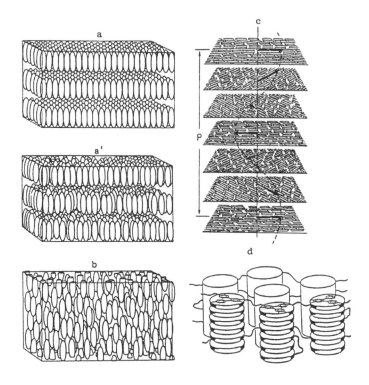

Figure 2.17 [S20]
Liquid Crystal Mesophases
(a) Smectic ordered, (a') Smectic unordered molecules in layers, (b)
Nematic, (c) Cholesteric, and (d) Discotic

polymer molecule can take. We have already seen that polymers that crystallize differ in properties from those that either do not crystallize or crystallize to a differing extent.

Normally, glassy polymers are brittle below their T_g's. Thermoset elastomers and rubbers are cross linked high polymers and rubbers which are formulated for applications where they are used at temperatures above their glass transition temperatures to provide chain mobility. When these materials are deformed, they return to their original shape after the applied force is removed. Lowering of the temperature below T_g can have tragic consequences as occurred in the "O" ring failure on the Challenger

Space Shuttle. On the other hand, many thermoplastics retain sufficient strength for use when their T_g's are not far above their use temperatures. Polycarbonate is a notable exception to this since it retains its strength at temperatures in excess of 130° below its T_g.

Crystallizable polymers are used in fiber applications since the drawing/extension of the fiber produces orientation in one direction which results in enhanced mechanical properties in that direction. This operation produces a high degree of symmetry and cohesive energy between the chains. For fibers as well as other applications, the importance of maintaining processing conditions that optimize the physical or use properties of the polymer cannot be overstated. This must also be done reproducibly.

The general performance/structure property relationships have been succinctly summarized [S21] and are illustrated in Table 2.2.

2.4   THERMODYNAMICS

2.4.1   Introduction/Elementary Thermodynamics

The laws which constitute thermodynamics are based on experimental evidence. The experimental evidence that led to the laws were performed on bodies of matter that contained very large numbers of atoms. Thermodynamics, therefore, provides us with a view of the average properties of large numbers of atoms and the mathematical relationships between pressure, volume, entropy, temperature, enthalpy, and internal energy without consideration of the mechanisms of how the final state was obtained.

Since thermodynamics provides the underpinnings for every branch of the physical sciences, it is appropriate that we consider concepts relevant to the topics in this monograph, particularly the chapter on miscibility.

2.4.2   Thermodynamic Classification

It is important to remember that thermodynamic quantities have real physical meaning and are applied successfully to a vari-

Table 2.2
General Property/Structure - Performance Relationships

Property	Increased Crystallinity	Increase Crosslinking	Increased Mol. Wt.	Increased Mol. Wt. Distribution	Addition of Polar Backbone Units	Addition of Backbone Stiffening Groups
Abrasion Resistance	+	+	+	–	+	–
Brittleness	–	M	+	+	+	+
Chemical Resistance	+	V	+	+	+	+
Hardness	+	+	+	+	+	+
T_g	+	+	+	–	+	+
Solubility	–	–	–	0	–	–
Tensile Strength	–	M	+	–	+	+
Toughness	–	–	+	–	+	–
Yield	+	+	+	+	+	+

+ = increase in property
0 = little or no effect
– = decrease in property
M = property passes through a maximum
V = variable results dependent on particular sample and temperature

80

ety of disciplines to understand, explain, and predict the way in which materials behave.

2.4.2.1 Internal Energy

Internal energy represents the total kinetic and potential energy of all the atoms or molecules in a material body or system. It depends only on the state of a substance.

In crystals for example, a large part of the internal energy is associated with the vibration of individual atoms in the lattice about their equilibrium positions in the x, y, and z directions. The Debye theory explains these vibrations as a result of elastic waves traveling within the crystal as opposed to individual atom vibrations. An increase of temperature causes the amplitudes of these elastic waves to increase, displacing them from their present positions with a corresponding increase in internal energy.

This is embodied in the first law of thermodynamics which states that the algebraic sum of all energy changes in an isolated system is zero. Thus energy may be converted from one form to another but cannot be created or destroyed. When a chemical system changes from one state to another the net transfer of energy to its surroundings must be balanced by a corresponding change in the internal energy of the system. But the energy transferred to the surroundings is the work done by the system. This is illustrated in Figure 2.18 and expressed as:

$$E = Q - W \tag{2.5}$$

where:

E = energy contained in the system
Q = heat absorbed by the system
W = work done on the system

For small changes in a system:

$$\Delta E = \Delta Q - \Delta W \tag{2.6}$$

and

$$dE = dQ - dW \tag{2.7}$$

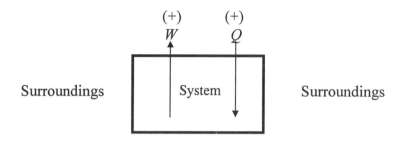

Figure 2.18
Representation of Transfer of Work and Heat Embodied in the
First Law of Thermodynamics

Since most chemical experiments are carried out at constant pressure,

$$dW = pdV \tag{2.8}$$

and

$$dE = dQ - PdV \tag{2.9}$$

where:

ΔE is a state function, i.e., path independent

ΔW and ΔQ are not state functions, i.e., are path dependent

2.4.2.2 Enthalpy

The enthalpy of a material at any stage is simply equal to the internal energy plus any pressure and corresponding volume effects. The term derives for the Greek (warm) and is sometimes referred to as the heat content. This is expressed as:

$$H = E + PV \tag{2.10}$$

2.4.2.3 Entropy

This important term comes from the Greek work meaning "change" and is represented by the letter S. It is a property, like internal energy that depends only on the state of a system and not on its previous history.

$$\Delta S = S_2 - S_1 = \int_1^2 \frac{dQ}{T}\bigg|_{rev} \tag{2.11}$$

where:

S_1 = entropy in state 1
S_2 = entropy in state 2
T = absolute temperature
dQ = increment of heat added to the system

The integration is assumed to be taken over a reversible path between the two equilibrium states of 1 and 2 and is independent of the way in which the system moved from state 1 to state 2. If the system moves from 1 to 2 as a result of an irreversible reaction, the entropy change remains equal to $(S_2 - S_1)$. However, the entropy change is given by Equation (2.11) only if the path is reversible. The integral for an irreversible path does not equal the entropy change, and in fact, is:

$$\Delta S = S_2 - S_1 > \int_1^2 \frac{dQ}{T}\bigg|_{irrev} \tag{2.12}$$

Differentiating Equations (2.11) and (2.12) yields:

$$dS = \frac{dQ}{T}\bigg|_{rev} \tag{2.13}$$

$$dS > \frac{dQ}{T}\bigg|_{irrev} \tag{2.14}$$

for an infinite small change in state.

2.4.2.4 Gibbs Free Energy

A chemical system is similar to a mechanical one and has a specific amount of energy associated with it. If this system can change by chemical reaction or otherwise to form a new system with lower energy, it will do so.

The Gibbs free energy is defined as:

$$G = E + PV - TS \tag{2.15}$$

and from Equation (2.10):

$$G = H - TS \tag{2.16}$$

2.4.2.5 Helmholtz Free Energy

For a reversible system, from above:

$$dH = TdS + VdP \tag{2.17}$$

$$= T\frac{dQ}{T} + VdP\Big|_{P=const.} \tag{2.18}$$

$$dH = dQ \tag{2.19}$$

Returning to Equation (2.7) which embodies the idea of maximum work,

$$dE = TdS - dW \tag{2.20}$$

the Helmholtz free energy can be defined as a state function such that

$$F = E - TS \tag{2.21}$$

2.4.2.6 Thermodynamic Interrelationships

It is instructive and useful to examine a few other relationships and their origins.

cf. Equation (2.9): $\qquad dE = dQ - PdV$

and Equation (2.13) $\qquad dS = \dfrac{dQ}{T}\Big|$

$$\boxed{dE = TdS - PdV} \quad \text{Internal Energy} \qquad (2.22)$$

Differentiating Equation (2.10):

$$dH = dE + PdV + VdP \qquad (2.23)$$

and substituting Equation (2.22) in Equation (2.23):

$$dH = TdS - PdV + PdV + VdP \qquad (2.24)$$

$$\boxed{dH = TdS + VdP} \quad \text{Enthalpy} \qquad (2.25)$$

Differentiating Equation (2.16):

$$dG = dH - TdS - SdT \qquad (2.26)$$

and substituting Equation (2.25) in (2.26):

$$dG = TdS + VdP - TdS - SdT \qquad (2.27)$$

$$\boxed{dG = VdP - SdT} \quad \text{Gibbs Free Energy} \qquad (2.28)$$

This is a most important thermodynamic relationship for understanding chemical reactions since it tells us how free energy, and the resulting equilibrium position varies with temperature and pressure.
Differentiating Equation (2.21):

$$dF = dE - TdS - SdT \qquad (2.29)$$

$$dF = TdS - PdV - TdS - SdT \qquad (2.30)$$

$$\boxed{dF = -PdV - SdT} \quad \text{Helmholtz Free Energy} \qquad (2.31)$$

Helmholtz free energy is thus a measure of the maximum amount of work that the system can perform on its surroundings.

2.5   PHASE TRANSITIONS

Molecules less complex than polymers exist as solids, liquids, and gases. These materials undergo transitions which are referred to as first order and exhibit sudden changes in volume or enthalpy.

Polymers, however, are more complex and because of their high molecular weight, partial crystallinity, and viscosity as liquids, these materials exhibit changes in their properties at transition temperatures at which they undergo phase changes from glassy (amorphous) to rubbery or from crystalline or semi-crystalline to melting.

These transitions and the particular phase in which a polymer exists at a given temperature determines its properties to a large extent. Even in the liquid state polymers exhibit viscoelasticity. In any of these

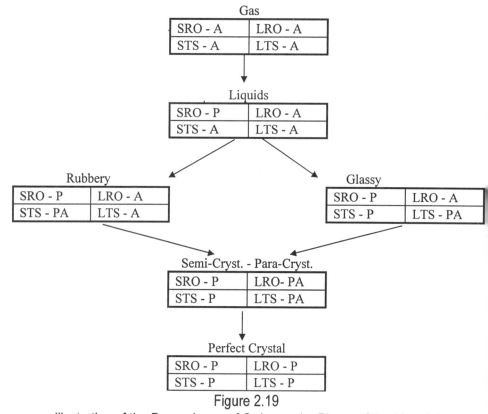

Figure 2.19
Illustration of the Dependence of Order on the Phase of the Material

states or phases, a polymer will exhibit various degrees of order and time dependence of stiffness.

Van Kreveln has presented a useful classification of phase states based on the degree of short-range order (SRO) and long-range order (LRO). Each state can be classified by a matrix of SRO, LRO, and time dependence of short time stiffness (STS) and long time stiffness (LTS). These concepts are illustrated in Figure 2.19 for non-oriented polymer systems since orientation of crystalline or semi-crystalline polymers has a large effect on properties. The presence of order is indicated by a "P" and the absence of order by an "A". Two possible phases or states is indicated by "PA".

2.5.1 Pure Assembly of One Species

Realizing that systems always tend to lower their free energies, for systems that undergo a transition at temperature, T_e or equilibrium temperature, the free energy of the system will vary as illustrated in Figure 2.20.

2.5.2 Definitions and Relationships to Thermodynamic Quantities

2.5.2.1 Heat Capacity

The specific heat capacity of a material is simply the change in

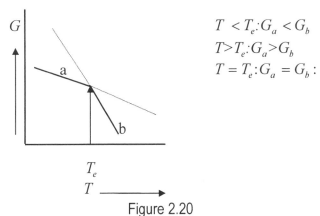

$$T < T_e : G_a < G_b$$
$$T > T_e : G_a > G_b$$
$$T = T_e : G_a = G_b :$$

Figure 2.20
Variation of Free Energy as a Function of Temperature

internal energy at constant volume or the change in enthalpy at constant pressure. Thus by definition from Equation (2.22):

$$\left(\frac{\partial E}{\partial T}\right)_V = T\frac{\partial S}{\partial T} = \frac{\partial Q}{\partial T} = C_V \tag{2.32}$$

$$\text{and} \quad \left(\frac{\partial H}{\partial T}\right)_P = T\frac{\partial S}{\partial T} = \frac{\partial Q}{\partial T} = C_P \tag{2.33}$$

For polymers, a plot of the specific heat as a function of temperature will result in transition curves similar to those in Figure 2.21 for polymeric materials.

2.5.2.2 Thermal Expansion

Next, we consider the variation of free energy as a function of temperature and pressure. From Equation (2.28):

From the P-V-T equation of state, the coefficient of thermal expansion is:

$$\alpha = \frac{1}{V}\frac{\partial V}{\partial T}\bigg|_P \tag{2.34}$$

The variation of the coefficient of thermal expansion at transition temperatures is illustrated in Figure 2.22.

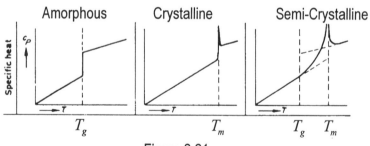

Figure 2.21
Behavior of Polymers at T_g and T_m

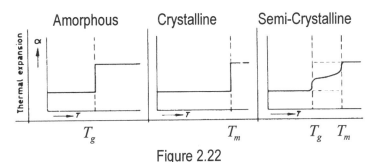

Figure 2.22
Behavior of Thermal Expansion at T_g and T_m

Similarly, from the P-V-T equation of state, the coefficient of compressibility is:

$$\beta = -\frac{1}{V}\frac{\partial V}{\partial P}\bigg|_T \tag{2.35}$$

2.5.2.3 Transition Orders
The order of a phase transformation is defined as the order of the derivative of the Gibbs free energy which exhibits a discontinuity.

- 1st Order
  From Equation (2.28);
  $$\left(\frac{\partial G}{\partial P}\right)_T = V \tag{2.36}$$

This is illustrated in Figure 2.23.

- 2nd Order
  From Equation (2.28);
  $$\left(\frac{\partial G}{\partial T}\right)_P = -S : \tag{2.37}$$

$$\text{and } \left(\frac{\partial^2 G}{\partial T^2}\right)_P = -\frac{\partial S}{\partial T}\bigg|_P = -\frac{C_P}{T} \tag{2.38}$$

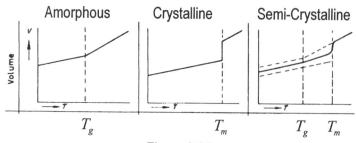

Figure 2.23
Behavior of Volume at T_g and T_m

$$\left.\frac{\partial H}{\partial T}\right|_P = T\left.\frac{\partial S}{\partial T}\right|_P \qquad (2.39)$$

Variation of the heat capacity is illustrated in Figure 2.21 in which a discontinuity is indicated at T_g.

Other properties of polymers vary during phase changes or changes in state but are not directly related to thermodynamic properties and will be considered later. Examples of these are the change in modulus and heat conductivity at T_g and T_m.

2.6   MOLECULAR WEIGHT CONCEPTS

As we have seen previously, polymers consist of molecules having different chain lengths. This is because polymerization reactions lead to polymers containing different numbers of repeat units whose numbers are governed by the mechanism of the polymerization reaction and the conditions under which it was carried out. Free radical polymerization reactions (vinyl chloride, e.g.), in general lead to narrower distributions than condensation reactions such as polyester formation.

These considerations are not trivial for polymers and molecular weight and molecular weight distribution control many properties and dictate performance. For example, melt viscosity, tensile strength, modulus, impact strength or toughness are dependent on the molecular weight of amorphous polymers and their molecular weight distribution (MWD). In contrast, density, specific heat capacity, and refractive index are essentially independent of the molecular weight at values above a critical molecular weight, required for chain entanglement ($\eta = \overline{M}^{3.4}$).

The molecular weight of a polymer is:

$$M_x = x \cdot M_o \qquad (2.40)$$

where: x = the degree of polymerization or number of repeat
            = DP (Degree of Polymerization)
          M_o = the molecular weight of the polymer repeat units

If W_x represents the weight fraction with a DP of x and N_x is the mole fraction with DP of x, then the molecular weight distribution is represented as illustrated in Figure 2.24.

The broadness of this curve represents the polydispersity of the particular polymer. The relationship between N_x and M_x is represented by Figure 2.25.

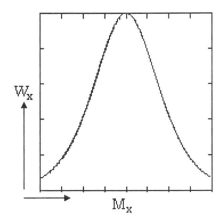

Figure 2.24
Variation of Molecular Weight as a Function of Weight Fraction of DP$_x$

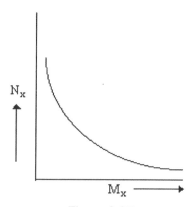

Figure 2.25

Variation of Molecular Weight as a Function of Mole Fraction of DP_x

If the polymer is monodisperse, it consists of a single molecular weight specie. This is generally not the case and we therefore have to describe the molecular weight in terms of an average value, $\overline{M}_w$ as given by Equation 2.41. The weight average molecular weight is determined from experiments in which each molecule or chain makes a contribution to the measured result. The weight average molecular weight is dependent upon the second power.

Consider an example where three molecules are present whose molecular weights are 150,000, 200,000, and 300,000.

$$\overline{M}_w = \sum_{x=1}^{\infty} W_x M_x = \frac{\sum_{x=1}^{\infty} N_x M_x^{2}}{\sum_{x=1}^{\infty} N_x M_x} \tag{2.41}$$

$$\overline{M}_w = \frac{150000(150000)}{600000} + \frac{200000(200000)}{600000} + \frac{300000(300000)}{600000} \tag{2.42}$$

$$= 233333$$

Any measurement that leads to the number of molecules, functional groups, or particles, that are present in a given weight of sample allows calculation of the number average molecular weight $\overline{M}_n$.

Thus for $\overline{M}_n$ we have:

$$\overline{M}_n = \sum_{x=1}^{\infty} N_x M_x = \frac{\text{total weight of sample}}{\text{no. of molecules } N_i} \tag{2.43}$$

$$= \frac{W}{\sum_{i=1}^{\infty} N_i} = \frac{\sum_{i=1}^{\infty} M_i N_i}{\sum_{i=1}^{\infty} N_i} \tag{2.44}$$

For the previous example, $\overline{M}_n$ is:

$$\overline{M}_n = \frac{1(100000)}{3} + \frac{1(200000)}{3} + \frac{1(300000)}{3} \tag{2.45}$$

$$= 200000$$

It should be noted that there are a number of ways of expressing the molecular weight and number averages, all of which are equivalent. $\overline{M}_n$ is a "colligative" property which is dependent on the number of particles present. $\overline{M}_n$ is independent of molecular size and is highly sensitive to small molecules present. Of course the ratio of $\overline{M}_w / \overline{M}_n$ is the polydispersity (PD) of the sample and is a measure of the narrowness/broadness of the distribution of the polymer molecules.

There are additional molecular weight averages which are used less frequently to characterize a polymer. $\overline{M}_z$ is the third moment or power average and is given by:

$$\overline{M}_z = \frac{\sum\limits_{x=1}^{\infty} N_x M_x^{3}}{\sum\limits_{x=1}^{\infty} N_x M_x^{2}} = \frac{\sum\limits_{i=1}^{\infty} N_i M_i^{3}}{\sum\limits_{i=1}^{\infty} N_i M_i^{2}} \qquad (2.46)$$

and

$$\overline{M}_{z+1} = \frac{\sum\limits_{x=1}^{\infty} N_x M_x^{4}}{\sum\limits_{x=1}^{\infty} N_x M_x^{3}} = \frac{\sum\limits_{i=1}^{\infty} N_i M_i^{4}}{\sum\limits_{i=1}^{\infty} N_i M_i^{3}} \qquad (2.47)$$

In the above example $\overline{M}_z$ is:

$$\overline{M}_z = \frac{1/3(1.0 \times 10^5)^3 + 1/3(2.0 \times 10^5)^3 + 1/3(3.0 \times 10^5)^3}{1/3(1.0 \times 10^5)^2 + 1/3(2.0 \times 10^5)^2 + 1/3(3.0 \times 10^5)^2} \qquad (2.48)$$

$$= 257143$$

The molecular weight of a polymer determined from viscosity measurement is M_v and is expressed as:

$$\overline{M}_v = \left[\frac{\sum\limits_{x=1}^{\infty} M_x^{1+a} N_x}{\sum\limits_{x=1}^{\infty} M_x N_x} \right]^{1/a} = \left[\frac{\sum\limits_{x=1}^{\infty} W_x M_x^{a}}{\sum\limits_{x=1}^{\infty} W_x} \right]^{1/a} \qquad (2.49)$$

When the exponent a in the Mark-Houwink equation is equal to 1, the average molecular weight, M_v, is equal to M_w. However, the exponent "a" varies typically from 0.5 to 0.8 and M_v is usually less than M_w.

Table 2.3 summarizes contemporary methods used for molecular weight determinations and the molecular weight ranges in which the techniques should be used. The relationship of the various molecular weight averages is shown in Figure 2.26.

Table 2.3
Typical Molecular Weight Determination Methods

Method	Type of Molecular Weight Average	Applicable Weight Range
Light scattering	$\overline{M}_w$	to ∞
Membrane osmometry	$\overline{M}_n$	2×10^4 to 2×10^6
Vapor phase osmometry	$\overline{M}_n$	to 40000
Electron and X-ray microscopy	$\overline{M}_{n,w,z}$	10^2 to ∞
Isopiestic method (isothermal distillation)	$\overline{M}_n$	to 20000
Ebulliometry	$\overline{M}_n$	to 40000
Cryoscopy	$\overline{M}_n$	to 50000
End-group analysis	$\overline{M}_n$	to 20000
Osmodialysis	$\overline{M}_n$	500-25000
Centrifugation Sedimentation equilibrium	M_z	to ∞
Archibald modification	$M_{z,w}$	to ∞
Trautman's method	M_w	to ∞
Sedimentation velocity	Gives a real M only for monodisperse systems	to ∞

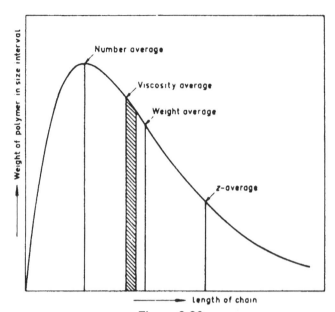

Figure 2.26
Molecular Weight Distributions

2.7   ABBREVIATIONS AND NOTATION

A	absence of order
C	number of components
C_V	heat capacity at constant volume
C_P	heat capacity at constant pressure
E	internal energy of a system
F	degrees of freedom
F	Helmholtz free energy
f_s	fraction of material in the solid phase
f_l	fraction of material in the liquid phase
G	Gibbs free energy
H	enthalpy
k	kinetic constant related to the rate of nucleation and growth

LRO	long range order
LTS	long time stiffness
M_o	molecular weight of polymer repeat units
$\overline{M}_w$	weight average molecular weight
$\overline{M}_n$	number average molecular weight
$\overline{M}_z$	z or power average molecular weight
$\overline{M}_v$	viscosity average molecular weight
n	integer
P	number of phases
P	presence of order
Q	heat absorbed by the system
S	entropy of a system
SRO	short range order
STS	short time stiffness
V_t	specific volume at time t
V_f	specific final volume
V_o	specific initial volume
W	work done on a system
x=DP	degree of polymerization or number of repeat units

2.8   SPECIFIC REFERENCES

1. Herman, K., Gerngross, O., and Abitz, W., *Z. Physik. Chem. B,* **10**, 271(1930).
2. Billmeyer, F.W., *Textbook of Polymer Science,* Interscience Publishers, a Division of John Wiley and Sons, New York, 1962.
3. Schlesinger, W., and Leeper, H.M., *J. Polymer Sci.* **11**, 203(1953).
4. Ranby, B.G., Moorhead, F.F., and Walter, N.M., *J. Polymer Sci.* **44**, 349(1960).
5. Geil, P.H., Symons, N.K.J., and Scott, R.G., *J. Appl. Phys.* **30**, 1516(1959).

6. Geil, P.H., *J. Polymer Sci.,* **44,** 449(1960).
7. Manley, R. St.J., *J. Polymer Sci.,* **47,** 509(1960).
8. Manley, R. St.J., *Nature* **189,**390(1961).
9. Ranby, B.G. and Noe, R.W., *J. Polymer Sci.* **51,** 337(1961).
10. Reneker, D.H. and Geil, P.H., *J. Appl. Phys.* **31,** 1916(1960).
11. Price, F.P. in *Growth and Perfection of Crystals,* R.H. Doremus, B.W. Roberts, and D. Turnbull, Eds., John Wiley and Sons, New York, 1958.
12. Avrami, M., *J. Chem. Phys.* **7,** 1103(1939): **8,** 212(1940).
13. Hoesman, R., *Polymer* **3,** 349(1962).
14. Natta, G., and Corradini, P., *J. Polymer Sci.* **20,** 251(1966).
15. Gaylord, N.G. and Mark, H., *Linear and Stereoregular Addition Polymers,* Interscience Publishers, John Wiley and Sons, New York, 1959.
16. Reinitzer, F., *Monatsh. Chem.* **9,** 421(1888).
17. Lehman, O., *Z. Phys. Chem.* **4,** 462(1889).
18. Finkelmann, H., in *Polymer Liquid Crystals,* Ciferri, A., Krigbaum, W.R., and Meyer, R.B., Eds., Academic Press, New York, 1982.
19. Ringsdorf, H. and Schneller, A., *Br. Polym. J.,* **13,**43(1981).
20. Platé, N.A. and Shibaev, V.P., *Comb-Shaped Polymers and Liquid Crystals,* Plenum Press, New York, 1987. Elsevier Scientific Publishing, Amsterdam, The Netherlands.
21. Seymour, R.B. and Carraher, Jr., C.E., *Polymer Chemistry,* Marcel Dekker, Inc., New York, 1987.

2.9   GENERAL REFERENCES

1. Coleman, M., Graf, J.F., and Painter, P.C., *Specific Interactions and the Miscibility of Polymer Blends*, Technomic Publishing Co., Lancaster, PA, 1991.
2. DiBenedetto, A.T., *The Structure and Properties of Materials*, McGraw-Hill, Inc., New York, N.Y., 1967.
3. Kingery, W. D., Bowen, H.K., Uhlmann, D.R., *Introduction to Ceramics*, 2nd Ed., John Wiley & Sons, New York, N.Y., 1976.

4. Moffatt, W.G., Pearsall, G.W., and Wulff, J., *Structure and Properties of Materials*, Vol. 1, Structure, John Wiley & Sons, New York, 1966.

5. Morawetz, H., *Polymers, the Origins and Growth of a Science,* John Wiley and Sons, New York, 1985.

6. Pauling, L., *Nature of the Chemical Bond*, 3d ed., Cornell University Press, Ithaca, N.Y., 1960.

7. Reed-Hill, R.E., *Physical Metallurgy Principles*, 2nd Ed., D. Van Nostrand Company, New York, N.Y., 1973.

8. Seymour, R.B., Carraher, C.E. Jr., *Polymer Chemistry*, Marcel Dekker, Inc., New York, N.Y., 1988.

9. Van Krevelen, D.W., *Properties of Polymers*, Elsevier, New York, N.Y., 1990.

10. Van Vlack, L.H., *Physical Ceramics for Engineers*, Addison-Wesley Publishing Co., Reading, Massachusetts, 1964.

11. Zachariasen, W.H., "The Atomic Arrangement in Glass", *J. Am. Chem. Soc.*, **54**, 3941(1932).

CHAPTER 3

Polymer Orientation and Rubber Elasticity

3.1   ORIENTATION IN POLYMERS

In Chapters 1 and 2 we have seen the different morphologies that materials can exhibit due to their structural configurations. References were also made as to the effects of the configurations on the use properties. We now consider further the effects of orientation and how it is measured.

Polymer molecules are anisotropic because of their long length. If all the chains are aligned to form oriented or extended chain crystal structures, we have perfect orientation. In most cases, however, the many crystallites and their orientation, relative to each other, may be random with the net result being the chain axes on the whole are randomly oriented. To orient the chains, drawing or stretching the polymer sample is carried out which usually results in the chain ends being oriented primarily in the direction of the draw. Most of this orientation results from orientation of the crystallites that already exists.

Drawing can result in some changes in crystallinity and orientation of the amorphous regions as well. It may at first seem contradictory to talk about orientation of amorphous regions in polymers but this can be considered as ordering on a two dimensional scale. This is similar to the ordering of a cross-linked elastomer by stretching which we will consider in more detail in the following sections.

If an amorphous polymer or cross-linked elastomer is cooled below its glass transition temperature and maintained there,

Table 3.1

Techniques Employed to Study and Characterize Orientation in Polymers

Short Range Interactions	Long-Range Interactions
Birefringence	Electron Diffraction
Raman Scattering	Wide Angle X-ray Scattering
Depolarized Light Scattering	Electron Microscopy
Rayleigh Scattering	Density
Bruillouin Scattering	Small Angle Neutron Scattering
NMR Relaxation	
Small Angle X-ray Scattering	

any ordering that occurred will be permanent unless it is again warmed above its T_g. Partial ordering and preservation of ordered regions occurs during molding or drawing operations.

3.2   DETECTION OF MOLECULAR ORIENTATION IN POLYMERS

There are a number of experimental tools that are used in detecting orientation and characterizing amorphous and crystalline regions in polymers. In these complex materials, the moments of distribution of the structural units must be determined for understanding the properties of the polymer. The popular techniques are listed in Table 3.1.

As examples of the type of information that might be obtained we will consider the use of X-ray diffraction and birefringence as techniques of detecting orientation. The material being examined can be in the form of a film or a fiber. Metal, ceramic, and polymer samples can be in the form of powders or films.

3.2.1   Detection of Orientation with X-rays

When X-rays are reflected, not from an array of atoms arranged in a solitary plane, but from atoms on a number of equally spaced parallel planes, such as exist in crystals, constructive interference occurs under restricted conditions. The law that governs this phenomenon is known as Bragg's Law and is represented

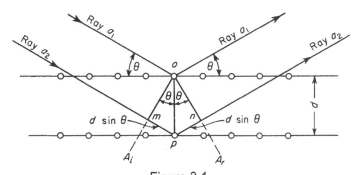

Figure 3.1
Illustration of Bragg's Law

in Figure 3.1.

The lattice spacing, or distance between planes, is d. The lines oA_i and oA_r are the wave fronts of incidence and reflection, respectively. Braggs Law is given by:

$$n\lambda = 2d\ \sin\theta \qquad (3.1)$$

where:

$n = 1, 2, 3,\ldots$

λ = wavelength of radiation in $\mathring{A}$

d = interplanar distance $\mathring{A}$

θ = angle of incidence of reflection of X-ray beam

Bragg's Law is satisfied if the distance *mpn* equals a multiple of a complete wavelength, i.e., it must equal λ or 2λ or 3λ, etc..

From the angles of the various diffraction peaks the type of unit cell and its dimensions can be determined. If the crystals in the sample are randomly oriented, diffraction rings on a photographic record will be observed. If the crystals are preferentially oriented by drawing the fiber or film, the rings will become arcs. The width of these arcs indicates the degree of orientation. This is illustrated in Figure 3.2. For perfect orientation the arcs become spots.

Amorphous materials do not diffract X-rays sharply as crystalline ones do, but some diffraction occurs even for liquids. The diffraction pattern is a diffuse halo in place of a sharp ring.

Figure 3.2
X-ray Diffraction Patterns as a Result of Orientation
of a Crystalline Material

This pattern does not change on stretching even though the molecules have become oriented (two dimensionally). This is illustrated in Figure 3.3.

3.2.2   Detection of Orientation with Birefringence

Birefringence refers to anisotropy of the refractive index of a material. This simply means that light does not travel with the same velocity in all directions. The refractive index of a material is determined by the electronic structure of its molecules. The long chain polymer molecules account for their anisotropic polarizabilities. Consequently, the refractive index along the chain axes is quite different from that perpendicular to the axes as illustrated in Figure 3.4 for perfectly oriented chains.

If we consider a film which is not oriented perfectly, the anisotropy in n can be measured quantitatively by techniques which make use of cross polarizers. One can find the refractive index along the sample axis, n_a, and the refractive index vertical to

Figure 3.3
X-ray Diffraction Patterns as a Result of Orientation
of an Amorphous Material

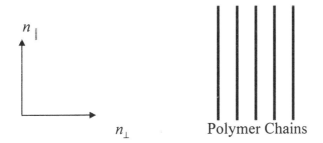

Figure 3.4
Effects on Birefringence as a Result of Polymer Orientation

the sample axis, n_v. The fractional orientation is then:

$$f = \frac{n_a - n_v}{n_\parallel - n_\perp} \qquad (3.2)$$

where:

$n_\parallel$ = refractive index along the chain axis

$n_\perp$ = refractive index perpendicular to the chain axis

For perfect alignment, f is 1 and for perfectly random orientation, $f = 0$.

3.3   EFFECT OF MOLECULAR ORIENTATION ON MECHANICAL PROPERTIES

Nearly all fibers and films are drawn or stretched to develop molecular orientation and enhance mechanical properties. Fibers are drawn continuously by passing the fiber over two sets of rolls which rotate at different speeds. The draw ratio is typically around five for commercial fibers.

This results in a reduction of cross sectional area for fibers and thickness for films. In the case of films, two sets of rollers operating at different speeds are used and clamps which stretch the sheet in the machine or running direction and simultaneously in the

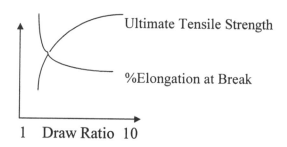

Figure 3.5
Effect of Draw Ratio on Mechanical Properties

transverse direction. This operation is known as biaxial orientation and produces a film strong in both directions. Temperatures during drawing or stretching operations are precisely and uniformly controlled to maintain exact thickness or diameter. Subsequent cooling or annealing operations are also closely controlled to minimize shrinkage due to the polymers' "memory", particularly when the product will be used above its T_g.

As draw ratios increase the initial modulus increases, the ultimate tensile strength increase and the percent elongation at break decrease. This is illustrated in Figure 3.5.

Beyond some maximum draw ratio, the fiber breaks. The effects are so dramatic that most undrawn fibers are useless because of poor strength. Usually some level of trade-off between an ultimate tensile strength and a desirable amount of elongation is imposed. As the draw ratio increases, the amount of orientation also increases to an upper limit as illustrated in Figure 3.6.

3.4   RUBBER ELASTICITY

This section is a prerequisite to understanding and interpreting many property estimations and providing the groundwork for detailed discussions of the mechanical properties of rubbers/elastomers. Polymers that display a high degree of rubber like elasticity consist of long chain molecules as discussed in the previous sections. In these types of materials the elastic behavior results directly from a decrease in entropy associated with the distortion of a chain molecule from its most probable confor-

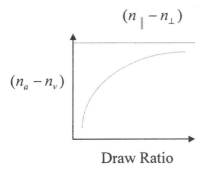

Figure 3.6
Effect of Draw Ratio on Fiber Orientation

mation. The types of materials that display these characteristics are amorphous materials whose usage is confined to temperatures above their T_g's.

In order to provide this elastic or recovery behavior the mobility of chains in elastomers must be low. This is accomplished by controlling a network of primary bonds or crosslinks between the polymer chains. If there are too few crosslinks and they are not spaced properly, desirable properties will not be attained.

The polymers generally consist of chains with low cohesive energies which are largely hydrocarbon or contain polar groups distributed randomly along the chain. The requirement of sites for crosslinking often leads to the choice of a diene as a monomer or comonomer for an elastomer.

A stretched elastomer should have a high tensile strength and modulus similar to crystalline plastics. Thus, rubbers in which crystallinity can develop on stretching such as natural rubber and its stereoregular counterparts, usually have more desirable properties than those with less regular structures. However, the addition of a reinforcing filler can improve properties and result in behavior similar to that obtained on the development of crystallinity. An example of this is reinforced styrene-butadiene rubber whose properties are nearly equivalent to natural rubber.

These materials stretch rapidly and exhibit elongations from 500-1000% with little loss of energy as heat. At these elongations they are expected to show excellent rebound with little permanent set.

3.4.1 Thermodynamics

The basic thermodynamic concepts were discussed in Chapter 2 and will be extended here to account for rubber elasticity.

When a sample of rubber is stretched, work is done on it and its free energy and internal energy are changed. By restricting the development to stretching in one direction, for simplicity, the elastic work W_{el} equals $f\,dl$, where f is the retractive force and dl the change in length. From Chapter 2:

$$\Delta F = \Delta H - T\Delta S \tag{3.3}$$

and
$$\Delta E = Q - W \tag{3.4}$$

For a reversible system, we have seen that:

$$dS = \frac{dQ}{T} \tag{3.5}$$

The retractive force is given by:

$$f = \left(\frac{\partial F}{\partial l}\right)_{T,P} = \left(\frac{\partial H}{\partial l}\right)_{T,P} - T\left(\frac{\partial S}{\partial l}\right)_{T,P} \tag{3.6}$$

For an ideal elastomer:

$$\left(\frac{\partial H}{\partial l}\right)_{T,P} = 0;\ f = -T\left(\frac{\partial S}{\partial l}\right)_{T,P} \tag{3.7}$$

Now the work done on stretching a rubber with force f and distance dl is:

$$dW = PdV - fdl \tag{3.8}$$

Differentiating and substituting:

$$dE = TdS - PdV + fdl \qquad (3.9)$$

Making the assumption of constant volume:

$$f = \left(\frac{\partial E}{\partial l}\right)_{T,V} - T\left(\frac{\partial S}{\partial l}\right)_{T,V} \qquad (3.10)$$

but $\qquad \left(\frac{\partial f}{\partial T}\right)_{P,\alpha} = -\left(\frac{\partial S}{\partial l}\right)_{T,V} \qquad (3.11)$

$$\therefore f = \left(\frac{\partial E}{\partial l}\right)_{T,V} + T\left(\frac{\partial f}{\partial T}\right)_{P,\alpha} \qquad (3.12)$$

Equation (3.12) is the "thermodynamic equation of state" for rubber, where $\alpha = \dfrac{l}{l_o}$ the ratio of length to unstretched length at T.

We have seen from Equation (2.34) in Chapter 2:

$$\frac{C_P}{T}\partial T = \partial S|_P \qquad (3.13)$$

and Equation (3.11) becomes:

$$\left(\frac{\partial f}{\partial T}\right)_{P,\alpha} = -\frac{C_P}{T}\left(\frac{\partial T}{\partial l}\right)_{P,adiabatic} \qquad (3.14)$$

or $\qquad \left(\frac{\partial T}{\partial f}\right)_{P,adiabatic} = -\frac{T}{C_P}\left(\frac{\partial l}{\partial T}\right)_{P,f} \qquad (3.15)$

which illustrates that rubbers contract on heating rather than expanding.

Changes in internal energy for a system are identified with changes in temperature and energy stored by bond bending and stretching. Changes in ΔS are due to differences in the polymers' conformations. Therefore, the concept of entropy as a measure of probability is useful. When a specimen is stretched, the polymer has the lowest entropy and will seek to maximize the entropy which thermodynamically is responsible for recovery.

The ideal rubber should respond to an external stress only by uncoiling, i.e., $\dfrac{\partial E}{\partial l} = 0$. Studies in which f, l, and T are varied confirm that for many amorphous, cross-linked materials above T_g, $\dfrac{\partial E}{\partial l}$ is very small. Below T_g, segments cannot deform in the time scale in which T_g was measured, therefore $\dfrac{\partial S}{\partial l} = 0$. In a glass, the external stress will result in bond bending and breaking. However, in a real system, stretching of a rubber involves overcoming energy barriers in polymer chains and the internal energy may be expected to depend slightly on the elongation. The contributions of the energy and entropy terms are shown in Figure 3.7 [S1,S2].

The entropy of a system in a given state (Ω_2) with respect to a reference state (Ω_1) is related to the ratio of the probability of that state (Ω_2) to that in the reference state (Ω_1) by:

$$\Delta S = k \ln \frac{W_2}{W_1} \qquad (3.16)$$

or $\qquad\qquad \Delta S = k \ln W \qquad\qquad (3.17)$

where k is the Boltzmann constant. If we stretch a piece of rubber whose polymer segments are part of a loose network, we move the ends of the segment to new positions as we do the entire specimen. Since these chains are part of a network, the chain ends cannot di-

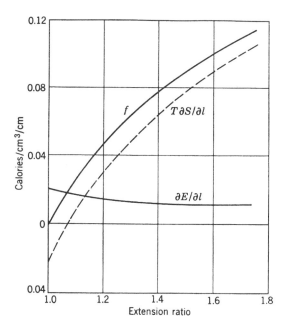

Figure 3.7 [S1]
Contributions of Energy and Entropy to the Retractive Force
in Natural Rubber

ffuse to more probable positions. The rubber can only relax if the external forces are removed. The probability of locating a polymer segment, $W(x,y,z)$, with one end fixed at the origin of a coordinate system and the other end at (x,y,z) is:

$$W(x,y,z) = \left(\frac{b^2}{p}\right)^{3/2} \exp\left[-b^2\left(x^2 + y^2 + z^2\right)\right]$$ (3.18)

where $b^2 = 3/2\ nl^2$ and n is the number of bonds of length l. This is limited to small deformations. This system is illustrated in Figure 3.8 [S3].

    A simple but useful model is one where the polymer chains can be represented as being freely jointed between atoms constituting the chain. If the chain has n links of length a, the average distance traveled, $\bar{r}_o$ in flight of any number of links will

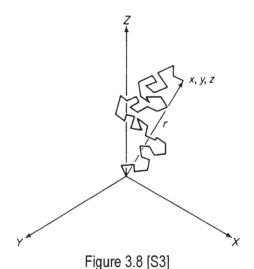

Figure 3.8 [S3]
Representation of Random Flight Movement of a Polymer Chain End
on the Cartesian Coordinate Axes

be zero since all starting directions are equally probable. However, the average square of the distance traveled $\bar{r}_o^2$ is not zero but increase linearly with the number of links.

$$\bar{r}_o^2 = nl^2 \tag{3.19}$$

A more realistic approach includes a constant bond angle characteristic of any chemical combination. If θ' is the bond angle ($109.5°$ for tetrahedral carbon) and $\theta = 180° - \theta'$ as shown in Figure 3.9 [S3].

From geometric considerations we know that:

$$r^2 = x^2 + y^2 + z^2 \tag{3.20}$$

Equation (3.19) indicates that $\bar{r}_o^2$ is the most probable value of r^2 even though the most probable value of r, as well as that of x, y, z, is zero. For a change in the overall sample dimensions of α so that: $x_2 = \alpha_x x$, $y = \alpha_y y$, and $z_2 = a_z z$, the ratio of the probabilities of the two states is:

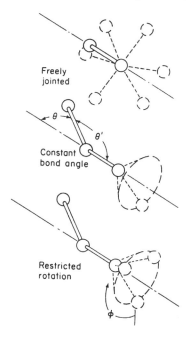

Figure 3.9 [S3]
Possible Polymer Chain Conformations

$$\ln\frac{\Omega_2}{\Omega_1} = -b^2\left[\left(\alpha_x^2-1\right)x^2 + \left(\alpha_y^2-1\right)y^2 + \left(\alpha_z^2-1\right)z^2\right] \tag{3.21}$$

The probability of a particular molecule i being contained in a particular volume is:

$$\Omega_i = \frac{V}{V_o} \tag{3.22}$$

and for a system containing υ molecules the probability is:

$$\Omega = \left(\frac{V}{V_o}\right)^{\upsilon} \tag{3.23}$$

Initially chain ends are randomly oriented in all directions, so x^2, y^2, and z^2 can each be replaced by their average value, $\bar{r}_o^2 / 3$:

$$\therefore \Delta S = S_2 - S_1 = k \ln \frac{\Omega_2}{\Omega_1} = -k \frac{3}{2nl^2} \frac{\bar{r}_o^2}{3} \left(\alpha_x^2 + \alpha_y^2 + \alpha_z^2 - 3 \right) \quad (3.24)$$

$$\text{or } \Delta S = -\frac{k}{2} \left(\alpha_x^2 + \alpha_y^2 + \alpha_z^2 - 3 \right) \quad (3.25)$$

For the total number of effective chains, υ_c,

$$\Delta S = -\frac{k}{2} \upsilon_c \left(\alpha_x^2 + \alpha_y^2 + \alpha_z^2 - 3 \right) \quad (3.26)$$

If the total volume remains constant:

$$\alpha_x \alpha_y \alpha_z = 1 \quad (3.27)$$

$$\alpha_x = \frac{1}{\alpha_y \alpha_z} \quad (3.28)$$

For uniaxial stretching, $\alpha_y = \alpha_z$

$$\therefore \alpha_x = \alpha = \frac{1}{\alpha_y^2} \quad (3.29)$$

$$\Delta S = -\frac{k\upsilon_c}{2} \left[\alpha^2 + \frac{2}{\alpha} - 3 \right] \quad (3.30)$$

A change in length l is:

$$l = x\alpha_x \text{ and } dl = x d\alpha_x \quad (3.31)$$

and:

$$\left(\frac{\partial S}{\partial l}\right)_{T,P} = \left(\frac{\partial \Delta S}{\partial l}\right)_{T,P} = \left(\frac{\partial \Delta S}{\partial \alpha}\right)_{T,P} \frac{1}{x} \tag{3.32}$$

$$\frac{\partial \Delta S}{\partial \alpha} = \frac{k\upsilon_c}{2}\left(2\alpha - \frac{2}{\alpha^2}\right) = -k\upsilon_c\left(\alpha - \frac{1}{\alpha^2}\right) \tag{3.33}$$

Then
$$f = \frac{Tk\upsilon_c}{x}\left(\alpha - \frac{1}{\alpha^2}\right) \tag{3.34}$$

If the retractive force per unit area is:

$$\sigma = \frac{f}{A_o} \tag{3.35}$$

where A_o is the original cross sectional area, and the original volume is V_o, where $V_o = l_o A_o$, then:

$$\sigma = \frac{kT\upsilon_c}{V_o}\left[\alpha_x - \left(\frac{1}{\alpha_x}\right)^2\right] = RTN\left(\alpha - \frac{1}{\alpha^2}\right) \tag{3.36}$$

where R is the gas constant per mole and N is the number of moles of polymer chains per unit volume.

Equation 3.36 has been confirmed for rubbery materials when viscous effects are negligible, e.g. by swelling to a new volume after immersion in a low molecular weight solvent. We must also account for polymer segments at chain ends that do not contribute to the modulus since they are not restrained at one end. The chain density ends per unit volume, $2\rho/M_n$ must be subtracted from the chain density, N. Deviation from Gaussian statistics at large deformations can be accounted for with an empirical constant C. This gives the "Mooney Rivlin" form of Equation (3.36).

The values of σ and α are based on the original unswollen cross section and the ratio of the original length to swollen, unstretched length, respectively.

$$\sigma = \left[\left(N - \frac{2\rho}{M_n} \right) RT + \frac{2C_2}{\alpha} \left(\alpha - \frac{1}{\alpha^2} \right) \left(\frac{1}{v_2} \right) \right]^{1/3}$$

(3.37)

and:

$$v_2 = \frac{V_o}{V_s}$$

(3.38)

3.5   ABBREVIATIONS AND NOTATION

b	constant
d	interplanar distance
k	Boltzmann's constant
l	length
n	integer
N	number of moles of polymer chains per unit volume
Q	heat
R	gas constant
T	temperature, Kelvin
v	ratio of original volume to solvent swollen volume
V	volume
W	work
A_o	original cross sectional area
α	thermal expansion coefficient
α	change in a polymer's dimensions
α	ratio of original polymer length to swollen polymer
λ	wavelength of radiation
θ	angle of incidence or reflection
n_a	refractive index along sample axis
n_v	refractive index vertical to sample axis
ΔF	free energy change
ΔH	enthalpy change
ΔS	entropy change

ΔE	internal energy change
T_g	glass transition temperature
Ω	probability of a given polymer state
$\bar{r}_o$	average distance traveled
$\bar{r}_o^2$	mean radius of gyration
M_n	number of average molecular weight
υ_c	number of effective chains
ρ	density
V_o	original volume
σ	force per unit area

3.6   SPECIFIC REFERENCES

1. Billmeyer, F.W., *Textbook of Polymer Science,* Courtesy of Interscience Publishers, a Division of John Wiley and Sons, New York, 1962.
2. Treloar, L.R.G., *The Physics of Rubber Elasticity*, 2nd ed., Clarendon Press, Oxford, 1958.
3. Rodriguez, F., *Principles of Polymer Systems*, McGraw-Hill Book Company, New York, 1970.

3.7   GENERAL REFERENCES

1. Billmeyer, F.W., *Textbook of Polymer Science,* Courtesy of Interscience Publishers, a Division of John Wiley and Sons, New York, 1962.
2. DiBenedetto, A.T., *The Structure and Properties of Materials*, McGraw-Hill, Inc., New York, N.Y., 1967.
3. Moffatt, W.G., Pearsall, G.W., and Wulff, J., *Structure and Properties of Materials*, Vol. 1, Structure, John Wiley & Sons, New York, 1966.
4. Morawetz, H., *Polymers, the Origins and Growth of a Science,* John Wiley and Sons, New York, 1985.

5. Pauling, L., *Nature of the Chemical Bond*, 3d ed., Cornell University Press, Ithaca, N.Y., 1960.
6. Seymour, R.B., Carraher, C.E. Jr., *Polymer Chemistry*, Marcel Dekker, Inc., New York, N.Y., 1988.
7. Van Krevelen, D.W., *Properties of Polymers*, Elsevier, New York, N.Y., 1990.

CHAPTER 4

Effective Use of Spreadsheets for Scientific Applications and Estimation of Materials Properties and Miscibility

A Guide to Using MATPROP$^©$

4.1 INTRODUCTION AND BACKGROUND

The spreadsheet has evolved from a simple concept of providing a means for carrying out repetitive calculations while changing the values of one or more variables and comparing the results produced from each variable change. Present day spreadsheets have perfected this approach and added statistical and mathematical modeling as well as elaborate graphics. Current spreadsheets have been modified to be more "user friendly" with many cosmetic enhancements.

The spreadsheet concept was first embodied in commercially obtainable software for an Apple II in 1979 in a program called Visicalc. This concept is very powerful and at the same time it is so simple that the authors, not realizing the potential of this software, did not fully protect it. Subsequently, it was copied and developed by a number of software companies, of which the most prominent of today are: Lotus 1-2-3 (Lotus Development Corp., now IBM), Excell (Microsoft Corporation), and Quattro® pro (originally developed by Borland), and later acquired by Novell and currently Corel.

These spreadsheets are available in DOS, MacIntosh, and the Windows™ (Microsoft Corporation) operating systems. When the spreadsheets were originally introduced, the operating systems did not have graphic interfaces nor the features of the spreadsheets of today. DOS, as the major system used to manipulate and communicate with IBM based PC's has been declining and will continue to do so until it is completely replaced with a graphic interface similar to the MacIntosh system. The graphic PC systems will use Microsoft Windows™ or another system as yet to be introduced and then accepted.

Originally, there were no "mice" and the menu was selected with the backslash key which attained the name "hot key". The term hot key stuck and this and other keys which are programmed to provide instant access to a feature or features which would normally require multiple key entries are still referred to in this manner. The backslash key brought up a menu whose choice was selected with the arrow and entry keys.

The memories of the Apple IIe (1979), MacIntosh (1983), and the IBM PC (1981) were severely limited and programs had to be small to run in their specified environments. The MS DOS (Microsoft Disk Operating System) was and is severely limited in its ability to handle larger memory requirements because of its basic design although at the time it was developed, it was quite adequate for programs then available. The memory requirements imposed by the graphic interfaces of today were not considered. In 1978 the Intel Corporation introduced a microprocessor capable of transferring 16 bits of data at a time and was capable of accessing one megabyte of memory. However, DOS was and is still limited to 640K of memory for application programs. The remaining 384K is generally reserved for hardware and operating systems use. DOS has subsequently been improved and this factor along with a "memory manager" allows the 384K or "upper memory" to be used and accessed in a more or less transparent manner to the user. The point is that these factors often result in less than optimum speed and performance since the program has to access instructions and data either from additional or "extended memory" or from the disk.

The basic spreadsheet consists of numbered rows and alphabetically listed columns or fields which define each cell by location. Examples of these might be cell A1, cell B3, etc.

Numerical cell entries can be thought of as vectors which can be multiplied, divided, summed, etc. to obtain whatever total is of interest. When a cell, e.g., A1 contains the entry "CH", e.g., the spreadsheet recognizes this entry as text and this cell cannot be numerically manipulated.

When a number is entered the spreadsheet recognizes this as an entry which can be numerically added, subtracted, raised to a power, or in other ways mathematically changed. Numerical operations are carried out using the standard operators familiar to users of Basic and FORTRAN programming languages, i.e., add(+), subtract(−), multiply(*), divide(/), etc. More complex operations begin with the @ symbol which tells the spreadsheet that an internally defined function is being invoked.

In the example of Table 4.1, cell C6 would contain the formula @sum(C2..C5); cell D6 contains the formula @sum(D2..D5); and cell E6 contains the formula C6/D6.

Table 4.1
Illustration of a Basic Spreadsheet

	A	B	C	D	E
1	Groups	No. of Groups	Group Wt.	Group Vol.	Density
2	CH	1	13.02	9.85	
3	CH_2	1	14.03	16.45	
4	CH_3	1	15.03	22.8	
5	Totals		42.08	49.10	
6					0.857

Cell contents of C5

@Sum(C2..C5)

Cell contents of D5

@Sum(D2..D5)

Cell contents of E6

+C5/+D5

Originally, spreadsheets were limited to a 256 x 256 matrix. This size has now expanded to 8192 x IV(236) with many other features added. As the size of spreadsheets was increased, it became necessary to provide more computer memory. Consequently, before the 80286 processor was introduced, Lotus, Intel, and Microsoft came up with the Expanded Memory Specification (EMS), also known as the LIM 3.2 specification. Windows™ 3.1 uses the DOS EMM386 utility to create expanded memory which is also applicable to older DOS programs that require EMS. This eliminates the need for special memory boards.

With the introduction of newer processors, viz., the 80286, the 80386, and the 80486 or higher, the protected mode operation of these processors are turned on via DOS by loading HIMEM.SYS at boot up from the CONFIG.SYS file. HIMEM is an extended memory manager that enables DOS to address memory above the conventional 1 megabyte limitation.

Windows has allowed the spreadsheet programs to take advantage of cut and paste to and from other applications, better graphics and a myriad of new features. Examples of these are statistical add "ons" such as regression analysis of data, data modeling, and optimization, among others. Optimization techniques employ linear programming techniques as the basis for the calculations. Each company offering spreadsheets will list individual features and their operation in the user guide and/or manual that accompanies the spreadsheet.

While it is true that the many editing features, drag and drop, e.g., viewing features such as zooming and tool bar usage, have greatly enhanced the ease of operation and use of the spreadsheet, the basic concepts and purposes remain much the same as they were when spreadsheets were first introduced. Software reviews are prone to "hype" new features, many of which are only small incremental improvements in appearance and result in little or no substantive improvements in overall performance of the program. On the contrary these "improvements" often require much greater hard disk memory and ultimately require newer more powerful hardware.

As competition for the market increased, the features contained in the programs greatly increased but contrary to what we might expect, the prices have dropped dramatically. The prices went from $300-$400 in 1990 to less than $50 today. However, to make up the difference in production costs and profits, free telephone technical support was discontinued and many manuals are now sub-standard in their explanations of the features and the detail that is presented to help the user. In most cases, manuals do not accompany the program and must be purchased separately.

On the upside, more details and explanation of the software operation are placed in the help menus. For many users, familiar with older versions, this is sufficient. But for many others this simply is not sufficient guidance and tutoring that leads to proficiency. Therefore, we have seen a plethora of second and third party products aimed at teaching people how to use the products for which the supplier should have furnished such information. In fact, it is not unusual for the user to pay as much or more for supplemental books and videos in order to use the software product. Perhaps the software producers have missed an opportunity to increase their profits and customer good-will.

4.2 BUSINESS VS. SCIENTIFIC

Primary focus has been on the use of the spreadsheet for business applications. This was to be expected since this was the use that was first visualized by the developers. All you had to do was change a single production cost, distribution cost, overhead cost, selling price, volume, or some other variable to obtain an immediate readout of how this affected a profit and loss statement or balance sheet on a month by month basis.

The scientific community very soon began to pick up on the native power inherent in the spreadsheet and began using it for scientific applications. The attributes of the format that attracts scientist/engineers are:

1. Easy and comprehensive viewing of entries and calculated results
2. Easy change in entry additions that do not require programming and compiling

3. Inherent ability to link calculations by specifying appropriate cell entries
4. Use of scientific formats including exponential, transcendental, and other functions required for scientific applications
5. Standard graphic and printing capabilities
6. Development of multipage or notebook type spreadsheet formats which allow easy access to related data

Even early features led to a symposium on Scientific Applications of Spreadsheets, sponsored by the AIChE at the Spring Meeting in Houston, Texas, April 1989. Applications illustrating the use for kinetics, process design, property estimation, and other applications were presented.

4.3   MATPROP© ESSENTIALS
What is MATPROP©?

MATPROP© is Windows™ based software program that consists of five related but separate modules as follows:
1. A plastics property database
2. A rubber/elastomers property database
3. A general material properties database
4. A solubility parameter database
5. A program that estimates the properties of polymers
6. A program that estimates the miscibility of polymers

A spreadsheet based format was chosen for MATPROP© for the reasons mentioned above but with a number of modifications to provide a simpler look and greater appeal. However, it should be noted that MATPROP© is not a spreadsheet and was not designed to provide the attributes and versatility of a spreadsheet since it is an application program that contains only the specific feature enumerated above.

4.4   MATPROP© COMPONENTS
When MATPROP© is started a screen similar to that shown

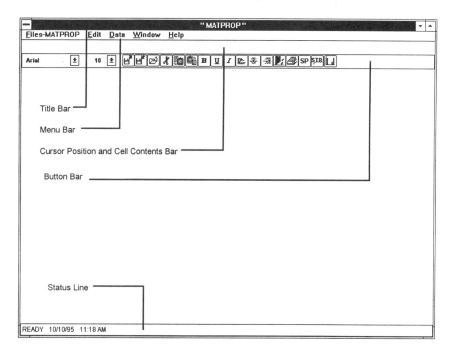

Figure 4.1
MATPROP© Components

in Figure 4.1 will open.

Each of the six separate modules, mentioned above, is a workbook that contains one or more worksheets. Access to these modules and their use is available through:

- The Menu Bar
  The menu bar offers the main level of commands. It uses pull down menus as in standard Window™ applications.
- The Cursor Position and Cell Contents Bar
  The cursor position and cell contents bar show the physical location (vector position) of the cursor. In some worksheets, entries can be made either on the worksheet itself, which is the simplest technique, or from this bar as is done in a standard spreadsheet.
- The Button Bar

READY 11/17/94 12:10 PM		Open workbook file
READY 11/17/94 12:15 PM		Save current workbook file
READY 11/17/94 12:16 PM		Close current workbook
READY 11/17/94 12:20 PM		Move the selected text to the clipboard
READY 11/17/94 12:21 PM		Copy the selected text to the clipboard
READY 11/23/94 10:54 AM		Paste data from clipboard to the selected range
READY 11/23/94 10:57 AM		Turn bold font on or off
READY 11/23/94 10:58 AM		Turn underlining on or off
READY 11/23/94 10:59 AM		Turn italic fonts on or off
READY 11/23/94 11:01 AM		Left justify text in selected block
READY 11/23/94 11:02 AM		Center text in selected block
READY 11/23/94 11:03 AM		Right justify text in selected block
READY 11/23/94 11:04 AM		Exit application
READY 11/23/94 11:06 AM		Print selected text
READY 11/23/94 11:07 AM		Transfer Solubility Parameters to MPCalc
READY 11/23/94 11:08 AM		Access Structural Graphic Files via Windows Paintbrush
READY 10/10/95 12:33 PM		Graph MPCalc Results

Figure 4.2
Button Bar Buttons and Corresponding Status Line Commands

The button bar contains buttons for often used commands and provides easier and faster access to the applications. In place of browsing through the menus, a command can be executed by one simple mouse click on the appropriate button. Figure 4.2 illustrates the button bar icons and the corresponding commands which are illustrated in the status line when the mouse cursor is placed on a button.

• The Status Bar

The status bar displays information about the current state of the program and any additional information that the user needs to be aware of based on cursor position.

4.5   FILES- MATPROP© MENU

The workbook modules can be opened, closed, saved, etc. by clicking on the icons on the button bar or by using the pull down menu. For many operations the use of the Button Bar will be easier and quicker to carry out the desired command.

The Menu Bar follows standard Windows format so that you should be comfortable with its appearance. The menu for the Files-MATPROP© section of the program is shown in Figure 4.3.

- Open - Displays a listing of MATPROP© databases and programs available (Figure 4.4).
- Close - Closes the current module without saving any work that may have been performed.
- Save - Saves the active file or workbook.
- Save As - Saves the active data or workbook to a different filename than the current one loaded. This command is not available from the Button Bar.
- Get Version - Loads data from alternate workbook data files (*.BDT files). Unlike FileOpen, this command will overwrite your current worksheet or workbook with the *.BDT data file. This command is used for retrieving different scenarios or versions of data of the same model.
- Put Version - Stores the data in alternate workbook data files (*.BDT) (Figure 4.5a).
- File|Put does not save the entire workbook but by default saves only the data entry cell. This is useful for saving different sce-

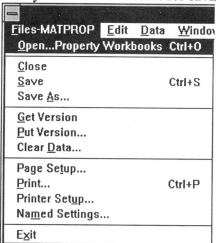

Figure 4.3
Files-MATPROP© Menu

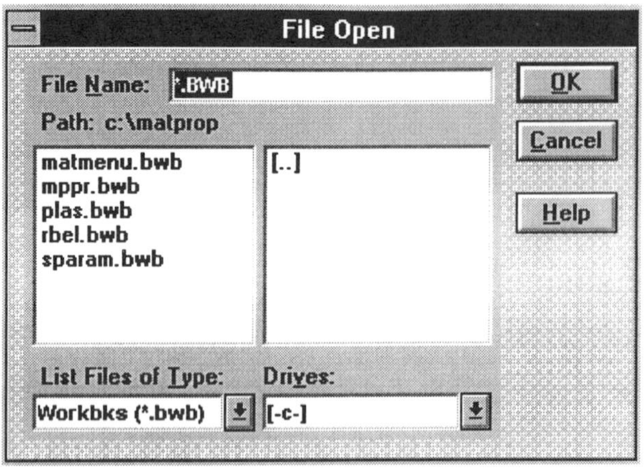

Figure 4.4
MATPROP© Databases and Programs Available

narios or version of data for the same model while minimizing file size and corresponding disk space required. If you assign a filename, a dialog box gives you additional choices (Figure 4.5b). This command is not available from the Button Bar.

- Put Version Options - These additional choices are shown in Figure 4.5b.
- Input cells - Saves only the data entry cells

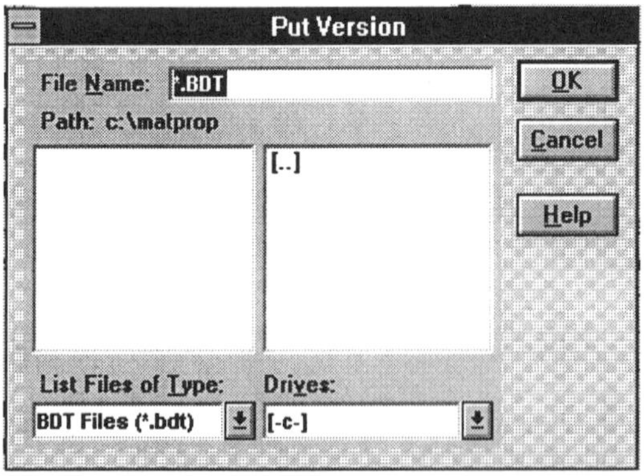

Figure 4.5a
Files-Put Dialog Box

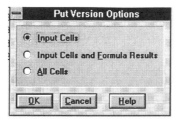

Figure 4.5b
Files Put Options Available

- Input cells and Formula Results - Saves input cell and calculated results
- All Cells - This command seems simple enough but built into this apparent simplicity is the power of MATPROP© to save all data in the entire workbook, even the hidden cells. This allows you to create your own versions of particular interest to you that are not protected and which you can use to modify and change the worksheet data as well as the appearance of the worksheet.
- Clear Data - Clears all data from input cells. This command is useful when you want to clear all entry data in preparation for entering new data. It functions the same as the delete key in deleting data from individual cells. This command is not available from the Button Bar.

  NOTE: If you are working with a database worksheet, choosing Input Cells and Formula Results will save only a blank worksheet(s).

  Although this menu item does not save any data it does provide you with a new worksheet in the typical spreadsheet format that you can use for building your own workbooks/worksheets that are subsets tailored to your own needs and interests.

- Page Setup - This command allow you to set up the layout for a printed page. Activating this command opens up a dialog box as shown below in Figure 4.6. Once you have specified the reg-

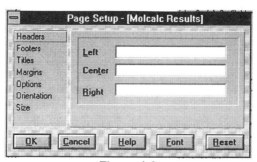

Figure 4.6
Files-Page Setup

ions you wish to print, such as, MATPROP© results, printing is then carried out automatically by clicking on the Print icon. This command is not available from the Button Bar.

- Print - Prints ranges/pages specified or listed under Page Setup.
- Printer Setup - This command opens up a dialog box which allows you to select a printer from those available to your system. Normally it will not be necessary to do this since the printer will have already been chosen from the control panel and available to other Window™ applications.
- Named Settings - This command allows you to create and name several different page setups so that different sections of your workbooks or modules can be printed simply by choosing the appropriate file name. If you choose this command, a dialog box opens up (Figure 4.7). This command is not available from the Button Bar.
- Exit - Exits the MATPROP© applications.

4.5.1   Additional Command Information

Additional information on each of the commands and supplemental information on the dialog boxes can be obtained by consulting the Help menu for a topic of interest. Help on general topics can be accessed from the contents menu or help on many specific topics can be obtained through the Help Search Index.

4.6   MATPROP© START UP

When MATPROP© is installed a MATPROP© program

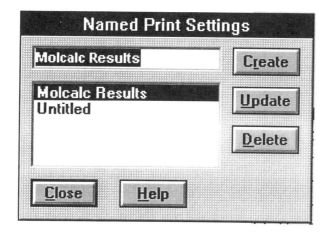

Figure 4.7
Files-Named Settings

group is installed in the Window™ program manager or as a "Matl Prop & DBs" group accessible from the Windows95/98™ Start menu. When MATPROP© is initiated by "double clicking" on the MATPROP© group icon or from the Windows95™ Start, Programs menu, the program initiates by opening the miscibility and property estimation module. No other changes are made to your hard disk or operating system. The opening screen that you will see is illustrated in Figure 4.8.

4.7   WORKING WITH MATPROP©

Data can be entered only in areas that have a white background. All other areas of the modules are protected and you will receive an error signal if data entry is attempted.

Molcalc is a worksheet of the mppr.bwb module that estimates 54 properties of materials. It can be used to estimate the properties of both simple organic compounds as well as more complicated materials such as polymers. In the case of polymers, the repeating structural groups are used to estimate properties. A graphic on screen display can be used to confirm the selection of the proper group for property estimation. This is particularly important in instances where there is ambiguity in the selection of groups based on chemical formulae.

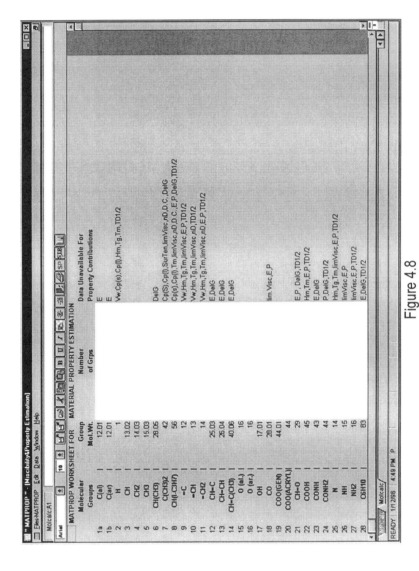

Figure 4.8
MATPROP© Opening Screen

4.8 CHOOSING GROUPS AND ESTIMATING PROPERTIES

The groups available are visible in the mppr.bwb workbook module. You can quickly explore the groups available by using:

- the "down arrow" cursor key
- the "page down" key
- the mouse and scroll bar

Once you have found the group or groups that constitute the structure you wish to estimate, simply enter the number of each group. The estimated properties will be displayed in the Calculated Molecular Properties section of the module.

⇒ Example 1 (Straight Forward Group Selection)
Poly(ethylene) (PE)

—(CH_2CH_2)—

Since the repeat structure of PE consists of two CH_2 groups, the group selection is unambiguous and you enter a two on the input line and the results or Calculated Molecular Properties will be displayed.

Additional Entries - As you scroll down the properties list you will see that there are several cells in the Calculated Molecular Properties section that require additional entries in order to take advantage of all the properties estimated.

Since PE has two groups that constitute the number of main chain groups, enter a "2" here. Enter the molecular weight of interest or the calculated critical molecular weight. For now, enter the calculated MW and accept "N_2" as the gas of interest and "None" as the thermal decomposition group, "T Decom Grp".

The Calculated Molecular Properties section should look like Table 4.2.

4.9 UNAVAILABLE PROPERTIES

Particular attention should be given to the last column of the Molcalc module which lists any property contributions which are not available for that particular group.

Table 4.2 Estimated Poly(ethylene) Properties

		StructuralGroup	
	CALCULATED	Ethylene	
	MOLECULAR	PolymerClass	
	PROPERTIES	Poly(ethylene)	
Mol.Wt.	28.06	g/mol.	
Density Am.	0.853	g/cm3.	
Density Gl.	0.885	g/cm3.	
Density Cr.	0.964	g/cm3.	
VdeWaalsVol	20.46	cm3/mol	
Eg	9.21E-03	cm3/mol*K,for glass,therm expan	
E1	2.05E-02	cm3/mol*K,forliquid,therm expan	
e1	7.29E-04	cm3/g*K,for liq,sp.therm.expan	
eg	3.28E-04	cm3/g*k,for glass,sp.therm expan	
alpha liq. cubic	6.22E-04	1/K,cubic therm.expan., liq.state	
alpha gl. cubic	2.90E-04	I/K, cubic therm.expan., gl.state	
alpha liq. linear	2.07E-04	1/K, linear coeff.liq.therm.expan.	
alpha gl. linear	9.68E-05	1/K, linear coeff., gl.therm.expan	
Specf Cp(s)	1.807	J/g*K, (298 deg. K)	
Specf Cp(l)	2.167	J/g*K, (298 deg. K)	
Hm	8000	J/mol.	
Tg	-120	C	
Tm	133	C	
SolPar(d)	8.022	Based on Vr,Cal1/2cm2/3disper.	
SolPar(p)	0.000	Based on Vr,Cal1/2cm2/3polar	
SolPar(h)	0.000	Based on Vr.Cal1/2cm2/3 H bond.	
SolPrCal/Cm	8.022	Cal1/2cm2/3, based on Vr	
SolParJ/Cm	16.413	J1/2/cm3/2	
SurTension	36.66	erg/cm2	
MChainGrps	2	Enter# of backboneatms/strucuni	
C(constant)	13.1	Ktheta1/2, x M, cm3/g	
K Theta	0.218	Un.Visc.Coeff.,cm3xmol1/2/g3/2	
M(Critical)	3558	critical molecular weight,g	
Refr. Index	1.50	refractive index, nD	
Dielc Cons	2.16	Dielectric Constant,D.C.	
Bulk Mod.(K)	4.57	(K)BulkModulus, GPa	
ShearMod.(G)	0.93	(G)Shear or Rigidity Modulus, GP	
Ten. Mod.(E)	2.62	(E)Tensile or Young'sModulus,GF	
PoissonRatio	0.40	Poisson's Ratio	
SndVel.Long.	2610.6	Longitudinal SoundVelocity, m/s	
SndVel.Shear	1045.7	Transverse Sound Velocity, m/s	
SndAbs,Long.	4.18	Long.SoundAbs.Coeff., dB/cm	
SndAbs,Shr.	20.89	Shear/Trans Snd.AbsCoeff,dB/cm	
ThermalCond	0.20	J/s*m*K	
SolPr(d)Cal	8.326	Based on Vg,cal1/2cm2/3	
SolPr(p)Cal	0.000	Based on Vg. cal1/2cm2/3	
SolPr(h)Cal	0.000	Based on Vg. cal1/2cm2/3	
SolPrCal/Cm	8.326	Cal1/2cm2/3, based on Vg	
E sub eta Inf.	26.8	E,Act.energy of visc flow,g KJ/mc	
M(Actual)	3558	Molecular Wt. @or>M(crit)	
T	298	Temp.of interest,degK	
A	9.17	Empirical Constant	
Log Eta Cr.Inf.	-3.680	Visc.@M(critical),T=inf.,Ns/m2	
Log Eta Cr.	1.04	Visc.@T of interest,Ns/m2	
LogNewtVisc.	1.036	Newt.Visc.@T,N*s/m2	
Gas	N2	Enter Gas of Interest from Table	
Log Perm.	-13.8	P,Coeff.of Perm,cm3*cm/cm2*s*P	
Delta H(A)@298	-44	Free Enthalpy of Poly., kJ/mol	
TempCorrFac(B)	204	K	
Delt Gsub f @T	16.8	Free Energy Change @T, kJ/mol	
TDecompGrp	None	Enter Poly.Grp.for Therm Decomp	
TD1/2	677	Temp.ofHalfDecomp, deg K	

134

For those property contributions that are listed as unavailable, the property estimations for that property will <u>not</u> be accurate. In some cases it may be possible to choose a different combination of groups to synthesize the structural unit from those available in order to obtain an approximation for that particular property of interest. Thus it may be possible to perform two or more estimates in order to approximate the widest possible listing of estimated properties. Table 4.3 shows a comparison of the calculate properties with literature values. In all cases, the comparisons are good.

Table 4.3
Comparison of Calculated Molecular Properties with Literature
Values for Poly(ethylene)

Structural Group	Ethylene	
Polymer Class	Poly(ethylene)	
CALCULATED MOLECULAR PROPERTIES	LITERATURE VALUES	

Properties		Literature Values	
Mol.Wt.	28.06	28.06	g/mol.
Density Am.	0.853	0.85	g/cm3.
Density Gl.	0.885	----	g/cm3.
Density Cr.	0.964	0.96-1.0	g/cm3.
VdeWaalsVol	20.46	20.5	cm3/mol
Eg	9.21E-03	6.7-10.1E-03	cm3/mol*K,glass,therm expan
E1	2.05E-02	2.1-2.69E-02	cm3/mol*K,liq.,therm expan
e1	7.29E-04	7.5-9.6E-04	cm3/g*K,liq,sp.therm.expan
eg	3.28E-04	2.4-3.6	cm3/g*k,for glass,sp.ther expan
alpha liq. cubic	6.22E-04	----	1/K,cubic therm.expan, liq.state
alpha gl. cubic	2.90E-04	----	1/K,cubic therm.expan., gl.state
alpha liq. linear	2.07E-04	----	1/K,linear coeff.liq.thermexpan.
alpha gl. linear	9.68E-05	----	1/K,linear coeff gl.therm.expan.
Specf Cp(s)	1.807	1.55-1.76	J/g*°K,
Specf Cp(l)	2.17	2.26	J/g*°K
Hm	8000	8220	J/mol.
Tg	-120	~ -125°	C
Tm	133	137.5°	C
SolPar(d)	8.022	8.05	Based on Vr,Cal1/2cm2/3disper.
SolPar(p)	0.000	0	Based on Vr,Cal1/2cm2/3polar
SolPar(h)	0.000	0	Basedon Vr.Cal1/2cm2/3Hbond.
SolPrCal/Cm	8.022	8.05	Cal 1/2cm2/3, based on Vr
SolParJ/Cm	16.413	15.8-17.1	J1/2/cm3/2

(continued)

Table 4.3 (continued)

Properties		Literature Values	
SurTension	36.66	35.7	erg/cm2
MChainGrps	2	2	Enter# of backbonatms/strucunit
K	13.1	13.1	molarlim.visc1/4xcm3/2/mol3/4
K Theta	0.218	0.22	lim.visc.#,cm3xmol1/2/g3/2
M(Critical)	3558	3500	critical molecular weight,g
Refr. Index	1.50	1.49	refractive index, nD
Dielc Cons	2.16	2.3	Dielectric Constant,D.C.
Bulk Mod.(K)	4.57	4.54	(K)BulkModulus, GPa
ShearMod.(G)	0.93	0.91	(G)Shear orRIgidityModulusGPa
Ten. Mod.(E)	2.62	2.55	(E)TensileYoung'sModulusGPa
PoissonRatio	0.40	0.41	Poisson's Ratio
SndVel.Long.	2610.5	2430	Longitudinal SoundVelocity, m/s
SndVel.Shear	1045.7	950	Transverse Sound Velocity, m/s
SndAbs,Long.	4.18	----	Long.SoundAbs.Coeff., dB/cm
SndAbs,Shr.	20.89	----	Shear/TransSnd.AbsCoeffdB/cm
ThermalCond	0.20	0.16-0.48	J/s*m*K
SolPr(d)Cal	8.326	---	Based on Vg,cal1/2cm2/3
SolPr(p)Cal	0.000	---	Based on Vg. cal1/2cm2/3
SolPr(h)Cal	0.000	---	Based on Vg. cal1/2cm2/3
SolPrCal/Cm	8.326	8.3	Cal1/2cm2/3, based on Vg
E sub eta Inf.	27	27	E,Actenergy,visc flow,g KJ/mol
M(Actual)	3558	input	Molecular Wt. @or>M(crit)
T	298		inputTemp.of interest,degK
A	9.17	----	Empirical Constant
Log Eta Cr.Inf.	-3.680	----	Visc@M(critical),T=inf.,Ns/m2
Log Eta Cr.	1.04	----	Visc.@T of interest,Ns/m2
LogNewtVisc.	1.036	----	Newt.Visc.@T,N*s/m2
Gas	N2	input	Enter Gas of Interest from Table
Log Perm.	-13.8	-13.2	PCoeffPerm,cm3*cm/cm2*s*Pa
Delta H(A)@298K	-90	-93.5gg	Free Enthalpy of Poly., kJ/mol
TempCorrFac(B)	204K	----	
Delt Gsub f @T	-44	----	Free Energy Change @T kJ/mol
TDecompGrp	None	----	Enter Poly.Grp.for TherDecmp
TD1/2	677	677	Temp.ofHalfDecomp, deg, K

$\Rightarrow$ Example 2 (Ambiguous Group Selection)
Poly(vinyl chloride) (PVC)

—(CH_2CHCl)—

In the case of PVC, it is possible to choose the groups of (A) or (B).

<u>A</u>

Group	Number of Groups Selected
CH_2	2
CH	1
Cl	1

	B
Group	Number of Groups Selected
CH_2	1
CHCl	1

In this instance, it is possible to choose the combination of groups in (*A*) or (*B*). The general rule is to choose the group or groups that most closely approximate that of the structure in which you are interested. The second choice of groups would therefore be the preferable choice. The results appear as shown in Table 4.4.

4.10   GROUP SELECTION ALTERNATIVES

Even in the case of poly(ethylene), it would have been possible to choose two CH groups and two H's or two C's and four H's. Either of these choices would have resulted in serious errors being introduced in the estimated properties.

> NOTE: The reasons why group construction using atomic or lower molecular weight species often contribute to inaccurate estimates are discussed in Chapter 5.

In the case of PVC, a majority of the estimated properties would not have been altered significantly. The most serious discrepancies would have occurred for the estimations of T_g and T_m. Comparisons of the estimated properties with those of literature values are shown in Table 4.5.

$\Rightarrow$ Example 3 (Double Bond Group Selection)
Poly(butadiene) (PB)

$-(CH_2CH=CHCH_2)-$

Based on the repeat structure, the groups to choose are: two $- CH_2$ groups and one $- CH = CH -$ group. The results of the estimations appear as shown in Table 4.6.

Table 4.4 Estimated Poly(vinyl chloride) Properties

	CALCULATED MOLECULAR PROPERTIES	StructuralGroup Vinyl Chloride PolymerClass Poly(vinyl chloride)
Mol.Wt.	62.51	g/mol.
Density Am.	1.398	g/cm3.
Density Gl.	1.383	g/cm3.
Density Cr.	1.580	g/cm3.
VdeWaalsVol	29.23	cm3/mol
Eg	1.32E-02	cm3/mol*K,for glass,therm expan
E1	2.92E-02	cm3/mol*K,forliquid,therm expan
e1	4.68E-04	cm3/g*K,for liq,sp.therm.expan
eg	2.10E-04	cm3/g*k,for glass,sp.therm.expa
alpha liq. cubic	6.54E-04	1/K,cubic therm.expan., liq.state
alpha gl. cubic	2.91E-04	I/K, cubic therm.expan., gl.state
alpha liq. linear	2.18E-04	1/K, linear coeff.liq.therm.expan.
alpha gl. linear	9.70E-05	1/K, linear coeff., gl.therm.expan.
Specf Cp(s)	1.089	J/g*K, (298 deg. K)
Specf Cp(l)	1.458	J/g*K, (298 deg. K)
Hm	11000	J/mol.
Tg	81	C
Tm	278	C
SolPar(d)	8.747	Based on Vr,Cal1/2cm2/3disper.
SolPar(p)	6.014	Based on Vr,Cal1/2cm2/3polar
SolPar(h)	1.462	Based on Vr.Cal1/2cm2/3 H bond.
SolPrCal/Cm	10.715	Cal1/2cm2/3, based on Vr
SolParJ/Cm	21.924	J1/2/cm3/2
SurTension	42.19	erg/cm2
MChainGrps	2	Enter# of backboneatms/strucuni
C(constant)	24.15	Ktheta1/2, x M, cm3/g
K Theta	0.149	Un.Visc.Coeff.,cm3xmol1/2/g3/2
M(Critical)	7586	critical molecular weight,g
Refr. Index	1.54	refractive index, nD
Dielc Cons	3.13	Dielectric Constant,D.C.
Bulk Mod.(K)	5.54	(K)BulkModulus, GPa
ShearMod.(G)	1.71	(G)Shear or Rigidity Modulus, GP
Ten. Mod.(E)	4.65	(E)Tensile or Young'sModulus,GF
PoissonRatio	0.36	Poisson's Ratio
SndVel.Long.	2364.5	Longitudinal SoundVelocity, m/s
SndVel.Shear	1105.0	Transverse Sound Velocity, m/s
SndAbs,Long.	2.41	Long.SoundAbs.Coeff., dB/cm
SndAbs,Shr.	12.06	Shear/Trans Snd.AbsCoeff,dB/cm
ThermalCond	0.18	J/s*m*K
SolPr(d)Cal	8.651	Based on Vg,cal1/2cm2/3
SolPr(p)Cal	5.947	Based on Vg. cal1/2cm2/3
SolPr(h)Cal	1.454	Based on Vg. cal1/2cm2/3
SolPrCal/Cm	10.598	Cal1/2cm2/3, based on Vg
E sub eta Inf.	85.1	E,Act.energy of visc flow,g KJ/mc
M(Actual)	7586	Molecular Wt. @or>M(crit)
T	298	Temp.of interest,degK
A	12.59	Empirical Constant
Log Eta Cr.Inf.	-8.637	Visc.@M(critical),T=inf.,Ns/m2
Log Eta Cr.	6.33	Visc.@T of interest,Ns/m2
LogNewtVisc.	6.330	Newt.Visc.@T,N*s/m2
Gas	N2	Enter Gas of Interest from Table
Log Perm.	-15.1	P,Coeff.of Perm,cm3*cm/cm2*s*P
Delta H(A)@298F	-73.7	Free Enthalpy of Poly., kJ/mol
TempCorrFac(B)	213	K
Delt Gsub f @T	-10.2	Free Energy Change @T, kJ/mol
TDecompGrp	None	Enter Poly.Grp.for Therm Decomp
TD1/2	544	Temp.ofHalfDecomp, deg K

Table 4.5
Comparison of Calculated Molecular Properties with Literature
Values for Poly(vinyl chloride)

Structural Group	Vinyl Chloride	
Polymer Class	Poly(vinyl chloride)	
CALCULATED MOLECULAR PROPERTIES	LITERATURE VALUES	

Properties		Literature Values	
Mol.Wt.	62.51	62.51	g/mol.
Density Am.	1.398	-----	g/cm3.
Density Gl.	1.383	1.385	g/cm3.
Density Cr.	1.580	1.52	g/cm3.
VdeWaalsVol	29.23	29.2	cm3/mol
Eg	1.32E-02	0.88-1.31E-02	cm3/mol*Kglass therm expan
E1	2.92E-02	2.62-3.25E-02	cm3/mol*K,forliq,ther expan
e1	4.68E-04	4.2-5.2E-02	cm3/g*K, liq,sp.therm.expan.
eg	2.10E-04	1.4-2.1E-02	cm3/g*k,glass,sp.ther expan
alpha liq. cubic	6.54E-04	----	1/K,cubictherm.expan., liq.state
alpha gl. cubic	2.91E-04	----	1/K,cubic therm.expan., gl.state
alpha liq. linear	2.18E-04	----	1/K,linear coeff.liq.ther.expan.
alpha gl. linear	9.70E-05	----	1/K,linear coeff.gl.therm.expan.
Specf Cp(s)	1.09	0.96-1.09	J/g*K,
Specf Cp(l)	1.46	1.22	J/g*K
Hm	11000	11000	J/mol.
Tg	81	81	C
Tm	278	273	C
SolPar(d)	8.747	----	Based on Vr,Cal1/2cm2/3disper.
SolPar(p)	6.014	----	Based on Vr,Cal1/2cm2/3polar
SolPar(h)	1.462	----	Based onVr.Cal1/2cm2/3Hbond.
SolPrCal/Cm	10.715	10.7	Cal1/2cm2/3, based on Vr
SolParJ/Cm	21.924	19.2-22.1	J1/2/cm3/2
SurTension	42.19	41.5	erg/cm2
MChainGrps	2	2	Enter# ofbackboneatms/strucunit
K	24.15	----	molarlimviscg1/4xcm3/2/mol3/4
K Theta	0.149	0.162	lim.visc.#,cm3xmol1/2/g3/2
M(Critical)	7586	6200	critical molecular weight,g
Refr. Index	1.54	1.54	refractive index, nD
Dielc Cons	3.1	3.1	Dielectric Constant,D.C.
Bulk Mod.(K)	5.54	5.5	(K)BulkModulus, GPa
ShearMod.(G)	1.71	1.81	(G)Shear, RigidityModulusGPa
Ten. Mod.(E)	4.65	4.9	(E)Tensile,Young'sModulusGPa
PoissonRatio	0.36	0.35	Poisson's Ratio
SndVel.Long.	2364.5	2376	Longitudinal SoundVelocity, m/s
SndVel.Shear	1105	1140	Transverse Sound Velocity, m/s
SndAbs,Long.	2.41	----	Long.SoundAbs.Coeff., dB/cm
SndAbs,Shr.	12.06	----	Shear/TransSndAbsCoeff,dB/cm
ThermalCond	0.18	0.17	/s*m*K

(continued)

Table 4.5 (continued)

Properties		Literature Values	
SolPr(d)Ca l	8.651	---	Based on Vg,cal1/2cm2/3
SolPr(p)Cal	5.947	---	Based on Vg. cal1/2cm2/3
SolPr(h)Cal	1.454	----	Based on Vg. cal1/2cm2/3
SolPrCal/Cm	10.598	----	Cal1/2cm2/3, based on Vg
E sub eta Inf.	85	85	E,Act.energy viscflow,g KJ/mol
M(Actual)	7586		input Mol. Wt. @or>M(crit)
T	298	input	Temp.of interest,degK
A	12.59	---	Empirical Constant
Log Eta Cr.Inf.	-8.637	---	Visc.@M(critical)T=inf.,Ns/m2
Log Eta Cr.	6.33	---	Visc.@T of interest,Ns/m2
LogNewtVisc.	6.330	4.6	Newt.Visc.@T,N*s/m2
Gas	N2	Input	Enter Gas of Interest from Table
Log Perm.	-15.1	-15.1	PCoeffofPercm3*cm/cm2*s*Pa
Delta H(A)@298K	-85.5gg	-71gg	Free Enthalpy of Poly., kJ/mol
TempCorrFac(B)	213	----	K
Delt Gsub f @T	-85.5	----	Free Energy Change @T kJ/mol
TDecompGrp	None	----	Enter Poly.Grp.for TherDecomp
TD1/2	544	543	Temp.ofHalfDecomp, deg., K

4.11   SPECIAL AND NAVIGATION KEYS

In addition to the standard techniques for moving to Molcalc results discussed above, Ctrl+R will immediately take you to the results section of Molcalc.

Information and use of other special keys can be found in the Help menu using the Basics-Navigation Keys, or Search choice.

4.12   CALCULATING FREE ENTHALPY OF FORMATION (ΔH_f^o)

The final value for a given material results from two separate calculations. This is because this value is the molar free enthalpy of formation for the imaginary gaseous polymer minus the enthalpy of the gaseous starting material (monomer).

For poly(ethylene), we have: $(CH_2CH_2) - (CH_2 = CH_2)$. We use 2 CH_2 groups for the polymer, as before, and 2 $(= CH_2)$ groups for the monomer. Thus the ΔH_f^o for PE is calculated and then the value of ΔH_f^o for the monomer is calculated. These values are then subtracted. For PE we have:

$$-44 - 46 = -90 \text{ kJ/mol} @ 298°K$$

Table 4.6 Estimated Poly(butadiene) Properties

	CALCULATED MOLECULAR PROPERTIES	StructuralGroup Butadiene PolymerClass Poly(butadiene)	
Mol.Wt.	54.1	g/mol.	
Density Am.	0.892	g/cm3.	
Density Gl.	0.910	g/cm3.	
Density Cr.	1.008	g/cm3.	
VdeWaalsVol	37.4	cm3/mol	
Eg	1.68E-02	cm3/mol*K,for glass,therm expar	
E1	3.74E-02	cm3/mol*K,forliquid,therm expan	
e1	6.91E-04	cm3/g*K,for liq,sp.therm.expan	
eg	3.11E-04	cm3/g*k,for glass,sp.therm expar	
alpha liq. cubic	6.17E-04	1/K,cubic therm.expan., liq.state	
alpha gl. cubic	2.83E-04	I/K, cubic therm.expan., gl.state	
alpha liq. linear	2.06E-04	1/K, linear coeff.liq.therm.expan.	
alpha gl. linear	9.44E-05	1/K, linear coeff., gl.therm.expan.	
Specf Cp(s)	1.627	J/g*K, (298 deg. K)	
Specf Cp(l)	1.915	J/g*K, (298 deg. K)	
Hm	9000	J/mol.	
Tg	-116	C	
Tm	30	C	
SolPar(d)	8.010	Based on Vr,Cal1/2cm2/3disper.	
SolPar(p)	0.000	Based on Vr,Cal1/2cm2/3polar	
SolPar(h)	0.000	Based on Vr.Cal1/2cm2/3 H bond.	
SolPrCal/Cm	8.010	Cal1/2cm2/3, based on Vr	
SolParJ/Cm	16.389	J1/2/cm3/2	
SurTension	35.39	erg/cm2	
MChainGrps	4	Enter# of backboneatms/strucuni	
C(constant)	22	Ktheta1/2, x M, cm3/g	
K Theta	0.165	Un.Visc.Coeff.,cm3xmol1/2/g3/2	
M(Critical)	6180	critical molecular weight,g	
Refr. Index	1.52	refractive index, nD	
Dielc Cons	2.29	Dielectric Constant,D.C.	
Bulk Mod.(K)	2.87	(K)BulkModulus, GPa	
ShearMod.(G)	0.62	(G)Shear or Rigidity Modulus, GP	
Ten. Mod.(E)	1.74	(E)Tensile or Young'sModulus,GF	
PoissonRatio	0.40	Poisson's Ratio	
SndVel.Long.	2035.4	Longitudinal SoundVelocity, m/s	
SndVel.Shear	834.2	Transverse Sound Velocity, m/s	
SndAbs,Long.	3.96	Long.SoundAbs.Coeff., dB/cm	
SndAbs,Shr.	19.81	Shear/Trans Snd.AbsCoeff,dB/cm	
ThermalCond	0.15	J/s*m*K	
SolPr(d)Cal	8.172	Based on Vg,cal1/2cm2/3	
SolPr(p)Cal	0.000	Based on Vg. cal1/2cm2/3	
SolPr(h)Cal	0.000	Based on Vg. cal1/2cm2/3	
SolPrCal/Cm	8.172	Cal1/2cm2/3, based on Vg	
E sub eta Inf.	25.9	E,Act.energy of visc flow,g KJ/mc	
M(Actual)	6180	Molecular Wt. @or>M(crit)	
T	298	Temp.of interest,degK	
A	8.62	Empirical Constant	
Log Eta Cr.Inf.	-3.599	Visc.@M(critical),T=inf.,Ns/m2	
Log Eta Cr.	0.95	Visc.@T of interest,Ns/m2	
LogNewtVisc.	0.948	Newt.Visc.@T,N*s/m2	
Gas	N2	Enter Gas of Interest from Table	
Log Perm.	-12.3	P,Coeff.of Perm,cm3*cm/cm2*s*P	
Delta H(A)@298F	32	Free Enthalpy of Poly., kJ/mol	
TempCorrFac(B)	280	K	
Delt Gsub f @T	115.4	Free Energy Change @T, kJ/mol	
TDecompGrp	None	Enter Poly.Grp.for Therm Decomp	
TD1/2	680	Temp.ofHalfDecomp, deg K	

The free energy of formation ΔG_f^o for the monomer is 63.9 kJ/mol @ 298°K and 82.0 kJ/mol @ 600°K.

For poly(butadiene), the free energy of formation of the imaginary gaseous poly(butadiene) and the gaseous 1,3-butadiene monomer is given by:

$(CH_2CH = CHCH_2) - (CH_2 = CHCH = CH_2)$. The ΔH_f^o and temperature correction factor (B) for the polymer are calculated by choosing two CH_2 groups and one $CH = CH$ group. In this case the choice of two $= CH_2$ groups will provide identical results for ΔH_f^o.

- Conjugated Double Bonds

    The presence of conjugated double bonds in a compound or structural unit, necessitates a separate entry in Molcalc. This is noted in line No. 58 of Molcalc. The ΔH_f^o value for the monomer is calculated by choosing two CH_2 groups and two $= CH$ groups.

    Since the monomer possesses one conjugated double bond, a "1" is entered in entry Cell D62. This gives a value of -104 kJ/mol for ΔH_f^o of the monomer. The final estimation becomes: 32 kJ/mol $-$ 104 kJ/mol $= -72$ kJ/mol as listed in Table 4.7 along with comparisons to literature values for many of the properties.

    Subtraction of the temperature correction factors for the polymer and monomer, respectively, yields $128T$. Thermodynamically, polymerization below a temperature of 562°K (72,000/128) is possible as we can calculate from Equation (19.35) in Chapter 19.

    The free energy change ΔG_f^o may be estimated in a similar fashion for the polymer and monomer at the temperatures of interest.

4.13   SPECIAL ENTRY GROUPS

- Permeation - For calculating the permeation coefficient for gases other than nitrogen (N_2), select the gas desired from those available in the Table of Gas Permeation Factors immediately to the right of the results (Calculated Molecular

Table 4.7
Comparison of Calculated Molecular Properties with Literature Values for Poly(butadiene)

Structural Group	Butadiene	
Polymer Class	Poly(butadiene)	
CALCULATED MOLECULAR PROPERTIES	LITERATURE VALUES	

Properties		Literature Values	
Mol.Wt.	54.1	54.1	g/mol.
Density Am .	0.892	0.89	g/cm3.
Density Gl.	0.910	---	g/cm3.
Density Cr .	1.008	1.01	g/cm3.
VdeWaalsVol	37.4	37.4	cm3/mol
Eg	1.68E-02	1.08-1.68E-02	cm3/mol*K,glass,ther expan
E1	3.74E-02	3.46-4.17E-02	cm3/mol*K,forliq,ther expan
e1	6.91E-04	6.4-7.7E-04	cm3/g*K, liq,sp.therm.expan
eg	3.11E-04	2.0E-04	cm3/g*k,for glass,sp.ther expan
alpha liq. cubic	6.17E-04	----	1/K,cubic ther.expan.liq.state
alpha gl. cubic	2.83E-04	----	1/K, cubic therm.expan., gl.state
alpha liq. linear	2.06E-04	----	1/K, linearcoeffliq.therm.expan.
alpha gl. linear	9.44E-05	----	1/K, linear coeffgl.therm.expan.
Specf Cp(s)	1.63	1.63	J/g*K,
Specf Cp(l)	1.92	1.89	J/g*K
Hm	9000	9000	J/mol.
Tg**	-116	-102	C
Tm **	30	11	C
SolPar(d)	8.010	8.1	Based on Vr,Cal1/2cm2/3disper.
SolPar(p)	0.000	0	Based on Vr,Cal1/2cm2/3polar
SolPar(h)	0.000	0	Based onVr.Cal1/2cm2/3Hbond.
SolPrCal/Cm	8.010	8.1	Cal1/2cm2/3, based on Vr
SolParJ/Cm	16.389	16.6-17.6	J1/2/cm3/2
SurTension	35.39	32.5	erg/cm2
MChainGrps	4	4	Enter# of backbonatms/strucunit
K	22	----	mollim.visc.g1/4xcm3/2/mol3/4
K Theta	0.165	0.13-0.185	lim.visc.#,cm3xmol1/2/g3/2
M(Critical)	6180	6000	critical molecular weight,g
Refr. Index	1.52	1.516	refractive index, nD
Dielc Cons	2.3	2.5	Dielectric Constant,D.C.
Bulk Mod.(K)	2.87	~2.2	(K)BulkModulus, GPa
ShearMod.(G)	0.62	~0.4	(G)Shear or Rigidity Mod, GPa
Ten. Mod.(E)	1.74	~1.3	(E)Ten, Young'sModulus,GPa
PoissonRatio	0.40	----	Poisson's Ratio
SndVel.Long.	2035.4	----	Longitudinal SoundVelocity, m/s
SndVel.Shear	834.2	----	Transverse Sound Velocity, m/s
SndAbs,Long.	3.96	----	Long. Snd. Abs. Coeff., dB/cm
SndAbs,Shr.	19.81	----	Shear/TransSnd.AbsCoef,dB/cm
ThermalCond	0.15	----	J/s*m*K

(continued)

Table 4.7 (continued)

Properties		Literature Values	
SolPr(d)Cal	8.172	8.1	Based on Vg,cal1/2cm2/3
SolPr(p)Cal	0.000	----	Based on Vg. cal1/2cm2/3
SolPr(h)Cal	0.000	----	Based on Vg. cal1/2cm2/3
SolPrCal/Cm	8.172	8.1	Cal1/2cm2/3, based on Vg
E sub eta Inf.	25.9	---	Act.energy of visc low,g kJ/mol
M(Actual)	6180		inputMol Wt. @or>M(crit)
T	298		input Temp.of interest,degK
A	8.62	----	Empirical Constant
Log Eta Cr.Inf.	-3.599	--- -	Visc.@M(crit),T=inf.,Ns/m2
Log Eta Cr.	0.95	---	Visc.@T of interest,Ns/m2
LogNewtVisc.	0.948	----	Newt.Visc.@T,N*s/m2
Gas	N2		Enter Gas of Interest from Table
Log Perm.	-12.3	-12.3	P,CoefofPer,cm3*cm/cm2*s*Pa
Delta H(A)@298K	-72gg	-73gg	Free Enthalpy of Poly., kJ/mol
TempCorrFac(B)	280	----	K
Delt Gsub f @T	-72	----	FreeEnergy Change @T, kJ/mol
TDecompGrp	None	----	Enter Poly.Grp.for TherDecomp
TD1/2	680	680	Temp.ofHalfDecomp, deg., K

**This property estimation is based on cis 1,4 poly(butadiene) which is reflected in the literature values.

Properties). These factors are listed in Table 4.8. MATPROP© will automatically supply the correct factor and complete the calculation.

As you enter a different gas into cell C119 of the results table, you will notice that the value of the corresponding permeation factor appears in the Permeation Factor Table.

NOTE: The gas selected must be entered exactly as it appears in the Table 4.8, i.e., the correct factor will not be used in the calculation unless case sensitivity is observed.

- Thermal Decomposition - Estimation of the thermal decomposition temperature for many complex structures of repeating groups represents a special situation. These groups exert a strong influence on temperatures of decomposition and their specialized structures preclude the use of constructing these structures from corresponding groups using additivity functions. Therefore, these groups must be entered as such and cannot be constituted from simpler groups to provide accurate decomposition temperatures.

Table 4.8
Permeation Factors for Various Gases

Gas	Perm. Factor
N2	0
CO	0.079
CH4	0.53
O2	0.58
He	1.18
H2	1.35
CO2	1.38
H2O	2.74
Factor	0

Properties, other than the temperature of decomposition, are estimated as outlined previously. To estimate the decomposition temperature of structures containing any of the structures listed in the Table "Poly.Grp.Ctn.to Temp. of Decomp." (Polymer Groups Contributing to the Temperature of Decomposition).

- Select group and type the exact name in the entry cell (C124) in the results Table.
- Enter the number of repeating groups of that structure in the entry column and row labeled Polymer Groups, For Decomp.(Cell D63)
- Enter other repeating groups that are not constituted by these specialized structures as before.
- After $TD_{1/2}$ is obtained, clear the number of groups entry and re-enter "None" in the results table in order to render subsequent property estimations accurate. The structures of these groups are shown in Table 4.9.

NOTE: A None entered in the TDecompGrp cell(C118) is the default setting that provides $TD_{1/2}$ estimations for other groups.

Table 4.9
Specialized Group Structures for Thermal Decomposition Estimates

Oxadiazole

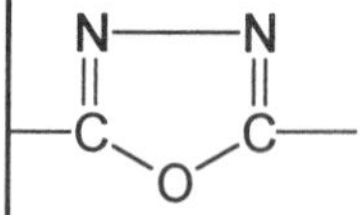

Thiadiazole

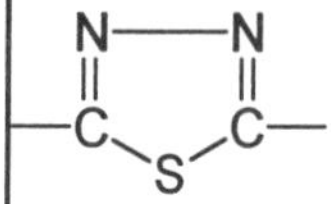

Triazole

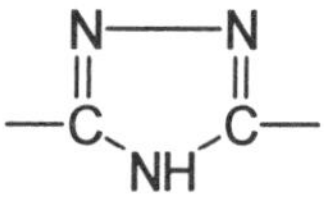

Benzoxazole

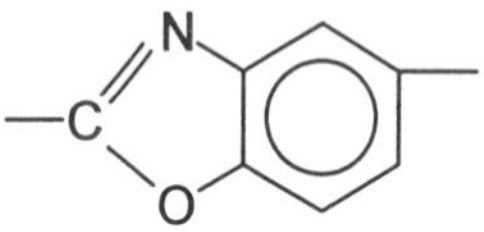

Benzthiazole

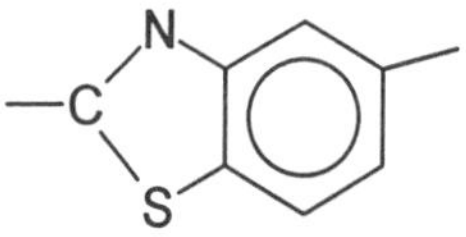

Benzimidazole

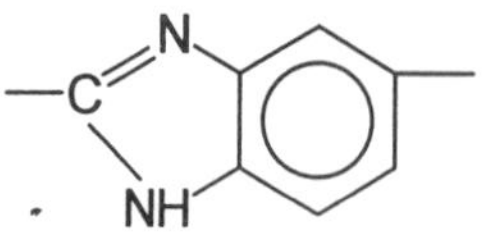

Quinoxaline

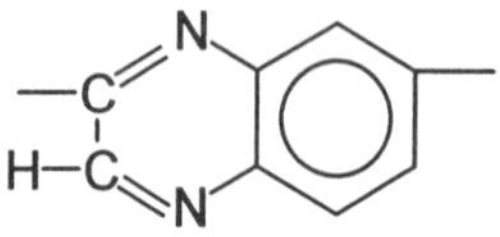

Phthalimide

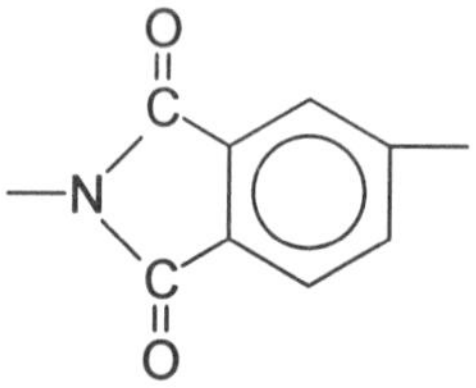

Diimidazobenzene

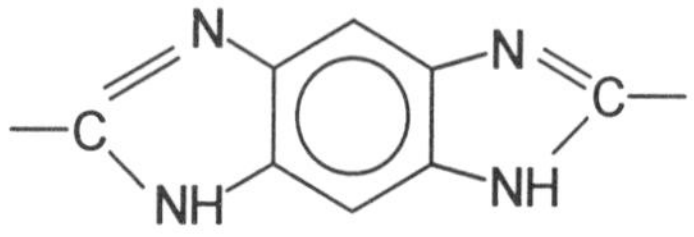

Pyromellitimide

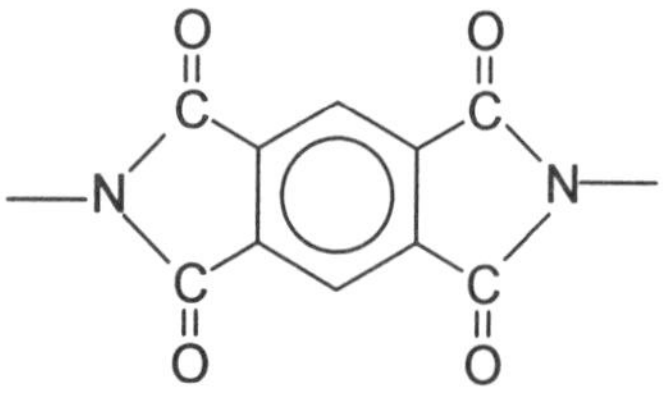

Tetrazopyrene

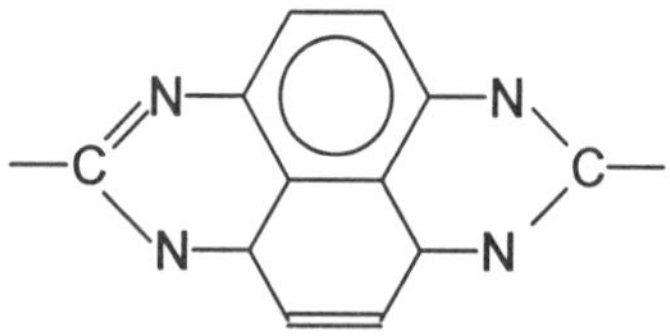

Benzimidopyrrolone

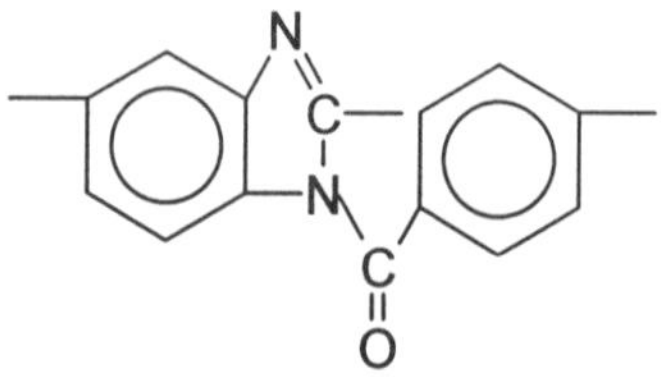

⇒   Example

Suppose we wish to calculate the $TD_{\frac{1}{2}}$ of poly(m-phenylene 2,5-oxadiazole)

Entering:

1 - m- C_6H_4

1 - Oxadiazole

These groups give a $TD_{\frac{1}{2}}$ of 798 °K vs. a literature value of 800 °K.

4.14   WORKING WITH THE EDIT MATPROP© MENU

The Edit pull down menus are shown in Figure 4.9. These edit commands follow normal Windows™ protocol.

- Cut - Deletes data and places it on the clipboard.
- Copy - Copies data from the selected cell or range to the clipboard.
- Paste - Copies data from the clipboard to the workbook page.
- Settings
- Directory - Establishes the directory that you choose to keep your files. Normally, this will be in the MATPROP© directory.
- GoTo - This command is used to jump to another cell or portion of the workbook. The cell containing the cursor will al ways be displayed in the upper left portion of the window immediately below the files menu on the menu bar. Enter the cell to which you wish to go in the Range entry cell of the GoTo dialog box. For example, if you wish to go to cell D14, simply enter D14 and click on the "GoTo" button every time you want to jump to this location.

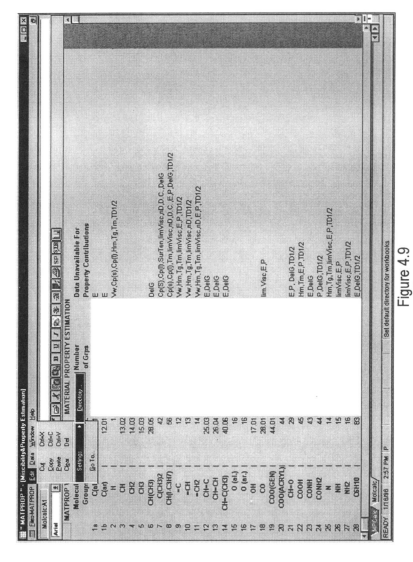

Figure 4.9
Edit MATPROP© Pull Down Menus

NOTE: - There are two entries that appear in the panel of the "GoTo" dialog box, viz., IO and /R. Although these symbols appear, they are <u>not</u> available for use and are used internally by MATPROP©.

4.15 WORKING WITH THE DATA QUERY MENU

The Data-Query command is used in conjunction with one

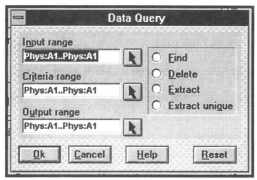

Figure 4.10a
The Query Menu Dialog Box Used to Establish Criteria in Searching
and/or Finding Specific Values of Interest

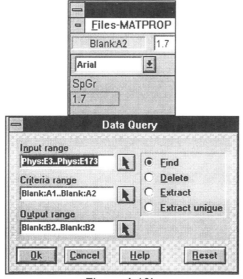

Figure 4.10b
Finding a Polymer Class of Specific Gravity of 1.7

of the database worksheets. The Query menu is useful for finding an item or a specific set of items for which you have established a set of criteria. When you use the Data menu, Query will be the only pull down item on the menu. Positioning the mouse cursor over Query and clicking will cause the dialog box shown in Figure 4.10a to appear.

The essentials of carrying out a query are establishing the following:

- Input Range
  This is the range that contains a database table.
- Criteria Range
- The range that contains your criteria are used to identify matching records in the table.
- Output Range
- The range where you want to copy the selected records based on the criteria you have established.
  Consider the following example. Suppose you wish to find those Polymer Class materials that have a SpGr of 1.7. Click on the blank page and in any cell, say A1, enter SpGr and in A2 enter 1.7.
- Activate the Query dialog box.
- *Click* on the arrow under input range and either enter Phys:E3..Phys:E173 or go to the physical property worksheet and highlight the range by holding down the mouse button and dragging the cursor over the desired range.
- *Enter* SpGr in cell A1 and 1.7 in cell A2.
- *Click* on the solid arrow under criteria range and highlight cells A1 and A2.
- *Click* on the button next to the range selected. This will return you to the dialog box menu.
- *Click* on the solid arrow under output range and highlight the output cells you have selected, e.g., B1 and B2.
- *Activate* the find radio button. The dialog box should appear as in Figure 4.10b.
- *Choose* OK.

The Query will automatically take you back to the Physical Properties worksheet to the polymer class having a specific gravity of 1.7. The ↓ will allow you to view every entry in the table with a value matching your criteria. If you are looking for those materials that possess a specific gravity > 1.7, *enter* a > 1.7 in cell A2. Activating the find command will again take you to the SpGr column in the physical properties worksheet, and find all materials matching your criteria.

MATPROP© stops at each one it finds. Simply use the down arrow (↓) to find every match in the table meeting the established criteria.

> NOTE: All blanks will be counted as a match and indicated by the program pausing each time you use the down arrow key.

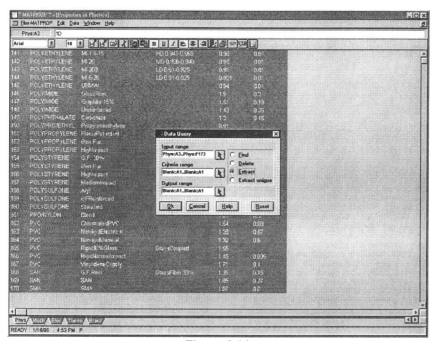

Figure 4.11
The Query Menu Dialog Box Used to Establish Criteria for Extracting
Data of Interest

- The Extract Command
  To illustrate the power and usefulness of the Data-Query command, the following additional example is included. Suppose you wish to extract all plastics in the database that are classified as "Blends" under the "Type" field in order to examine and/or work with them separately.
- *Activate* the Query dialog box.
- *Choose* the entire table as the input range either by entering the range or highlighting as Figure 4.11 shows.
- Again using the blank worksheet of the plastics workbook, enter the precise titles as they appear in the worksheet from which you intend to extract.
- Under Type, *type* Blend and select this as your criteria range (D1..D2) as shown in Figure 4.12.
- Again enter the precise titles as they appear in the worksheet from which you intend to extract. In most cases it will prove easier to use the copy command or icon to transfer the titles or heading to the blank worksheet.

  NOTE: Remember there is case and position sensitivity.

- For the Output range, specify a range (including field names) sufficient to accommodate all the results of the extraction. See Figure 4.12 as an example.
- *Choose*, Extract and OK. Your blank page should now look like Figure 4.13.
- The Delete Command.
  Since all worksheets are protected and cannot be modified except for those cells that are clearly designated as input cells, the Delete Command is not operable and therefore will not have any effect on the original worksheet.

  NOTE: The delete command can be used in conjunction with the EditPut command to modify your worksheets created with the *.BDT extension (Consult the previous description of Put command and Help menu).

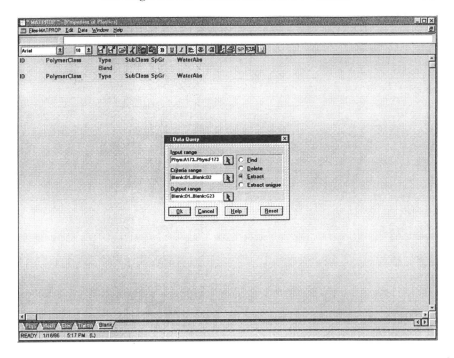

Figure 4.12
The Query Menu Dialog Box in Use to Specify the Criteria Range
as Type Blend (D1..D2)

4.16 WORKING WITH THE WINDOW MATPROP© MENU

The Window pull down menu allows you to select the ways in which you can work with your worksheets. This will be particularly advantageous in those instances when you wish to scroll a worksheet repeatedly or may be moving between work-books. The Window pull down menu is shown in Figure 4.14a.

- Titles - When you click on Titles, the dialog box shown in Figure 4.14b appears. If you are using one of the database workbooks, e.g., the properties of plastics, by setting the cursor at an appropriate cell, the desired field titles can be maintained as the worksheet is scrolled vertically, horizontally, or both. By placing the cursor in cell (B4) and choosing both, the headings or field titles will be maintained as you scroll down the sheet. Scrolling from left to right will maintain the ID number.

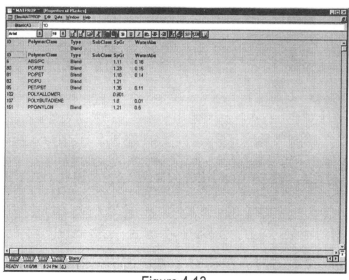

Figure 4.13
Results of Carrying Out an Extraction from the Data Query Menu
for All "Blend Type" Materials

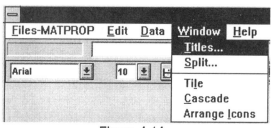

Figure 4.14a
Window- MATPROP© Pull Down Menu

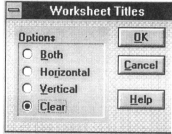

Figure 4.14b
The Windows Menu Titles Dialog Box Used for Locking Titles
as a Worksheet is Scrolled

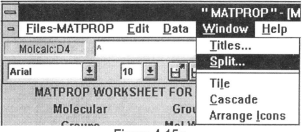

Figure 4.15a
Window MATPROP© Pull Down Menu

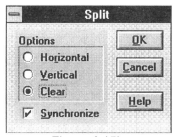

Figure 4.15b
The Window Menu Split Dialog Box Used for Producing Multiple
Views of a Worksheet

- Split - When you *click* on Split, Figure 4.15a, the dialog box shown in Figure 4.15b appears. You can simultaneously view different portions of the same worksheet. This allows you, e.g., to change your selection of groups while viewing the effects of this change on the results as shown in Figure 4.16.

4.16.1 Worksheets as a Result of the Windows™-Tile Command and Active Workbooks Minimized as Shown by the Lower Icons

- Tile Command- Tiles the worksheets as shown in Figure 4.17a and 4.17b. These figures are included only to illustrate the difference in appearance of the worksheets and the minimized icons between Windows™ 3.2 and Windows95/98™.
- Cascade Command - Cascades multiple worksheets as shown in Figure 4.18.
- Arrange Icons Command - When worksheets are minimized by using standard Window procedures by clicking on the minimize button, the Arrange Icons command arranges icons along

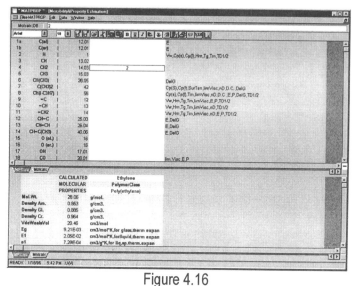

Figure 4.16

Split Screen Available as a Result of Invoking the Window - Split Horizontal Command

the lower left portion of the main window as shown in Figures 4.17 and 4.18.

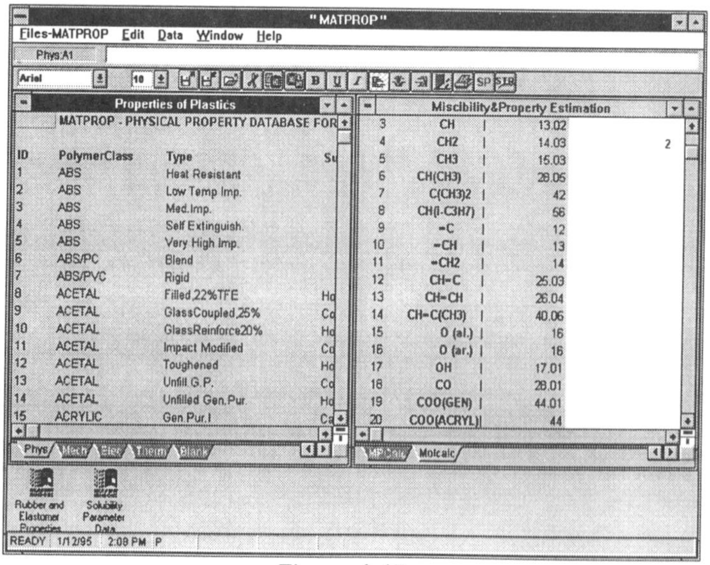

Figure 4.17a

Worksheet as a Result of the Window 3.1 Tile Command

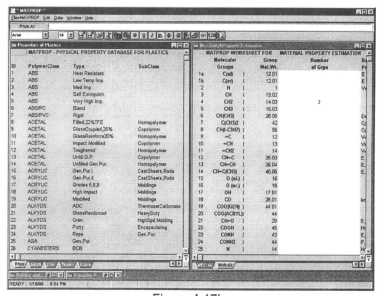

Figure 4.17b.
Worksheet as a Result of the Window(WIN95/98) Tile Command

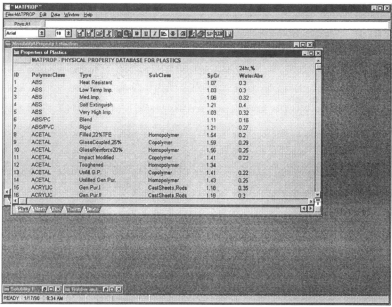

Figure 4.18
Worksheets as a Result of the Window - Cascade Command and Active
Workbooks Minimized as Shown by the Lower Icons

Figure 4.19
Help MATPROP© Pull Down Menu

4.17 WORKING WITH THE HELP MATPROP© MENU

The Help pull down menu allows you to obtain on line help about many of the topics discussed above as well as other topics (Figure 4.19). This feature has been provided to assist in answering many questions that may arise in the use of MATPROP©.

However, this is a rapid help option and is no substitute for the detail and examples contained in the users guide.

- Content - The Content command will allow you to go from general topics and descriptions to more specific topics.
- Search Index - The search index displays a complete alphabetical listing of topics. You can obtain information on other aspects and subtopics by selecting Display for the topic of interest.
- About MATPROP© - This menu item provides quick access to the supplier and the version installed. This information is also covered elsewhere. The MATPROP© start-up screen lists the program version and your system requirements are covered under this topic in the Search Index.

4.18 ESTIMATING POLYMER MISCIBILITIES USING MPCALC AND ENTERING DATA

MPCalc is also a worksheet of the mppr.bwb that is used to estimate polymer miscibilities from the solubility parameters estimated in Molcalc.

The tabs on the bottom of the worksheets labeled MPCalc and Molcalc, are used to easily switch between worksheets by left button clicking on the tab of the worksheet of interest as illustrated in Figure 4.20a. The opening screen of MPCalc is shown in Figure 4.20b. The opening screen has been designed to provide a view of essential input and readout areas.

Input Cells - As in Molcalc, the input cells have white backgrounds.

4.18.1 Molcalc And MPCalc

Entering Data in MPCalc from Molcalc - The results section of Molcalc has been placed on the side of the entry table for the solubility parameters so that the estimated solubility parameters

Left Click on MPCalc to Choose the Worksheet

Figure 4.20a
MPCalc Worksheet Tab

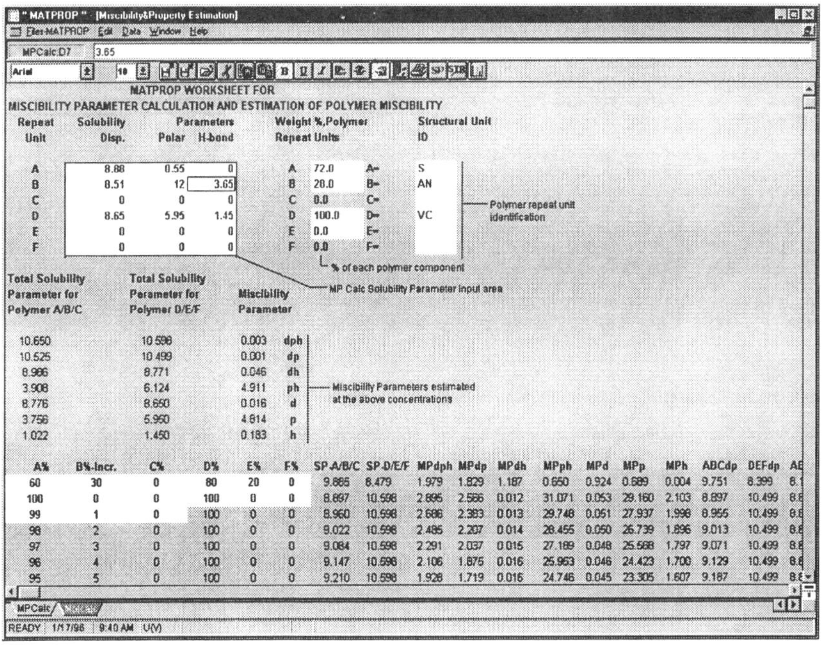

Figure 4.20b
MPCalc Opening Screen of MATPROP©

can be entered immediately after estimation based on entry of the proper molecular groups. For example, Figure 4.21 shows the estimated molecular properties of poly(styrene) and the entry table for solubility parameters.

The estimated solubility parameters for poly(styrene) have been entered in the component A row, as illustrated in Figure 4.21.

> NOTE: It should be noted that in the case of glassy polymers such as poly(styrene), the solubility parameters estimated using the glassy volumes are used. The dispersive force shown is indicated by the cursor. Incorrect miscibilities would be indicated if the values based on V_r were used in this instance.

Once the values for the two polymer systems, i.e., ABC and DEF have been entered into the table you can *click* on the SP icon to automatically transfer the values from Molcalc into the input box of MPCalc.

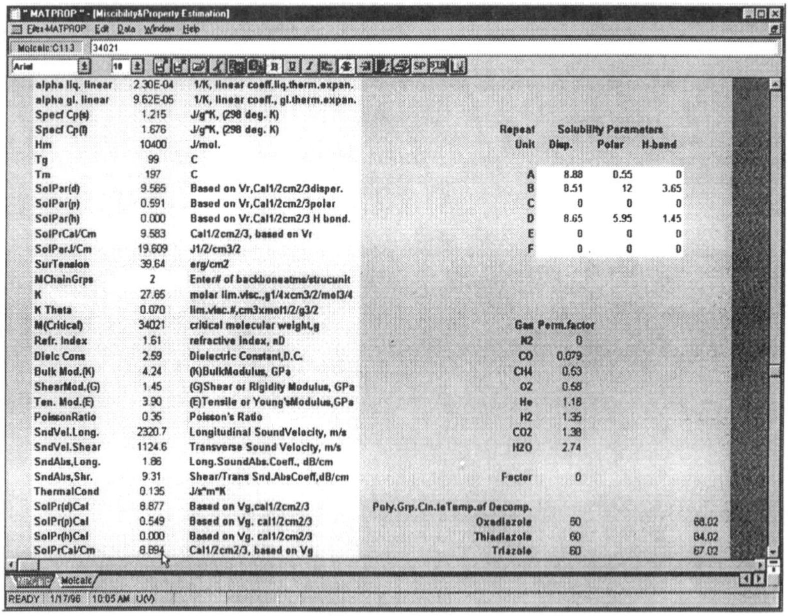

Figure 4.21
Results of Estimation of Molecular Properties for Polystyrene and Entry
into the Table for Component A

4.18.2 MPCalc

The solubility parameters may also be entered directly into the input box of MPCalc for the individual repeat units of A, B, C, D, E, and F.

Once the solubility parameters have been entered, MPCalc estimates the miscibility parameters (MP's). There are two types of Miscibility Parameters that are estimated, viz., static and dynamic.

- Static - A static analysis is one in which the MP's are desired at only one or more specific concentrations of two or more polymers.

In this instance, after the solubility parameters have been entered, enter the weight percent of each polymer component. The resulting miscibility parameters are listed in the Miscibility Parameter column as illustrated in Figure 4.20b.

> NOTE: Since this is a guide in how to use the program, Chapter 21, Miscibility, should be consulted for understanding and interpretation of the quantitative relationships between miscibility parameters and polymer miscibilities. However, generally one looks for regions where the MP is equal to or less than 0.1.

- Dynamic - A dynamic analysis is one in which the MP's are examined as a function of changing concentration of one of the components.

The first input row in the dynamic results table generated by MPCalc (Figure 4.22) can be used to enter individual polymer repeat unit concentrations for static analysis and comparison with the results of the dynamic analysis.

Initial concentrations of A and C can be entered along with the increment to be varied. MPCalc uses an increment of 1%. Obviously all three columns must sum to 100%. For increments greater than 1%, or for concentrations of *A* below 100%, the last usable entry in the MPCalc results will occur when the concentration of A reaches 0, i.e., all three columns must again sum to 100%. This is illustrated in Figure 4.23.

MPCalc provides all possible combinations of miscibility

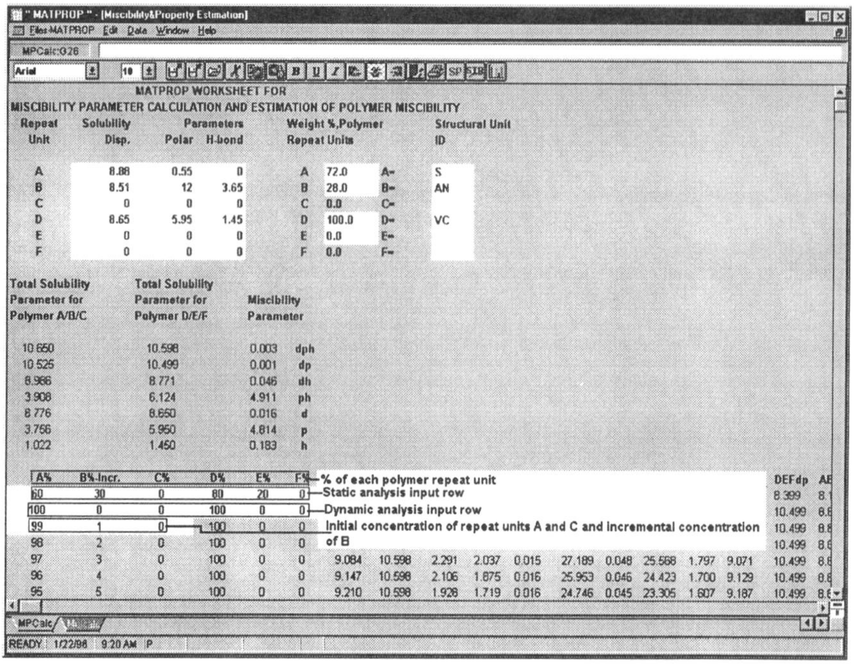

Figure 4.22
Identification and Input Areas of MPCalc for Static and Dynamic Results
Where the MP is Equal to or Less Than 0.1.

parameters for use. The most used combination is MP_{dph}.
However, others frequently consulted are: MP_{dp}, MP_{dh}, and MP_{ph}.

4.19 GRAPHING MISCIBILITY RESULTS OF MPCALC

Since MATPROP© is a Window™ based program, it is possible to copy data from the worksheets into other applications. Thus, in this case you can copy data from MPCalc and paste it into other Window™ applications.

If the MP data is copied into one of your favorite graphing programs, you can generate graphs using the data of specific interest to you. An easier way is to use the graphing capabilities MATPROP© itself. This is accomplished by *clicking* on the graphics icon on the toolbar (See section 4.4) while you are using the MPCalc worksheet.

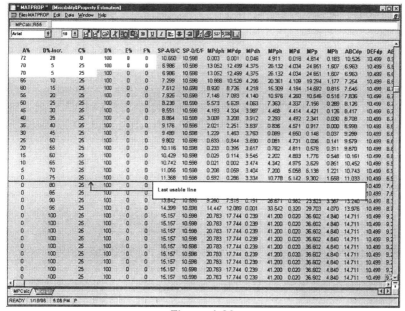

Figure 4.23
MPCalc Illustrating Inputs For Differing Concentrations of A and C and B
Component Increments Greater Than 1%

4.20   PRINTING GRAPH RESULTS

File|Print - This command causes a "pop-up" dialog box to appear which allows selection or changing of the printer. If you will be using the default printer originally specified for use in Windows™ and printing a half page of your graph.

- In portrait mode, you do not need to activate this command. For printing your graph in full page mode, *click* on setup and choose *landscape*.

- File|PageSetup - This menu command should be used to specify how you wish the printed graph to appear. If the graph is to be printed in the portrait mode it will be necessary to specify *half page*. For the graph to be printed in full page you must change the setting to *landscape* here as well as in the setup under File|Print.

After you have specified any other parameters that allow you to control the appearance of your graph, choose *OK*. Your graph will appear similar to the sample results included with MATPROP© which are shown in Figure 4.24.

4.21   DATABASE WORKBOOKS OF MATPROP©

MATPROP© contains four databases:

- A database of plastics (plas.bwb)
- A database of rubbers and elastomers (rbel.bwb)
- A database of general material properties(genprop.bwb)
- A database of solubility parameters (sparam.bwb)

4.21.1 Plastics Database

This workbook consists of five modules or worksheets, four of which contain comprehensive listings of plastics.

- Physical Properties
- Mechanical Properties
- Electrical Properties
- Thermal Properties

A blank worksheet has been provided so that you can copy, paste, and construct your own individual database tailored to your

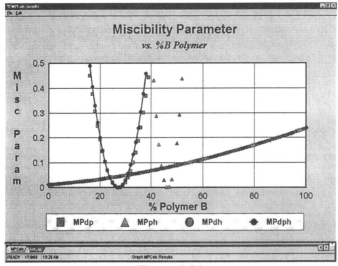

Figure 4.24
Plots of Miscibility Parameter Data from MPCalc for the SAN-PVC System
as a Function of AN Concentration

specific interests should you wish to do so. This blank sheet can also be used to conduct queries as explained previously.

4.21.2 Rubber and Elastomers Database

This workbook consists of six modules or worksheets five of which contain comprehensive listings of rubbers and elastomers. The sixth is again a blank page for copying and querying any of the worksheets.

- Physical Properties
- Mechanical Properties
- Electrical Properties
- Thermal Properties
- Environmental Properties

4.21.3 General Material Properties Database

This workbook consists of thirteen worksheets which list properties of a wide range of metals, ceramics, non-metals, and plastics. These allow comparisons of materials for properties of interest. The database consists of the following worksheets:

1. Density of Materials
2. Ultimate Tensile Strength of Materials
3. Tensile Yield Strength of Materials
4. Metals Elongation, %
5. Specific Strength of Materials
6. Materials Modulus of Elasticity in Tension
7. Hardness of Materials
8. Service Temperature of Non-Metallics
9. Specific Heat of Materials
10. Thermal Conductivity of Materials
11. Coefficient of Thermal Expansion of Materials
12. Electrical Resistivity of Materials
13. Blank Page for copying and querying the worksheets

4.21.4 Solubility Parameter Database

A module or worksheet has been provided that gives a listing of the estimated individual solubility parameters of 51 different polymers for use in estimating polymer miscibilities. Selections can be copied and pasted into the blank page along with additional estimations from Molcalc to construct your own database of those materials of interest.

For the sake of space, all structural groups and polymer classifications have been abbreviated. The precise polymer can be ascertained by activating or *clicking* on the structure (STR) button, as shown in Figure 4.25.

This activates the paintbrush program in Windows™ and gives you direct access to the MATPROP© structural files by scrolling through your active window programs (alt + tab). The polymer files are contained in poly1.bmp and poly2.bmp. The molstrc.bmp graphics sheet is used in conjunction with Molcalc.

The remaining two *.bmp files are used internally by MATPROP© and will not be useful to you. The names of each polymer and the structure are then available with the identification number as contained in the database table as illustrated in Figure 4.26.

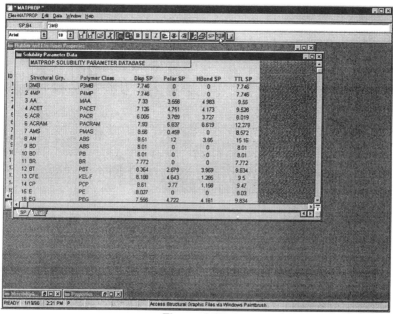

Figure 4.25
Activating the "STR" Button to Access the Structural Graphic Files

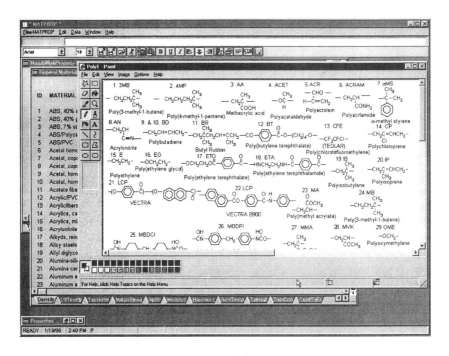

Figure 4.26
Using the Polymer Graphic Files to Identify and Establish the Structure
of a Specific Polymer

167

CHAPTER 5

Polymer Property Estimation and Estimated Properties of Selected Polymers

5.1 INTRODUCTION

In this chapter, we introduce the types of molar properties and the basis for materials properties estimation that is used by Molcalc of the MATPROP© software for estimating properties. Also included is a tabular listing of the molar properties estimated, their formulas, the associated structural group property, and the structural group symbol as it appears in the calculated properties section of Molcalc.

For ready reference, we have included property estimates and their units for a series of different types of polymers along with the structural groups used for estimation. This series has been included to serve as examples of the proper structural group selection in order to obtain "best" property estimates.

In subsequent chapters, detail will be provided for the basis of appropriately grouped properties along with comparisons to literature values when available.

5.2 BASIS OF MOLECULAR PROPERTY ESTIMATION

5.2.1 Types of Molar Properties

5.2.1.1 Colligative Properties

Colligative properties are properties that are special in the sense that they have identical values per mole of substance irregardless of the structure of the material. Thus ideal gases and solutions exhibit these properties. Examples of colligative or osmotic properties are: osmotic pressure, vapor pressure reduction, boiling point elevation, and freezing point depression.

5.2.1.2 Additive Properties

Additive properties are displayed by those materials whose molar properties are equal to the sum of values of the constituent atoms. On a rigorous basis, only the molar mass is strictly additive. Other molar quantities that are more interesting such as volume, refraction, etc. can be approximated using the additivity principle and is discussed below. Limitations and examples of limitations of using additivity are given throughout this text.

5.2.1.3 Constitutive Properties

Constitutive properties are a function of the structure of the molecule without consideration for colligative or additive properties. Examples are energy absorption of light at specific wavelengths or magnetic resonance absorption at a given field strength and frequency.

5.3   ADDITIVITY AND PROPERTIES FROM GROUP CONTRIBUTIONS

The use of the principle of additivity has been a powerful tool in the "semi empirical" approach for estimating properties of materials, particularly those of polymers. The additivity principle simply means that the molar properties of a polymer may be estimated by summing the atomic, group, or bond contribution. This is expressed mathematically as:

$$F = \sum_i n_i F_i \qquad (5.1)$$

where:

F = A molar property
n_i = Number of contributing components of type i
F_i = Numerical contribution of the component i

$$\text{and} \qquad F = \sum_{i=1}^{n} \sum_{j=1}^{n} w_{ij} F_{ij} \qquad (5.2)$$

where:

w_{ij} = weight factor taking into account the relative importance of contacts between groups i and j

F_{ij} = property contributions of F_{ij}

This approach is applicable to polymers because of their repeating structures. End groups play only a minor role, at least for high molecular weight polymers and the basis for additivity is the sum of contributions per mole of repeating unit or structural group.

There are three additive methods which are used to compute the total additivities of the structural unit contributions of a material. These are:

1. Atomic Contributions - This is the simplest system of additivities which assumes that a particular property of interest may be estimated from the contributions of the atoms of which it is composed.
2. Group Contribution - This is a more complex system for estimating properties from additivities since the property contributions of the groups as a whole are considered and group volumes, valence structures, and other relevant characteristics are inherent in the group selected.
3. Bond Contributions - The use of bond contributions is a refinement of the group contribution technique. The effects of bond contributions must be used in conjunction with the group additivity method since this technique is the only way of accounting for differences between various atoms such as single, double, or triple bonded carbons, single or double bonded carbon-oxygen or carbon-nitrogen, for example.

The estimation of properties based on additivities assumes that the contribution of each atom or group will be precisely identical regardless of its environment. We have seen in previous chapters that a polymer molecule can interact both intramolecularly and intermolecularly. The strict use of atomic contributions in most cases is much too simplistic to result in any meaningful property estimations. Therefore, group contributions must be used in order to provide good approximations of experimentally measured properties.

In general the closer one comes to the repeating structural unit(s) or group(s) which comprise the polymer, the better the estimated properties will be. This is the reason that selection of the largest group which more closely approximated that of the repeating structural groups of the polymer was recommended, cf. section 4.10 of Chapter 4.

In addition to the effects of bond type mentioned above, there are also the effects of the surrounding groups which influence estimated properties. Estimations based on additivity principles assume that the contributions of a CH_2 group, for example, are identical for each property estimated whether or not a CH_3, OH, or other group is attached. Differing adjacent or surrounding groups will obviously influence both intra- and intermolecular interactions and thus cause the estimated properties to differ from the actual properties. The extent of this difference or error will vary for the specific property being estimated but on the whole will be on the order of $\pm 15\%$.

It is "always" preferable to use the actual or experimentally measured properties of materials whenever possible. The key considerations in being able to estimate properties of materials is the ease of use of the tools for doing so, and in this case the formulas and calculation tools are incorporated into the software. Implicit in software that estimates polymer properties is that for one to become an effective user of the software, substantial investments in time or resources are not required.

The value of using the Molcalc software of MATPROP© for estimating polymer properties is:

1. Properties of interest can be estimated that are not easily obtainable in the literature.

2. Properties can be estimated that have not been experimentally measured.
3. "What if?" questions can be asked to determine the probable effects of changing a structure and the corresponding property changes.
4. Property changes as a result of changing the molecular architecture can be used to draw inferences regarding the structural influence on properties of interest.
5. A central "database" for estimating polymer properties from group contributions aids in making valid comparisons between polymeric materials especially in the case of homologous series.

It can be seen that many factors can be responsible for the introduction of estimation errors. The volume contribution of a group is used in a number of estimations and unfortunately it has a large effect upon the estimated property. This means that any error introduced in estimating the total volume of a complex group or molecule will play a correspondingly large role in error contribution of the properties requiring a value of this component.

A secondary problem in estimating properties of polymeric and organic materials is that in many cases the group contribution to a particular property of interest is not known and, therefore, the property cannot be estimated. In tabulating the group contributions and developing the software, every effort was made to:

1. Utilize only essential groups which can be used as building blocks to synthesize more complex structures.
2. Adjust group contributions, in some instances, to provide more accurate property estimations for a large number of different classes of polymers and their resultant properties.
3. Provide sufficient types of groups that will permit the properties to be estimated for a large number of different polymers. In a number of instances this was done by "synthesizing" larger groups from smaller ones and testing property estimations by comparison of

estimated properties to actual properties and subsequently refining the contributions of the groups.

Although a number of investigators [G1-G5] have studied and reported on techniques and results for group contributions for specific group properties, it is Van Krevelen who made a major contribution in this area. In fact, Van Krevelen [G3-G4] is the author who has organized and assembled data from independent measurements and provided correlations. It is through his efforts and the published information that serves the basis for this software. We have attempted to improve upon the estimations discussed above and provide a Windows™ based program that is easy to use and performs complicated calculations reproducibly and accurately.

The additive molar functions estimated along with their symbols, formulas, names, and pioneering investigators are listed in Table 5.1. Subsequent example polymer property tables list the estimated properties along with their dimensions.

The following section in this chapter illustrates examples of different classes of polymers, the groups selected to estimate their properties, and the resultant estimated properties.

In following chapters, detailed explanations of these properties will be discussed and the formulas for property estimations will be given along with discussions of the factors and issues that can be expected to lead to differing estimates based on structure, morphology, or other molecular or bulk property considerations.

5.4   EXAMPLES OF ESTIMATED POLYMER PROPERTIES FROM SELECTED GROUPS

The essence of the advantage of Molcalc is that it can be used to estimate the properties of whatever polymer/material that is of interest. This eliminates the need for comprehensive tables of polymers since these can be generated as needed. The polymers in this section have been selected to serve as representative examples of a number of different classes of polymers that illustrate the repeating group selected to obtain those properties.

Table 5.1
Additive Molar Structural Unit Functions

Name	Symbol	Formula	Structural Group Property	Structural Group Symbol	Pioneering Investigator
Molar Mass	M	$\sum N_i M_i$	Mass of Structural Group	M_i	Dalton(1801) Berzelius(1810)
Molar Number of Backbone Atoms	Z	$\sum N_i Z_i$	No. of Backbone Atoms/Str. Grp.	Z_i	Weyland(1970)
Molar Van der Waals Volume	V_w	$\sum_i N_i V_{w,i}$	Hard Volume/Str. Grp.	$V_{w,i}$	Bondi(1968)
Molar Glass Transition	Y_g	$M \cdot T_g$	Glass Transition Temperature	T_g	Van Krevelen-Hoftyzer(1976)
Molar Melt Transition	Y_m	$M \cdot T_m$	Melt Transition Temperature	T_m	Van Krevelen-Hoftyzer(1976)
Molar Unit Volume	V	M / ρ	Density	ρ	Traube(1895)
Molar Heat Capacity	C_p	$M \cdot c_p$	Specific Heat Capacity	c_p	Satoh(1948) Shaw(1969)
Molar Cohesive Energy	E_{coh}	$M \cdot e_{coh}$	Cohesive Energy Density	e_{coh}	Bunn(1955)
Molar Attraction	F	$V \cdot \delta$	Solubility Parameter	δ	Small(1953)
Molar Parachor	P_s	$V \cdot \gamma^{1/4}$	Surface Tension	γ	Sugden(1924)
Molar Elastic Wave Velocity	U	$V \cdot u^{1/3}$	Sound Wave Velocity	u	Rao(1940) Schuyer(1958)
Molar Intrinsic Viscosity	J	$K_q^{1/2} \cdot M - 4.2Z$	Intrinsic Viscosity Parameter	K_θ	Van Krevelen-Hoftyzer(1967/76)

(continued)

Name	Symbol	Formula	Structural Group Property	Structural Group Symbol	Pioneering Investigator
Molar Viscosity Temp. Gradient	H_η	$M \cdot E_h^{1/3}$	Activation Energy of Viscous Flow	$E_\eta(\infty)$	Van Krevelen-Hoftyzer (1975)
Molar Polarisation	P	$V \cdot \dfrac{\varepsilon - 1}{\varepsilon + 2}$	Dielectric Constant	ε	Mosotti(1850) Clausius(1879)
Molar Optical Refraction	R	$V \cdot (n-1)$	Index of Light Refraction	n	Gladstone-Dale (1858)
Molar Free Energy of Formation	ΔG_f^0	$A - BT$	Heat of Formation Entropy of Formation	$\dfrac{A(\Delta H_f^o)}{B}$	Van Krevelen-Chermin(1950)
Molar Thermal Decomposition	$Y_{d,1/2}$	$M \cdot T_{d,1/2}$	Temperature of Half-Way Decomposition	$T_{d,1/2}$	Van Krevelen (1988)

It should be remembered (cf. Chapter 4) that not all estimated properties displayed are accurate or usable as is. The "usability" depends on whether or not contributions are available for that particular group or property. This can be determined by consulting the "Unavailable Properties" column. In addition the final value for ΔG must be obtained using the techniques described previously (cf. Chapter 4) and further illustrated and discussed in Chapter 19.

In the following chapters of this monograph, limitations, sources of inaccuracies and basis for the estimations will be discussed and illustrated.

5.5 CALCULATED MOLECULAR PROPERTIES OF REPRESENTATIVE POLYMER CLASSES

The following is a tabulation of the example polymers contained in this section in the estimated property tables. A separate chapter appendix of abbreviations and notation is included to clarify the terminology used in property Tables 5.3-5.25. Property entries are not present for those cases where data are unavailable even though Molcalc will automatically perform the calculation. The user must consult the unavailable property column to determine whether or not the calculation of interest is valid.

For estimating the enthalpy of formation of the polymer, the value for the monomer must be subtracted as described in Chapter 4, to obtain a valid total ΔH_f^o. The values listed in the tables reflect the calculation for the polymer only. Comparisons of the total ΔH_f^o values with literature values are contained in Chapter 19.

In many cases word descriptions rather than symbols are used in the estimated molecular property tables in order to conform to the software requirements of Molcalc.

Table 5.2
Listing of Example Polymers

Table	Class	Polymer	Unavailable Properties
5.3	Polyolefins	Poly(ethylene)	----
5.4	Polyolefins	Poly(propylene)	----
5.5	Polyolefins	Poly(isobutylene)	----
5.6	Vinyls	Poly(vinyl chloride)	----
5.7	Vinyls	Poly(vinyl fluoride)	$E_\eta(\infty)$
5.8	Vinyls	Poly(chlorotrifluoroethylene)	$K_\Theta, M_{cr}, E_\eta(\infty)$
5.9	Vinyls	Poly(vinyl acetate)	----
5.10	Unsaturated Group Polymers	Poly(butadiene)	----
5.11	Unsaturated Group Polymers	Poly(isoprene)	----
5.12	Aromatic Polymers	Poly(styrene)	----
5.13	Aromatic Polymers	Poly(α-methyl styrene)	----
5.14	Aromatic Polymers	Polycarbonate	ΔH, ΔG
5.15	Aromatic Polymers	Poly(ethylene terephthalate)	----
5.16	Aromatic Polymers	Poly(butylene terephthalate)	----
5.17	Ether Group Polymers	Penton	----
5.18	Ether Group Polymers	Poly(methylene oxide)	----
5.19	Ether Group Polymers	Poly(vinyl methyl ether)	----
5.20	Ether Group Polymers	Poly(phenylene oxide)	$K_\Theta, M_{cr}, E_\eta(\infty)$ $\Delta H, \Delta G$,
5.21	Polyamides	Poly(hexamthylene adipamide)	ΔH, ΔG
5.22	Polyamides	Poly(acrylamide)	$P, TD_{1/2}$
5.23	Nitrile Polymers	Poly(acrylonitrile)	$C_p(l), E_\eta(\infty)$,
5.24	Acrylate Polymers	Poly(methyl methacrylate)	----
5.25	Acrylate Polymers With a Saturated Ring	Poly(cyclohexyl methacrylate)	----

Table 5.3 Estimated Properties of Poly(ethylene)

		StructuralGroup
CALCULATED		Ethylene
MOLECULAR		PolymerClass
PROPERTIES		Poly(ethylene)
Mol.Wt.	28.06	g/mol.
Density Am.	0.853	g/cm3.
Density Gl.	0.885	g/cm3.
Density Cr.	0.964	g/cm3.
VdeWaalsVo l	20.46	cm3/mol
Eg	9.21E-03	cm3/mol*K,for glass,therm expan
E1	2.05E-02	cm3/mol*K,forliquid,therm expan
e1	7.29E-04	cm3/g*K,for liq,sp.therm.expan.
eg	3.28E-04	cm3/g*k,for glass,sp.therm expan
alpha liq. cubic	6.22E-04	1/K,cubic therm.expan., liq.state
alpha gl. cubic	2.90E-04	1/K, cubic therm.expan., gl.state
alpha liq. linear	2.07E-04	1/K, linear coeff.liq.therm.expan.
alpha gl. linear	9.68E-05	1/K, linear coeff., gl.therm.expan.
Specf Cp(s)	1.807	J/g*K,
Specf Cp(l)	2.167	J/g*K
Hm	8000	J/mol.
Tg	-120	C
Tm	133	C
SolPar(d)	8.022	Based on Vr,Cal1/2cm2/3disper.
SolPar(p)	0.000	Based on Vr,Cal1/2cm2/3polar
SolPar(h)	0.000	Based on Vr.Cal1/2cm2/3 H bond.
SolPrCal/Cm	8.022	Cal1/2cm2/3, based on Vr
SolParJ/Cm	16.413	J1/2/cm3/2
SurTension	36.66	erg/cm2
MChainGrps	2	Enter# of backboneatms/strucunit
C(constant)	13.1	molar lim.visc.,g1/4xcm3/2/mol3/4
K Theta	0.218	lim.visc.#,cm3xmol1/2/g3/2
M(Critical)	3558	critical molecular weight,g
Refr. Index	1.50	refractive index, nD
Dielc Cons	2.16	Dielectric Constant,D.C.
Bulk Mod.(K)	4.57	(K)BulkModulus, GPa
ShearMod.(G)	0.93	(G)Shear or Rigidity Modulus, GPa
Ten. Mod.(E)	2.62	(E)Tensile or Young'sModulus,GPa
PoissonRatio	0.40	Poisson's Ratio
SndVel.Long.	2610.5	Longitudinal SoundVelocity, m/s
SndVel.Shear	1045.7	Transverse Sound Velocity, m/s
SndAbs,Long.	4.18	Long.SoundAbs.Coeff., dB/cm
SndAbs,Shr.	20.89	Shear/Trans Snd.AbsCoeff,dB/cm
ThermalCond	0.20	J/s*m*K
SolPr(d)Cal	8.326	Based on Vg,cal1/2cm2/3
SolPr(p)Cal	0.000	Based on Vg. cal1/2cm2/3
SolPr(h)Cal	0.000	Based on Vg. cal1/2cm2/3
SolPrCal/Cm	8.326	Cal1/2cm2/3, based on Vg
E sub eta Inf.	27	E,Act.energy of visc flow,g KJ/mol
M(Actual)	3558	Molecular Wt. @or>M(crit)
T	298	Temp.of interest,degK
A	9.17	Empirical Constant
Log Eta Cr.Inf.	-3.680	Visc.@M(critical),T=inf.,Ns/m2
Log Eta Cr.	1.04	Visc.@T of interest,Ns/m2
LogNewtVisc.	1.036	Newt.Visc.@T,N*s/m2
Gas	N2	Enter Gas of Interest from Table
Log Perm.	-13.8	P,Coeff.of Perm,cm3*cm/cm2*s*Pa
Delta H(A)@298K	-44	Free Enthalpy of Poly., kJ/mol
TempCorrFac(B)	204	K
Delt Gsub f @T	16.8	Free Energy Change @T, kJ/mol
TDecompGrp	None	Enter Poly.Grp.for Therm Decomp
TD1/2	677	Temp.ofHalfDecomp, deg, K

Structure

Repeat Structural Groups Selected

$-(CH_2CH_2)-$

$2 - CH_2$

179

Table 5.4 Estimated Properties of Poly(propylene)

		Structural Group
CALCULATED		Propylene
MOLECULAR		**PolymerClass**
PROPERTIES		Poly(propylene)
Mol.Wt.	42.08	g/mol.
Density Am.	0.857	g/cm3.
Density Gl.	0.855	g/cm3.
Density Cr.	0.968	g/cm3.
VdeWaalsVol	30.68	cm3/mol
Eg	1.38E-02	cm3/mol*K,for glass,therm expan
E1	3.07E-02	cm3/mol*K,forliquid,therm expan
e1	7.29E-04	cm3/g*K,for liq,sp.therm.expan
eg	3.28E-04	cm3/g*k,for glass,sp.therm expan
alpha liq. cubic	6.25E-04	1/K,cubic therm.expan., liq.state
alpha gl. cubic	2.81E-04	1/K, cubic therm.expan., gl.state
alpha liq. linear	2.08E-04	1/K, linear coeff.liq.therm.expan.
alpha gl. linear	9.35E-05	1/K, linear coeff., gl.therm.expan.
Specf Cp(s)	1.707	J/g*K,
Specf Cp(l)	2.097	J/g*K
Hm	8500	J/mol.
Tg	-18	C
Tm	181	C
SolPar(d)	7.665	Based on Vr,Cal1/2cm2/3disper.
SolPar(p)	0.000	Based on Vr,Cal1/2cm2/3polar
SolPar(h)	0.000	Based on Vr.Cal1/2cm2/3 H bond.
SolPrCal/Cm	7.665	Cal1/2cm2/3, based on Vr
SolParJ/Cm	15.682	J1/2/cm3/2
SurTension	31.98	erg/cm2
MChainGrps	2	Enter# of backboneatms/strucunit
C(constant)	15.45	molar lim.visc.,g1/4xcm3/2/mol3/4
K Theta	0.135	lim.visc.#,cm3xmol1/2/g3/2
M(Critical)	9300	critical molecular weight,g
Refr. Index	1.48	refractive index, nD
Dielc Cons	2.16	Dielectric Constant,D.C.
Bulk Mod.(K)	4.33	(K)BulkModulus, GPa
ShearMod.(G)	1.49	(G)Shear or Rigidity Modulus, GPa
Ten. Mod.(E)	4.02	(E)Tensile or Young'sModulus,GPa
PoissonRatio	0.35	Poisson's Ratio
SndVel.Long.	2715.1	Longitudinal SoundVelocity, m/s
SndVel.Shear	1320.0	Transverse Sound Velocity, m/s
SndAbs,Long.	1.81	Long.SoundAbs.Coeff., dB/cm
SndAbs,Shr.	9.04	Shear/Trans Snd.AbsCoeff,dB/cm
ThermalCond	0.20	J/s*m*K
SolPr(d)Cal	7.649	Based on Vg,cal1/2cm2/3
SolPr(p)Cal	0.000	Based on Vg. cal1/2cm2/3
SolPr(h)Cal	0.000	Based on Vg. cal1/2cm2/3
SolPrCal/Cm	7.649	Cal1/2cm2/3, based on Vg
E sub eta Inf.	43.	E,Act.energy of visc flow,g KJ/mol
M(Actual)	9300	Molecular Wt. @or>M(crit)
T	298	Temp.of interest,degK
A	8.92	Empirical Constant
Log Eta Cr.Inf.	-5.098	Visc.@M(critical),T=inf.,Ns/m2
Log Eta Cr.	2.55	Visc.@T of interest,Ns/m2
LogNewtVisc.	2.550	Newt.Visc.@T,N*s/m2
Gas	N2	Enter Gas of Interest from Table
Log Perm	-13.2	P,Coeff.of Perm,cm3*cm/cm2*s*Pa
Delta H(A)@298K	-70.7	Free Enthalpy of Poly., kJ/mol
TempCorrFac(B)	317	K
Delt Gsub f @T	23.8	Free Energy Change @T, kJ/mol
TDecompGrp	None	Enter Poly.Grp.for Therm Decomp
TD1/2	665	Temp.ofHalfDecomp, deg.,K

Structure

$-(CHCH_2)-$
CH_3

Repeat Structural Groups Selected

$1 - CH_2$

$1 - CH$

$1 - CH_3$

Table 5.5 Estimated Properties of Poly(isobutylene)

		StructuralGroup
CALCULATED		Isobutylene
MOLECULAR		PolymerClass
PROPERTIES		Poly(isobutylene) Butyl Rubber
Mol.Wt.	56.1	g/mol.
Density Am .	0.840	g/cm3.
Density Gl.	0.822	g/cm3.
Density Cr.	0.949	g/cm3.
VdeWaalsVol	40.9	cm3/mol
Eg	1.84E-02	cm3/mol*K,for glass,therm expan
E1	4.09E-02	cm3/mol*K,forliquid,therm expan
e1	7.29E-04	cm3/g*K,for liq,sp.therm.expan
eg	3.28E-04	cm3/g*k,for glass,sp.therm expan
alpha liq. cubic	6.12E-04	1/K,cubic therm.expan., liq.state
alpha gl. cubic	2.70E-04	1/K, cubic therm.expan., gl.state
alpha liq. linear	2.04E-04	1/K, linear coeff.liq.therm.expan.
alpha gl. linear	8.99E-05	1/K, linear coeff., gl.therm.expan.
Specf Cp(s)	1.664	J/g*K,
Specf Cp(l)	1.989	J/g*K
Hm	11350	J/mol.
Tg	-83	C
Tm	117	C
SolPar(d)	7.609	Based on Vr,Cal1/2cm2/3disper.
SolPar(p)	0.000	Based on Vr,Cal1/2cm2/3polar
SolPar(h)	0.000	Based on Vr.Cal1/2cm2/3 H bond.
SolPrCal/Cm	7.609	Cal1/2cm2/3, based on Vr
SolParJ/Cm	15.569	J1/2/cm3/2
SurTension	27.30	erg/cm2
MChainGrps	2	Enter# of backboneatms/strucunit
C(constant)	18.85	molar lim.visc.,g1/4xcm3/2/mol3/4
K Theta	0.113	lim.visc.#,cm3xmol1/2/g3/2
M(Critical)	13258	critical molecular weight,g
Refr. Index	1.49	refractive index, nD
Dielc Cons	2.15	Dielectric Constant,D.C.
Bulk Mod.(K)	3.70	(K)BulkModulus, GPa
ShearMod.(G)	1.29	(G)Shear or Rigidity Modulus, GPa
Ten. Mod.(E)	3.46	(E)Tensile or Young'sModulus,GPa
PoissonRatio	0.34	Poisson's Ratio
SndVel.Long.	2540.2	Longitudinal SoundVelocity, m/s
SndVel.Shear	1238.8	Transverse Sound Velocity, m/s
SndAbs,Long.	1.76	Long.SoundAbs.Coeff., dB/cm
SndAbs,Shr.	8.80	Shear/Trans Snd.AbsCoeff,dB/cm
ThermalCond	0.18	J/s*m*K
SolPr(d)Cal	7.448	Based on Vg,cal1/2cm2/3
SolPr(p)Cal	0.000	Based on Vg. cal1/2cm2/3
SolPr(h)Cal	0.000	Based on Vg. cal1/2cm2/3
SolPrCal/Cm	7.448	Cal1/2cm2/3, based on Vg
E sub eta Inf.	48.08	E,Act.energy of visc flow,g KJ/mol
M(Actual)	13258	Molecular Wt. @or>M(crit)
T	298	Temp.of interest,degK
A	13.27	Empirical Constant
Log Eta Cr.Inf.	-5.487	Visc.@M(critical),T=inf.,Ns/m2
Log Eta Cr.	2.97	Visc.@T of interest,Ns/m2
LogNewtVisc .	2.965	Newt.Visc.@T,N*s/m2
Gas	N2	Enter Gas of Interest from Table
Log Perm.	-13.3	P,Coeff.of Perm,cm3*cm/cm2*s*Pa
Delta H(A)@298K	-94	Free Enthalpy of Poly., kJ/mol
TempCorrFac(B)	432	K
Delt Gsub f @T	34.7	Free Energy Change @T, kJ/mol
TDecompGrp	None	Enter Poly.Grp.for Therm Decomp
TD1/2	622	Temp.ofHalfDecomp, deg., K

Structure

Repeat Structural Groups Selected

CH_3
$-(CH_2C)-$
CH_3

$1 - C(al)$

$1 - CH_2 \; ; \; 2 - CH_3$

181

Table 5.6 Estimated Properties of Poly(vinyl chloride)

CALCULATED MOLECULAR PROPERTIES		StructuralGroup
		Vinyl Chloride
		PolymerClass
		Poly(vinyl chloride)
Mol.Wt.	62.51	g/mol.
Density Am.	1.398	g/cm3.
Density Gl.	1.383	g/cm3.
Density Cr.	1.580	g/cm3.
VdeWaalsVol	29.23	cm3/mol
Eg	1.32E-02	cm3/mol*K,for glass,therm expan
E1	2.92E-02	cm3/mol*K,forliquid,therm expan
e1	4.68E-04	cm3/g*K,for liq,sp.therm.expan.
eg	2.10E-04	cm3/g*k,for glass,sp.therm expan
alpha liq. cubic	6.54E-04	1/K,cubic therm.expan., liq.state
alpha gl. cubic	2.91E-04	1/K, cubic therm.expan., gl.state
alpha liq. linear	2.18E-04	1/K, linear coeff.liq.therm.expan.
alpha gl. linear	9.70E-05	1/K, linear coeff., gl.therm.expan.
Specf Cp(s)	1.089	J/g*K,
Specf Cp(l)	1.458	J/g*K
Hm	11000	J/mol.
Tg	81	C
Tm	278	C
SolPar(d)	8.747	Based on Vr,Cal1/2cm2/3disper.
SolPar(p)	6.014	Based on Vr,Cal1/2cm2/3polar
SolPar(h)	1.462	Based on Vr.Cal1/2cm2/3 H bond.
SolPrCal/Cm	10.715	Cal1/2cm2/3, based on Vr
SolParJ/Cm	21.924	J1/2/cm3/2
SurTension	42.19	erg/cm2
MChainGrps	2	Enter# of backboneatms/strucunit
C(constant)	24.15	molar lim.visc.,g1/4xcm3/2/mol3/4
K Theta	0.149	lim.visc.#,cm3xmol1/2/g3/2
M(Critical)	7586	critical molecular weight,g
Refr. Index	1.54	refractive index, nD
Dielc Cons	3.13	Dielectric Constant,D.C.
Bulk Mod.(K)	5.54	(K)BulkModulus, GPa
ShearMod.(G)	1.71	(G)Shear or Rigidity Modulus, GPa
Ten. Mod.(E)	4.65	(E)Tensile or Young'sModulus,GPa
PoissonRatio	0.36	Poisson's Ratio
SndVel.Long.	2364.5	Longitudinal SoundVelocity, m/s
SndVel.Shear	1105.0	Transverse Sound Velocity, m/s
SndAbs,Long .	2.41	Long.SoundAbs.Coeff., dB/cm
SndAbs,Shr.	12.06	Shear/Trans Snd.AbsCoeff,dB/cm
ThermalCond	0.18	J/s*m*K
SolPr(d)Cal	8.651	Based on Vg,cal1/2cm2/3
SolPr(p)Cal	5.947	Based on Vg. cal1/2cm2/3
SolPr(h)Cal	1.454	Based on Vg. cal1/2cm2/3
SolPrCal/Cm	10.598	Cal1/2cm2/3, based on Vg
E sub eta Inf.	85	E,Act.energy of visc flow,g KJ/mol
M(Actual)	7586	Molecular Wt. @or>M(crit)
T	12.59	Empirical Constant
Log Eta Cr.Inf.	-8.637	Visc.@M(critical),T=inf.,Ns/m2
Log Eta Cr.	6.33	Visc.@T of interest,Ns/m2
LogNewtVisc	6.330	Newt.Visc.@T,N*s/m2
Gas	N2	Enter Gas of Interest from Table
Log Perm.	-15.1	P,Coeff.of Perm,cm3*cm/cm2*s*Pa
Delta H(A)@298K	-73.7	Free Enthalpy of Poly., kJ/mol
TempCorrFac(B)	213	K
Delt Gsub f @T	-10.2	Free Energy Change @T, kJ/mol
TDecompGrp	None	Enter Poly.Grp.for Therm Decomp
TD1/2	544	Temp.ofHalfDecomp, deg., K

Structure	Repeat Structural Groups Selected
	$1 - CH_2$
	$1 - CHCl$

$$-(CHCH_2)-$$
$$\ \ |$$
$$Cl$$

182

Table 5.7 Estimated Properties of Poly(vinyl fluoride)

		StructuralGroup
CALCULATED		Vinyl Fluoride
MOLECULAR		PolymerClass
PROPERTIES		Poly(vinyl fluoride)
Mol.Wt.	46.05	g/mol.
Density Am.	1.269	g/cm3.
Density Gl.	1.272	g/cm3.
Density Cr.	1.434	g/cm3.
VdeWaalsVol	23.23	cm3/mol
Eg	1.05E-02	cm3/mol*K,for glass,therm expan
E1	2.32E-02	cm3/mol*K,forliquid,therm expan
e1	5.04E-04	cm3/g*K,for liq,sp.therm.expan
eg	2.27E-04	cm3/g*k,for glass,sp.therm expan
alpha liq. cubic	6.40E-04	1/K,cubic therm.expan., liq.state
alpha gl. cubic	2.89E-04	1/K, cubic therm.expan., gl.state
alpha liq. linear	2.13E-04	1/K, linear coeff.liq.therm.expan.
alpha gl. linear	9.63E-05	1/K, linear coeff., gl.therm.expan.
Specf Cp(s)	1.354	J/g*K,
Specf Cp(l)	1.571	J/g*K
Hm	7500	J/mol.
Tg	-18	C
Tm	192	C
SolPar(d)	6.530	Based on Vr,Cal1/2cm2/3disper.
SolPar(p)	0.000	Based on Vr,Cal1/2cm2/3polar
SolPar(h)		0.000 Based on Vr.Cal1/2cm2/3 H bond.
SolPrCal/Cm	6.530	Cal1/2cm2/3, based on Vr
SolParJ/Cm	13.361	J1/2/cm3/2
SurTension	32.75	erg/cm2
MChainGrps	2	Enter# of backboneatms/strucunit
C(constant)	24.15	molar lim.visc.,g1/4xcm3/2/mol3/4
K Theta	0.275	lim.visc.#,cm3xmol1/2/g3/2
M(Critical)	2234	critical molecular weight,g
Refr. Index	1.42	refractive index, nD
Dielc Cons	8.05	Dielectric Constant,D.C.
Bulk Mod.(K)	3.55	(K)BulkModulus, GPa
ShearMod.(G)	0.75	(G)Shear or Rigidity Modulus, GPa
Ten. Mod.(E)	2.11	(E)Tensile or Young'sModulus,GPa
PoissonRatio	0.40	Poisson's Ratio
SndVel.Long.	1894.7	Longitudinal SoundVelocity, m/s
SndVel.Shear	771.0	Transverse Sound Velocity, m/s
SndAbs,Long.	4.03	Long.SoundAbs.Coeff., dB/cm
SndAbs,Shr.	20.15	Shear/Trans Snd.AbsCoeff,dB/cm
ThermalCond	0.16	J/s*m*K
SolPr(d)Cal	6.548	Based on Vg,cal1/2cm2/3
SolPr(p)Cal	0.000	Based on Vg. cal1/2cm2/3
SolPr(h)Cal	0.000	Based on Vg. cal1/2cm2/3
SolPrCal/Cm	6.548	Cal1/2cm2/3, based on Vg
E sub eta Inf .	0.8	E,Act.energy of visc flow,g KJ/mol
M(Actual)	2234	Molecular Wt. @or>M(crit)
T	298	Temp.of interest,degK
A	0.16	Empirical Constant
Log Eta Cr.Inf.	-1.464	Visc.@M(critical),T=inf.,Ns/m2
Log Eta Cr.	-1.33	Visc.@T of interest,Ns/m2
LogNewtVisc.	-1.331	Newt.Visc.@T,N*s/m2
Gas	N2	Enter Gas of Interest from Table
Log Perm.	-15.9	P,Coeff.of Perm,cm3*cm/cm2*s*Pa
Delta H(A)@298K	-219.7	Free Enthalpy of Poly., kJ/mol
TempCorrFac(B)	216	K
Delt Gsub f @T	-155.3	Free Energy Change @T, kJ/mol
TDecompGrp	None	Enter Poly.Grp.for Therm Decomp
TD1/2	662	Temp.ofHalfDecomp, deg., K

Structure	Repeat Structural Groups Selected

—(CHCH$_2$)—
    |
    F
	1 – CH$_2$
	 1 – CHF

Table 5.8 Estimated Properties of Poly(chlorotrifluoroethylene)

CALCULATED MOLECULAR PROPERTIES		StructuralGroup Chloro trifluoroethylene PolymerClass Poly(chlorotrifluoroethylene)
Mol.Wt.	116.48	g/mol.
Density Am.	2.012	g/cm3.
Density Gl.	1.885	g/cm3.
Density Cr.	2.273	g/cm3.
VdeWaalsVol	35.96	cm3/mol
Eg	1.62E-02	cm3/mol*K,for glass,therm expan
E1	3.60E-02	cm3/mol*K,forliquid,therm expan
e1	3.09E-04	cm3/g*K,for liq,sp.therm.expan
eg	1.39E-04	cm3/g*k,for glass,sp.therm expan
alpha liq. cubic	6.21E-04	1/K,cubic therm.expan., liq.state
alpha gl. cubic	2.62E-04	1/K, cubic therm.expan., gl.state
alpha liq. linear	2.07E-04	1/K, linear coeff.liq.therm.expan.
alpha gl. linear	8.73E-05	1/K, linear coeff., gl.therm.expan.
Specf Cp(s)	0.890	J/g*K,
Specf Cp(l)	1.010	J/g*K
Hm	7900	J/mol.
Tg	70	C
Tm	253	C
SolPar(d)	6.036	Based on Vr,Cal1/2cm2/3disper.
SolPar(p)	4.643	Based on Vr,Cal1/2cm2/3polar
SolPar(h)	1.285	Based on Vr.Cal1/2cm2/3 H bond.
SolPrCal/Cm	7.722	Cal1/2cm2/3, based on Vr
SolParJ/Cm	15.800	J1/2/cm3/2
SurTension	27.10	erg/cm2
MChainGrps	2	Enter# of backboneatms/strucunit
C(constant)	23.65	molar lim.visc.,g1/4xcm3/2/mol3/4
K Theta	----	lim.visc.#,cm3xmol1/2/g3/2
M(Critical)	----	critical molecular weight,g
Refr. Index	1.44	refractive index, nD
Dielc Cons	2.41	Dielectric Constant,D.C.
Bulk Mod.(K)	3.02	(K)BulkModulus, GPa
ShearMod.(G)	0.62	(G)Shear or Rigidity Modulus, GPa
Ten. Mod.(E)	1.75	(E)Tensile or Young'sModulus,GPa
PoissonRatio	0.40	Poisson's Ratio
SndVel.Long.	1382.4	Longitudinal SoundVelocity, m/s
SndVel.Shear	556.1	Transverse Sound Velocity, m/s
SndAbs,Long.	4.14	Long.SoundAbs.Coeff., dB/cm
SndAbs,Shr.	20.70	Shear/Trans Snd.AbsCoeff,dB/cm
ThermalCond	0.12	J/s*m*K
SolPr(d)Cal	5.655	Based on Vg,cal1/2cm2/3
SolPr(p)Cal	4.350	Based on Vg. cal1/2cm2/3
SolPr(h)Cal	1.243	Based on Vg. cal1/2cm2/3
SolPrCal/Cm	7.242	Cal1/2cm2/3, based on Vg
E sub eta Inf.	----	E,Act.energy of visc flow,g KJ/mol
M(Actual)	----	Molecular Wt. @or>M(crit)
T	298	Temp.of interest,degK
A	----	Empirical Constant
Log Eta Cr.Inf.	----	Visc.@M(critical),T=inf.,Ns/m2
Log Eta Cr. -	----	Visc.@T of interest,Ns/m2
LogNewtVisc.	-0.883	Newt.Visc.@T,N*s/m2
Gas	N2	Enter Gas of Interest from Table
Log Perm.	-14.5	P,Coeff.of Perm,cm3*cm/cm2*s*Pa
Delta H(A)@298K	-594	Free Enthalpy of Poly., kJ/mol
TempCorrFac(B)	253	K
Delt Gsub f @T	----	Free Energy Change @T, kJ/mol
TDecompGrp	None	Enter Poly.Grp.for Therm Decomp
TD1/2	650	Temp.ofHalfDecomp, deg., K

Structure

—(CF$_2$CFCl)—

Repeat Structural Groups Selected

2 – C(al)

3 – F

1 – Cl

184

Table 5.9 Estimated Properties of Poly(vinyl acetate)

		StructuralGroup
CALCULATED		Vinyl Acetate
MOLECULAR		PolymerClass
PROPERTIES		Poly(vinyl acetate)
Mol.Wt.	86.09	g/mol.
Density Am.	1.168	g/cm3.
Density Gl.	1.192	g/cm3.
Density Cr.	1.320	g/cm3.
VdeWaalsVol	45.88	cm3/mol
Eg	2.06E-02	cm3/mol*K,for glass,therm expan
E1	4.59E-02	cm3/mol*K,forliquid,therm expan
e1	5.33E-04	cm3/g*K,for liq,sp.therm.expan
eg	2.40E-04	cm3/g*k,for glass,sp.therm expan
alpha liq. cubic	6.23E-04	1/K,cubic therm.expan., liq.state
alpha gl. cubic	2.86E-04	1/K, cubic therm.expan., gl.state
alpha liq. linear	2.08E-04	1/K, linear coeff.liq.therm.expan.
alpha gl. linear	9.53E-05	1/K, linear coeff., gl.therm.expan.
Specf Cp(s)	1.369	J/g*K,
Specf Cp(l)	1.780	J/g*K
Hm	6000	J/mol.
Tg	14	C
Tm	239	C
SolPar(d)	7.693	Based on Vr,Cal1/2cm2/3disper.
SolPar(p)	3.250	Based on Vr,Cal1/2cm2/3polar
SolPar(h)	4.763	Based on Vr.Cal1/2cm2/3 H bond.
SolPrCal/Cm	9.614	Cal1/2cm2/3, based on Vr
SolParJ/Cm	19.670	J1/2cm3/2
SurTension	40.20	erg/cm2
MChainGrps	2	Enter# of backboneatms/strucunit
C(constant)	24.45	molar lim.visc.,g1/4xcm3/2/mol3/4
K Theta	0.081	lim.visc.#,cm3xmol1/2/g3/2
M(Critical)	25977	critical molecular weight,g
Refr. Index	1.47	refractive index, nD
Dielc Cons	3.32	Dielectric Constant,D.C.
Bulk Mod.(K)	3.64	(K)BulkModulus, GPa
ShearMod.(G)	1.13	(G)Shear or Rigidity Modulus, GPa
Ten. Mod.(E)	3.06	(E)Tensile or Young'sModulus,GPa
PoissonRatio	0.36	Poisson's Ratio
SndVel.Long .	2098.7	Longitudinal SoundVelocity, m/s
SndVel.Shea r	981.8	Transverse Sound Velocity, m/s
SndAbs,Long.	2.40	Long.SoundAbs.Coeff., dB/cm
SndAbs,Shr.	11.98	Shear/Trans Snd.AbsCoeff,dB/cm
ThermalCond	0.17	J/s*m*K
SolPr(d)Cal	7.853	Based on Vg,cal1/2cm2/3
SolPr(p)Cal	3.317	Based on Vg. cal1/2cm2/3
SolPr(h)Cal	4.813	Based on Vg. cal1/2cm2/3
SolPrCal/Cm	9.789	Cal1/2cm2/3, based on Vg
E sub eta Inf.	63.2	E,Act.energy of visc flow,g KJ/mol
M(Actual)	25977	Molecular Wt. @or>M(crit)
T	298	Temp.of interest,degK
A	11.52	Empirical Constant
Log Eta Cr.Inf.	-6.776	Visc.@M(critical),T=inf.,Ns/m2
Log Eta Cr.	4.34	Visc.@T of interest,Ns/m2
LogNewtVisc.	2.179	Newt.Visc.@T,N*s/m2
Gas	N2	Enter Gas of Interest from Table
Log Perm.	-14.1	P,Coeff.of Perm,cm3*cm/cm2*s*Pa
Delta H(A)@298K	-407.7	Free Enthalpy of Poly., kJ/mol
TempCorrFac(B)	433	K
Delt Gsub f @T	-278.7	Free Energy Change @T, kJ/mol
TDecompGrp	None	Enter Poly.Grp.for Therm Decomp
TD1/2	544	Temp.ofHalfDecomp, deg., K

Structure	Repeat Structural Groups Selected	
$-(CH_2CH)-$ $\quad\quad OCCH_3$ $\quad\quad\overset{\|}{O}$	$1 - CH_2$	$1 - CH$
	$1 - CH_3$	$1 - COO$

185

Table 5.10 Estimated Properties of Poly(butadiene)

		StructuralGroup
CALCULATED		Butadiene
MOLECULAR		PolymerClass
PROPERTIES		Poly(butadiene) Rubber
Mol.Wt.	54.1	g/mol.
Density Am.	0.892	g/cm3.
Density Gl.	0.910	g/cm3.
Density Cr.	1.008	g/cm3.
VdeWaalsVol	37.4	cm3/mol
Eg	1.68E-02	cm3/mol*K,for glass,therm expan
E1	3.74E-02	cm3/mol*K,forliquid,therm expan
e1	6.91E-04	cm3/g*K,for liq,sp.therm.expan
eg	3.11E-04	cm3/g*k,for glass,sp.therm expan
alpha liq. cubic	6.17E-04	1/K,cubic therm.expan., liq.state
alpha gl. cubic	2.83E-04	1/K, cubic therm.expan., gl.state
alpha liq. linear	2.06E-04	1/K, linear coeff.liq.therm.expan.
alpha gl. linear	9.44E-05	1/K, linear coeff., gl.therm.expan.
Specf Cp(s)	1.627	J/g*K,
Specf Cp(l)	1.915	J/g*K
Hm	9000	J/mol.
Tg	-116	C
Tm	30	C
SolPar(d)	8.010	Based on Vr,Cal1/2cm2/3disper.
SolPar(p)	0.000	Based on Vr,Cal1/2cm2/3polar
SolPar(h)	0.000	Based on Vr.Cal1/2cm2/3 H bond.
SolPrCal/Cm	8.010	Cal1/2cm2/3, based on Vr
SolParJ/Cm	16.389	J1/2/cm3/2
SurTension	35.39	erg/cm2
MChainGrps	4	Enter# of backboneatms/strucunit
C(constant)	22	molar lim.visc.,g1/4xcm3/2/mol3/4
K Theta	0.165	lim.visc.#,cm3xmol1/2/g3/2
M(Critical)	6180	critical molecular weight,g
Refr. Index	1.52	refractive index, nD
Dielc Cons	2.29	Dielectric Constant,D.C.
Bulk Mod.(K)	2.87	(K)BulkModulus, GPa
ShearMod.(G)	0.62	(G)Shear or Rigidity Modulus, GPa
Ten. Mod.(E)	1.74	(E)Tensile or Young'sModulus,GPa
PoissonRatio	0.40	Poisson's Ratio
SndVel.Long.	2035.4	Longitudinal SoundVelocity, m/s
SndVel.Shear	834.2	Transverse Sound Velocity, m/s
SndAbs,Long .	3.96	Long.SoundAbs.Coeff., dB/cm
SndAbs,Shr.	19.81	Shear/Trans Snd.AbsCoeff,dB/cm
ThermalCond	0.15	J/s*m*K
SolPr(d)Cal	8.172	Based on Vg,cal1/2cm2/3
SolPr(p)Cal	0.000	Based on Vg. cal1/2cm2/3
SolPr(h)Cal	0.000	Based on Vg. cal1/2cm2/3
SolPrCal/Cm	8.172	Cal1/2cm2/3, based on Vg
E sub eta Inf.	25.9	E,Act.energy of visc flow,a KJ/mol
M(Actual)	6180	Molecular Wt. @or>M(crit)
T	298	Temp.of interest,degK
A	8.62	Empirical Constant
Log Eta Cr.Inf.	-3.599	Visc.@M(critical),T=inf.,Ns/m2
Log Eta Cr.	0.95	Visc.@T of interest,Ns/m2
LogNewtVisc.	0.948	Newt.Visc.@T,N*s/m2
Gas	N2	Enter Gas of Interest from Table
Log Perm.	-12.3	P,Coeff.of Perm,cm3*cm/cm2*s*Pa
Delta H(A)@298K	32	Free Enthalpy of Poly., kJ/mol
TempCorrFac(B)	280	K
Delt Gsub f @T	115.4	Free Energy Change @T, kJ/mol
TDecompGrp	None	Enter Poly.Grp.for Therm Decomp
TD1/2	680	Temp.ofHalfDecomp, deg., K

Structure

$$-(CH_2CH=CHCH_2)-$$

Repeat Structural Groups Selected

$2 - CH_2$

$1 - CH = CH$

186

Table 5.11 Estimated Properties of Poly(isoprene)

		StructuralGroup
CALCULATED		Isoprene
MOLECULAR		PolymerClass
PROPERTIES		Poly(cis-1,4-isoprene) Natural Rubber
Mol.Wt.	68.12	g/mol.
Density Am.	0.900	g/cm3.
Density Gl:	0.907	g/cm3.
Density Cr.	1.017	g/cm3.
VdeWaalsVol	47.61	cm3/mol
Eg	2.14E-02	cm3/mol*K,for glass,therm expan
E1	4.76E-02	cm3/mol*K,forliquid,therm expan
e1	6.99E-04	cm3/g*K,for liq,sp.therm.expan
eg	3.15E-04	cm3/g*k, for glass,sp.therm expan
alpha liq. cubic	6.29E-04	1/K,cubic therm.expan., liq.state
alpha gl. cubic	2.85E-04	1/K, cubic therm.expan., gl.state
alpha liq. linear	2.10E-04	1/K, linear coeff.liq.therm.expan.
alpha gl. linear	9.51E-05	1/K, linear coeff., gl.therm.expan.
Specf Cp(s)	1.626	J/g*K,
Specf Cp(l)	1.982	J/g*K
Hm	9100	J/mol.
Tg	-67	C
Tm	37	C
SolPar(d)	7.916	Based on Vr,Cal1/2cm2/3disper.
SolPar(p)	0.000	Based on Vr,Cal1/2cm2/3polar
SolPar(h)	0.000	Based on Vr.Cal1/2cm2/3 H bond.
SolPrCal/Cm	7.916	Cal1/2cm2/3, based on Vr
SolParJ/Cm	16.196	J1/2/cm3/2
SurTension	36.03	erg/cm2
MChainGrps	4	Enter# of backboneatms/strucunit
C(constant)	24.4	molar lim.visc.,g1/4xcm3/2/mol3/4
K Theta	0.128	lim.visc.#,cm3xmol1/2/g3/2
M(Critical)	10267	critical molecular weight,g
Refr. Index	1.52	refractive index, nD
Dielc Cons	2.26	Dielectric Constant,D.C.
Bulk Mod.(K)	3.11	(K)BulkModulus, GPa
ShearMod.(G)	0.51	(G)Shear or Rigidity Modulus, GPa
Ten. Mod.(E)	1.46	(E)Tensile or Young'sModulus,GPa
PoissonRatio	0.42	Poisson's Ratio
SndVel.Long.	2053.0	Longitudinal SoundVelocity, m/s
SndVel.Shear	755.4	Transverse Sound Velocity, m/s
SndAbs,Long.	4.87	Long.SoundAbs.Coeff., dB/cm
SndAbs,Shr.	24.34	Shear/Trans Snd.AbsCoeff,dB/cm
ThermalCond	0.15	J/s*m*K
SolPr(d)Cal	7.979	Based on Vg,cal1/2cm2/3
SolPr(p)Cal	0.000	Based on Vg. cal1/2cm2/3
SolPr(h)Cal	0.000	Based on Vg. cal1/2cm2/3
SolPrCal/Cm	7.979	Cal1/2cm2/3, based on Vg
E sub eta Inf.	26.3	E,Act.energy of visc flow,g KJ/mol
M(Actual)	10267	Molecular Wt. @or>M(crit)
T	298	Temp.of interest,degK
A	6.70	Empirical Constant
Log Eta Cr.Inf.	-3.633	Visc.@M(critical),T=inf.,Ns/m2
Log Eta Cr.	0.98	Visc.@T of interest,Ns/m2
LogNewtVisc.	0.985	Newt.Visc.@T,N*s/m2
Gas	N2	Enter Gas of Interest from Table
Log Perm.	-12.3	P,Coeff.of Perm,cm3*cm/cm2*s*Pa
Delta H(A)@298K	-2	Free Enthalpy of Poly., kJ/mol
TempCorrFac(B)	387	K
Delt Gsub f @T	113.3	Free Energy Change @T, kJ/mol
TDecompGrp	None	Enter Poly.Grp.for Therm Decomp
TD1/2	595	Temp.ofHalfDecomp, deg., K

Structure

Repeat Structural Groups Selected

$$-(CH_2\underset{CH=C}{}\overset{CH_2)-}{\underset{CH_3}{}}$$

$2 - CH_2$

$1 - CH_3$

$1 - C = CH$

187

Table 5.12 Estimated Properties of Poly(styrene)

CALCULATED MOLECULAR PROPERTIES		StructuralGroup Styrene PolymerClass Poly(styrene)
Mol.Wt.	104.15	g/mol.
Density Am.	1.145	g/cm3.
Density Gl.	1.063	g/cm3.
Density Cr.	1.294	g/cm3.
VdeWaalsVol	62.85	cm3/mol
Eg	2.83E-02	cm3/mol*K,for glass,therm expan
E1	6.29E-02	cm3/mol*K,forliquid,therm expan
e1	6.03E-04	cm3/g*K,for liq,sp.therm.expan
eg	2.72E-04	cm3/g*k,for glass,sp.therm expan
alpha liq. cubic	6.91E-04	1/K,cubic therm.expan., liq.state
alpha gl. cubic	2.89E-04	1/K, cubic therm.expan., gl.state
alpha liq. linear	2.30E-04	1/K, linear coeff.liq.therm.expan.
alpha gl. linear	9.62E-05	1/K, linear coeff., gl.therm.expan.
Specf Cp(s)	1.215	J/g*K,
Specf Cp(l)	1.676	J/g*K
Hm	10400	J/mol.
Tg	99	C
Tm	197	C
SolPar(d)	9.565	Based on Vr,Cal1/2cm2/3disper.
SolPar(p)	0.591	Based on Vr,Cal1/2cm2/3polar
SolPar(h)	0.000	Based on Vr.Cal1/2cm2/3 H bond.
SolPrCal/Cm	9.583	Cal1/2cm2/3, based on Vr
SolParJ/Cm	19.609	J1/2/cm3/2
SurTension	39.64	erg/cm2
MChainGrps	2	Enter# of backboneatms/strucunit
C(constant)	30.15	molar lim.visc.,g1/4xcm3/2/mol3/4
K Theta	0.084	lim.visc.#,cm3xmol1/2/g3/2
M(Critical)	24064	critical molecular weight,g
Refr. Index	1.61	refractive index, nD
Dielc Cons	2.59	Dielectric Constant,D.C.
Bulk Mod.(K)	4.16	(K)BulkModulus, GPa
ShearMod.(G)	1.39	(G)Shear or Rigidity Modulus, GPa
Ten. Mod.(E)	3.75	(E)Tensile or Young'sModulus,GPa
PoissonRatio	0.35	Poisson's Ratio
SndVel.Long.	2292.3	Longitudinal SoundVelocity, m/s
SndVel.Shear	1101.7	Transverse Sound Velocity, m/s
SndAbs,Long .	1.99	Long.SoundAbs.Coeff., dB/cm
SndAbs,Shr.	9.96	Shear/Trans Snd.AbsCoeff,dB/cm
ThermalCond	0.16	J/s*m*K
SolPr(d)Cal	8.877	Based on Vg,cal1/2cm2/3
SolPr(p)Cal	0.549	Based on Vg. cal1/2cm2/3
SolPr(h)Cal	0.000	Based on Vg. cal1/2cm2/3
SolPrCal/Cm	8.894	Cal1/2cm2/3, based on Vg
E sub eta Inf.	57.5	E,Act.energy of visc flow,g KJ/mol
M(Actual)	24064	Molecular Wt. @or>M(crit)
T	298	Temp.of interest,degK
A	8.10	Empirical Constant
Log Eta Cr.Inf.	-6.288	Visc.@M(critical),T=inf.,Ns/m2
Log Eta Cr.	3.82	Visc.@T of interest,Ns/m2
LogNewtVisc .	3.82	Newt.Visc.@T,N*s/m2
Gas	N2	Enter Gas of Interest from Table
Log Perm.	-13.4	P,Coeff.of Perm,cm3*cm/cm2*s*Pa
Delta H(A)@298K	62.3	Free Enthalpy of Poly., kJ/mol
TempCorrFac(B)	389	K
Delt Gsub f @T	178.2	Free Energy Change @T, kJ/mol
TDecompGrp	None	Enter Poly.Grp.for Therm Decomp
TD1/2	634	Temp.ofHalfDecomp, deg., K

Structure	Repeat Structural Groups Selected
—(CHCH₂)—	1 – CH 1 – CH₂ 1 – C₆H₅

188

Table 5.13 Estimated Properties of Poly(α methyl styrene)

CALCULATED MOLECULAR PROPERTIES		StructuralGroup
		α methyl styrene
		PolymerClass
		Poly(α methyl styrene)
Mol.Wt.	118.17	g/mol.
Density Am.	1.088	g/cm3.
Density Gl.	1.010	g/cm3.
Density Cr.	1.229	g/cm3.
VdeWaalsVol	73.07	cm3/mol
Eg	3.29E-02	cm3/mol*K,for glass,therm expan
E1	7.31E-02	cm3/mol*K,forliquid,therm expan
e1	6.18E-04	cm3/g*K,for liq,sp.therm.expan
eg	2.78E-04	cm3/g*k,for glass,sp.therm expan
alpha liq. cubic	6.73E-04	1/K,cubic therm.expan., liq.state
alpha gl. cubic	2.81E-04	1/K, cubic therm.expan., gl.state
alpha liq. linear	2.24E-04	1/K, linear coeff.liq.therm.expan.
alpha gl. linear	9.36E-05	1/K, linear coeff., gl.therm.expan.
Specf Cp(s)	1.253	J/g*K,
Specf Cp(l)	1.675	J/g*K
Hm	13250	J/mol.
Tg	168	C
Tm	216	C
SolPar(d)	9.221	Based on Vr,Cal1/2cm2/3disper.
SolPar(p)	0.495	Based on Vr,Cal1/2cm2/3polar
SolPar(h)	0.000	Based on Vr.Cal1/2cm2/3 H bond.
SolPrCal/Cm	9.235	Cal1/2cm2/3, based on Vr
SolParJ/Cm	18.895	J1/2/cm3/2
SurTension	35.10	erg/cm2
MChainGrps	2	Enter# of backboneatms/strucunit
C(constant)	33.55	molar lim.visc.,g1/4xcm3/2/mol3/4
K Theta	0.081	lim.visc.#,cm3xmol1/2/g3/2
M(Critical)	26010	critical molecular weight,g
Refr. Index	1.59	refractive index, nD
Dielc Cons	2.54	Dielectric Constant,D.C.
Bulk Mod.(K)	3.83	(K)BulkModulus, GPa
ShearMod.(G)	1.29	(G)Shear or Rigidity Modulus, GPa
Ten. Mod.(E)	3.48	(E)Tensile or Young'sModulus,GPa
PoissonRatio	0.35	Poisson's Ratio
SndVel.Long.	2259.2	Longitudinal SoundVelocity, m/s
SndVel.Shea r	1090.0	Transverse Sound Velocity, m/s
SndAbs,Long.	1.93	Long.SoundAbs.Coeff., dB/cm
SndAbs,Shr.	9.66	Shear/Trans Snd.AbsCoeff,dB/cm
ThermalCond	0.15	J/s*m*K
SolPr(d)Cal	8.560	Based on Vg,cal1/2cm2/3
SolPr(p)Cal	0.459	Based on Vg. cal1/2cm2/3
SolPr(h)Cal	0.000	Based on Vg. cal1/2cm2/3
SolPrCal/Cm	8.572	Cal1/2cm2/3, based on Vg
E sub eta Inf.	58.2	E,Act.energy of visc flow,g KJ/mol
M(Actual)	26010	Molecular Wt. @or>M(crit)
T	298	Temp.of interest,degK
A	6.91	Empirical Constant
Log Eta Cr.Inf.	-6.349	Visc.@M(critical),T=inf.,Ns/m2
Log Eta Cr.	3.89	Visc.@T of interest,Ns/m2
LogNewtVisc.	3.885	Newt.Visc.@T,N*s/m2
Gas	N2	Enter Gas of Interest from Table
Log Perm.	-13.5	P,Coeff.of Perm,cm3*cm/cm2*s*Pa
Delta H(A)@298K	39	Free Enthalpy of Poly., kJ/mol
TempCorrFac(B)	504	K
Delt Gsub f @T	189.2	Free Energy Change @T, kJ/mol
TDecompGrp	None	Enter Poly.Grp.for Therm Decomp
TD1/2	563	Temp.ofHalfDecomp, deg., K

Structure	Repeat Structural Groups Selected	
CH3	1 – C(ar)	1 – CH3
–(CCH2)–	1 – CH2	1 – C6H5

189

Table 5.14 Estimated Properties of Poly(carbonate)

		StructuralGroup
CALCULATED		2,2 propanebis(4-phenyl carbonate)
MOLECULAR		PolymerClass
PROPERTIES		Poly(carbonate)
Mol.Wt.	254.26	g/mol.
Density Am.	1.197	g/cm3.
Density Gl.	1.208	g/cm3.
Density Cr.	1.353	g/cm3.
VdeWaalsVol	136.21	cm3/mol
Eg	6.13E-02	cm3/mol*K,for glass,therm expan
E1	1.36E-01	cm3/mol*K,forliquid,therm expan
e1	5.36E-04	cm3/g*K,for liq,sp.therm.expan
eg	2.41E-04	cm3/g*k,for glass,sp.therm expan
alpha liq. cubic	6.41E-04	1/K,cubic therm.expan., liq.state
alpha gl. cubic	2.91E-04	1/K, cubic therm.expan.,gl.state
alpha liq. linear	2.14E-04	1/K, linear coeff.liq.therm.expan.
alpha gl. linear	9.71E-05	1/K, linear coeff., gl.therm.expan.
Specf Cp(s)	1.134	J/g*K,
Specf Cp(l)	1.605	J/g*K
Hm	31350	J/mol.
Tg	148	C
Tm	231	C
SolPar(d)	8.746	Based on Vr,Cal1/2cm2/3disper.
SolPar(p)	2.110	Based on Vr,Cal1/2cm2/3polar
SolPar(h)	3.354	Based on Vr.Cal1/2cm2/3 H bond.
SolPrCal/Cm	9.602	Cal1/2cm2/3, based on Vr
SolParJ/Cm	19.646	J1/2/cm3/2
SurTension	44.36	erg/cm2
MChainGrps	12	Enter# of backboneatms/strucunit
C(constant)	118.6	molar lim.visc.,g1/4xcm3/2/mol3/4
K Theta	0.218	lim.visc.#,cm3xmol1/2/g3/2
M(Critical)	3570	critical molecular weight,g
Refr. Index	1.59	refractive index, nD
Dielc Cons	3.05	Dielectric Constant,D.C.
Bulk Mod.(K)	4.60	(K)BulkModulus, GPa
ShearMod.(G)	1.18	(G)Shear or Rigidity Modulus, GPa
Ten. Mod.(E)	3.27	(E)Tensile or Young'sModulus,GPa
PoissonRatio	0.38	Poisson's Ratio
SndVel.Long.	2271.1	Longitudinal SoundVelocity, m/s
SndVel.Shear	993.5	Transverse Sound Velocity, m/s
SndAbs,Long.	3.27	Long.SoundAbs.Coeff., dB/cm
SndAbs,Shr.	16.33	Shear/Trans Snd.AbsCoeff,dB/cm
ThermalCond	0.15	J/s*m*K
SolPr(d)Cal	8.827	Based on Vg,cal1/2cm2/3
SolPr(p)Cal	2.130	Based on Vg. cal1/2cm2/3
SolPr(h)Cal	3.370	Based on Vg. cal1/2cm2/3
SolPrCal/Cm	9.685	Cal1/2cm2/3, based on Vg
E sub eta Inf.	84.8	E,Act.energy of visc flow,g KJ/mol
M(Actual)	3570	Molecular Wt. @or>M(crit)
T	298	Temp.of interest,degK
A	10.55	Empirical Constant
Log Eta Cr.Inf.	-8.607	Visc.@M(critical),T=inf.,Ns/m2
Log Eta Cr.	6.30	Visc.@T of interest,Ns/m2
LogNewtVisc.	6.297	Newt.Visc.@T,N*s/m2
Gas	N2	Enter Gas of Interest from Table
Log Perm.	-13.5	P,Coeff.of Perm,cm3*cm/cm2*s*Pa
Delta H(A)@298K	----	Free Enthalpy of Poly., kJ/mol
TempCorrFac(B)	690	K
Delt Gsub f @T	----	Free Energy Change @T, kJ/mol
TDecompGrp	None	Enter Poly.Grp.for Therm Decomp
TD1/2	602	Temp.ofHalfDecomp, deg., K

Structure

Repeat Structural Groups Selected

1 – C(ar)   2 – CH$_3$

2 – p – C$_6$H$_4$ 1 – OCOO

190

Table 5.15 Estimated Properties of Poly(ethylene terephthalate)

CALCULATED MOLECULAR PROPERTIES		StructuralGroup
		Ethylene Terephthalate
		PolymerClass
		PET
Mol.Wt.	192.17	g/mol.
Density Am.	1.302	g/cm3.
Density Gl.	1.342	g/cm3.
Density Cr.	1.471	g/cm3.
VdeWaalsVol	94.18	cm3/mol
Eg	4.24E-02	cm3/mol*K,for glass,therm expan
E1	9.42E-02	cm3/mol*K,forliquid,therm expan
e1	4.90E-04	cm3/g*K,for liq,sp.therm.expan
eg	2.21E-04	cm3/g*k,for glass,sp.therm expan
alpha liq. cubic	6.38E-04	1/K,cubic therm.expan., liq.state
alpha gl. cubic	2.96E-04	1/K, cubic therm.expan., gl.state
alpha liq. linear	2.13E-04	1/K, linear coeff.liq.therm.expan.
alpha gl. linear	9.87E-05	1/K, linear coeff., gl.therm.expan.
Specf Cp(s)	1.153	J/g*K,
Specf Cp(l)	1.581	J/g*K
Hm	23000	J/mol.
Tg	64	C
Tm	249	C
SolPar(d)	8.576	Based on Vr,Cal1/2cm2/3disper.
SolPar(p)	3.266	Based on Vr,Cal1/2cm2/3polar
SolPar(h)	4.760	Based on Vr.Cal1/2cm2/3 H bond.
SolPrCal/Cm	10.338	Cal1/2cm2/3, based on Vr
SolParJ/Cm	21.152	J1/2/cm3/2
SurTension	42.49	erg/cm2
MChainGrps	10	Enter# of backboneatms/strucunit
C(constant)	81	molar lim.visc.,g1/4xcm3/2/mol3/4
K Theta	0.178	lim.visc.#,cm3xmol1/2/g3/2
M(Critical)	5354	critical molecular weight,g
Refr. Index	1.57	refractive index, nD
Dielc Cons	3.51	Dielectric Constant,D.C.
Bulk Mod.(K)	4.49	(K)BulkModulus, GPa
ShearMod.(G)	1.09	(G)Shear or Rigidity Modulus, GPa
Ten. Mod.(E)	3.06	(E)Tensile or Young'sModulus,GPa
PoissonRatio	0.39	Poisson's Ratio
SndVel.Long.	2139.3	Longitudinal SoundVelocity, m/s
SndVel.Shear	920.7	Transverse Sound Velocity, m/s
SndAbs,Long.	3.45	Long.SoundAbs.Coeff., dB/cm
SndAbs,Shr.	17.26	Shear/Trans Snd.AbsCoeff,dB/cm
ThermalCond	0.16	J/s*m*K
SolPr(d)Cal	8.840	Based on Vg,cal1/2cm2/3
SolPr(p)Cal	3.366	Based on Vg. cal1/2cm2/3
SolPr(h)Cal	4.833	Based on Vg. cal1/2cm2/3
SolPrCal/Cm	10.622	Cal1/2cm2/3, based on Vg
E sub eta Inf.	47.10	E,Act.energy of visc flow,g KJ/mol
M(Actual)	5354	Molecular Wt. @or>M(crit)
T	298	Temp.of interest,degK
A	7.32	Empirical Constant
Log Eta Cr.Inf.	-5.404	Visc.@M(critical),T=inf.,Ns/m2
Log Eta Cr.	2.88	Visc.@T of interest,Ns/m2
LogNewtVisc.	2.876	Newt.Visc.@T,N*s/m2
Gas	N2	Enter Gas of Interest from Table
Log Perm.	-13.0	P,Coeff.of Perm,cm3*cm/cm2*s*Pa
Delta H(A)@298K	-618	Free Enthalpy of Poly., kJ/mol
TempCorrFac(B)	616	K
Delt Gsub f @T	----	Free Energy Change @T, kJ/mol
TDecompGrp	None	Enter Poly.Grp.for Therm Decomp
TD1/2	721	Temp.ofHalfDecomp, deg., K

Structure

$-(OCH_2CH_2O-\overset{O}{\underset{\|}{C}}-\langle\bigcirc\rangle-\overset{O}{\underset{\|}{C}})-$

Repeat Structural Groups Selected

$2 - CH_2$

$1 - p - C_6H_4$

$2 - COO$

191

Table 5.16 Estimated Properties of Poly(butylene terephthalate)

		StructuralGroup
CALCULATED		Butylene terephthalate
MOLECULAR		PolymerClass
PROPERTIES		Poly(butylene terephthalate)
Mol.Wt.	220.23	g/mol.
Density Am.	1.220	g/cm3.
Density Gl.	1.259	g/cm3.
Density Cr.	1.379	g/cm3.
VdeWaalsVol	114.64	cm3/mol
Eg	5.16E-02	cm3/mol*K,for glass,therm expan
E1	1.15E-01	cm3/mol*K,forliquid,therm expan
e1	5.21E-04	cm3/g*K,for liq,sp.therm.expan
eg	2.34E-04	cm3/g*k,for glass,sp.therm expan
alpha liq. cubic	6.35E-04	1/K,cubic therm.expan., liq.state
alpha gl. cubic	2.95E-04	1/K, cubic therm.expan., gl.state
alpha liq. linear	2.12E-04	1/K, linear coeff.liq.therm.expan.
alpha gl. linear	9.83E-05	1/K, linear coeff., gl.therm.expan.
Specf Cp(s)	1.236	J/g*K,
Specf Cp(l)	1.656	J/g*K
Hm	31000	J/mol.
Tg	41	C
Tm	235	C
SolPar(d)	8.475	Based on Vr,Cal1/2cm2/3disper.
SolPar(p)	2.670	Based on Vr,Cal1/2cm2/3polar
SolPar(h)	4.304	Based on Vr.Cal1/2cm2/3 H bond.
SolPrCal/Cm	9.873	Cal1/2cm2/3, based on Vr
SolParJ/Cm	20.202	J1/2/cm3/2
SurTension	41.08	erg/cm2
MChainGrps	12	Enter# of backboneatms/strucunit
C(constant)	94.1	molar lim.visc.,g1/4xcm3/2/mol3/4
K Theta	0.183	lim.visc.#,cm3xmol1/2/g3/2
M(Critical)	5070	critical molecular weight,g
Refr. Index	1.55	refractive index, nD
Dielc Cons	3.32	Dielectric Constant,D.C.
Bulk Mod.(K)	4.57	(K)BulkModulus, GPa
ShearMod.(G)	1.08	(G)Shear or Rigidity Modulus, GPa
Ten. Mod.(E)	3.01	(E)Tensile or Young'sModulus,GPa
PoissonRatio	0.39	Poisson's Ratio
SndVel.Long.	2220.1	Longitudinal SoundVelocity, m/s
SndVel.Shear	942.7	Transverse Sound Velocity, m/s
SndAbs,Long.	3.60	Long.SoundAbs.Coeff., dB/cm
SndAbs,Shr.	18.0	Shear/Trans Snd.AbsCoeff,dB/cm
ThermalCond	0.17	J/s*m*K
SolPr(d)Cal	8.747	Based on Vg,cal1/2cm2/3
SolPr(p)Cal	2.756	Based on Vg. cal1/2cm2/3
SolPr(h)Cal	4.373	Based on Vg. cal1/2cm2/3
SolPrCal/Cm2/3	10.160	Cal1/2cm2/3, based on Vg
E sub eta Inf.	44.1	E,Act.energy of visc flow,g KJ/mol
M(Actual)	5071	Molecular Wt. @or>M(crit)
T	298	Temp.of interest,degK
A	7.36	Empirical Constant
Log Eta Cr.Inf.	-5.147	Visc.@M(critical),T=inf.,Ns/m2
Log Eta Cr.	2.60	Visc.@T of interest,Ns/m2
LogNewtVisc .	2.603	Newt.Visc.@T,N*s/m2
Gas	N2	Enter Gas of Interest from Table
Log Perm.	-13.2	P,Coeff.of Perm,cm3*cm/cm2*s*Pa
Delta H(A)@298K	-662	Free Enthalpy of Poly., kJ/mol
TempCorrFac(B)	820	K
Delt Gsub f @T	----	Free Energy Change @T, kJ/mol
TDecompGrp	None	Enter Poly.Grp.for Therm Decomp
TD1/2	716	Temp.ofHalfDecomp, deg., K

Structure	Repeat Structural Groups Selected
	$4 - CH_2$
	$1 - p - C_6H_4$
	$2 - COO$

Table 5.17 Estimated Properties of Poly(3,3-bischloromethyl oxacyclobutane)

		StructuralGroup
CALCULATED		3,3-bis(chloromethyl-oxacyclobutane)
MOLECULAR		PolymerClass
PROPERTIES		PENTON
Mol.Wt.	155.05	g/mol.
Density Am.	1.338	g/cm3.
Density Gl.	1.316	g/cm3.
Density Cr.	1.512	g/cm3.
VdeWaalsVol	74.15	cm3/mol
Eg	3.34E-02	cm3/mol*K,for glass,therm expan
E1	7.42E-02	cm3/mol*K,forliquid,therm expan
e1	4.78E-04	cm3/g*K,for liq,sp.therm.expan
eg	2.15E-04	cm3/g*k,for glass,sp.therm expan
alpha liq. cubic	6.40E-04	1/K,cubic therm.expan., liq.state
alpha gl. cubic	2.83E-04	1/K, cubic therm.expan., gl.state
alpha liq. linear	2.13E-04	1/K, linear coeff.liq.therm.expan.
alpha gl. linear	9.44E-05	1/K, linear coeff., gl.therm.expan.
Specf Cp(s)	1.152	J/g*K,
Specf Cp(l)	1.575	J/g*K
Hm	24650	J/mol.
Tg	2	C
Tm	193	C
SolPar(d)	8.480	Based on Vr,Cal1/2cm2/3disper.
SolPar(p)	4.938	Based on Vr,Cal1/2cm2/3polar
SolPar(h)	2.799	Based on Vr.Cal1/2cm2/3 H bond.
SolPrCal/Cm	10.204	Cal1/2cm2/3, based on Vr
SolParJ/Cm	20.878	J1/2/cm3/2
SurTension	39.01	erg/cm2
MChainGrps	4	Enter# of backboneatms/strucunit
C(constant)	53.8	molar lim.visc.,g1/4xcm3/2/mol3/4
K Theta	0.120	lim.visc.#,cm3xmol1/2/g3/2
M(Critical)	11659	critical molecular weight,g
Refr. Index	1.55	refractive index, nD
Dielc Cons	3.14	Dielectric Constant,D.C.
Bulk Mod.(K)	6.27	(K)BulkModulus, GPa
ShearMod.(G)	1.34	(G)Shear or Rigidity Modulus, GPa
Ten. Mod.(E)	3.76	(E)Tensile or Young'sModulus,GPa
PoissonRatio	0.40	Poisson's Ratio
SndVel.Long.	2453.4	Longitudinal SoundVelocity, m/s
SndVel.Shear	1001.5	Transverse Sound Velocity, m/s
SndAbs,Long.	4.0	Long.SoundAbs.Coeff., dB/cm
SndAbs,Shr.	20.00	Shear/Trans Snd.AbsCoeff,dB/cm
ThermalCond	0.19	J/s*m*K
SolPr(d)Cal	8.340	Based on Vg,cal1/2cm2/3
SolPr(p)Cal	4.856	Based on Vg. cal1/2cm2/3
SolPr(h)Cal	2.776	Based on Vg. cal1/2cm2/3
SolPrCal/Cm	10.042	Cal1/2cm2/3, based on Vg
E sub eta Inf.	67.7	E,Act.energy of visc flow,g KJ/mol
M(Actual)	11659	Molecular Wt. @or>M(crit)
T	298	Temp.of interest,degK
A	12.91	Empirical Constant
Log Eta Cr.Inf.	-7.156	Visc.@M(critical),T=inf.,Ns/m2
Log Eta Cr.	4.75	Visc.@T of interest,Ns/m2
LogNewtVisc.	4.748	Newt.Visc.@T,N*s/m2
Gas	N2	Enter Gas of Interest from Table
Log Perm.	-14.7	P,Coeff.of Perm,cm3*cm/cm2*s*Pa
Delta H(A)@298K	-286	Free Enthalpy of Poly., kJ/mol
TempCorrFac(B)	600	K
Delt Gsub f @T	----	Free Energy Change @T, kJ/mol
TDecompGrp	None	Enter Poly.Grp.for Therm Decomp
TD1/2	542	Temp.ofHalfDecomp, deg., K

Structure	Repeat Structural Groups Selected	

CH₂Cl

−(OCH₂CCH₂)−

CH₂Cl

4 – CH₂	2 – Cl
1 – C(al)	1 – O(al)

193

Table 5.18 Estimated Properties of Poly(methylene oxide)

CALCULATED MOLECULAR PROPERTIES		StructuralGroup
		Methylene oxide
		PolymerClass
		DELRIN
Mol.Wt.	30.03	g/mol.
Density Am.	1.204	g/cm3.
Density Gl.	1.162	g/cm3.
Density Cr.	1.360	g/cm3.
VdeWaalsVol	15.73	cm3/mol
Eg	7.08E-03	cm3/mol*K,for glass,therm expan
E1	1.57E-02	cm3/mol*K,forliquid,therm expan
e1	5.24E-04	cm3/g*K,for liq,sp.therm.expan
eg	2.36E-04	cm3/g*k,for glass,sp.therm.expan
alpha liq. cubic	6.30E-04	1/K,cubic therm.expan., liq.state
alpha gl. cubic	2.74E-04	1/K, cubic therm.expan., gl.state
alpha liq. linear	2.10E-04	1/K, linear coeff.liq.therm.expan.
alpha gl. linear	9.13E-05	1/K, linear coeff., gl.therm.expan.
Specf Cp(s)	1.404	J/g*K,
Specf Cp(l)	2.198	J/g*K
Hm	7500	J/mol.
Tg	-85	C
Tm	183	C
SolPar(d)	7.248	Based on Vr,Cal1/2cm2/3disper.
SolPar(p)	7.836	Based on Vr,Cal1/2cm2/3polar
SolPar(h)	5.359	Based on Vr.Cal1/2cm2/3 H bond.
SolPrCal/Cm	11.944	Cal1/2cm2/3, based on Vr
SolParJ/Cm	24.437	J1/2/cm3/2
SurTension	37.57	erg/cm2
MChainGrps	2	Enter# of backboneatms/strucunit
C(constant)	10.85	molar lim.visc.,g1/4xcm3/2/mol3/4
K Theta	0.131	lim.visc.#,cm3xmol1/2/g3/2
M(Critical)	9917	critical molecular weight,g
Refr. Index	1.49	refractive index, nD
Dielc Cons	2.84	Dielectric Constant,D.C.
Bulk Mod.(K)	6.53	(K)BulkModulus, GPa
ShearMod.(G)	1.49	(G)Shear or Rigidity Modulus, GPa
Ten. Mod.(E)	4.15	(E)Tensile or Young'sModulus,GPa
PoissonRatio	0.39	Poisson's Ratio
SndVel.Long.	2659.6	Longitudinal SoundVelocity, m/s
SndVel.Shear	1112.6	Transverse Sound Velocity, m/s
SndAbs,Long..	3.76	Long.SoundAbs.Coeff., dB/cm
SndAbs,Shr.	18.78	Shear/Trans Snd.AbsCoeff,dB/cm
ThermalCond	0.22	J/s*m*K
SolPr(d)Cal	6.996	Based on Vg,cal1/2cm2/3
SolPr(p)Cal	7.563	Based on Vg. cal1/2cm2/3
SolPr(h)Cal	5.265	Based on Vg. cal1/2cm2/3
SolPrCal/Cm	11.570	Cal1/2cm2/3, based on Vg
E sub eta Inf.	26.9	E,Act.energy of visc flow,g KJ/mol
M(Actual)	9917	Molecular Wt. @or>M(crit)
T	298	Temp.of interest,degK
A	7.49	Empirical Constant
Log Eta Cr.Inf.	-3.688	Visc.@M(critical),T=inf.,Ns/m2
Log Eta Cr.	1.04	Visc.@T of interest,Ns/m2
LogNewtVisc.	1.044	Newt.Visc.@T,N*s/m2
Gas	N2	Enter Gas of Interest from Table
Log Perm.	-14.2	P,Coeff.of Perm,cm3*cm/cm2*s*Pa
Delta H(A)@298K	-142	Free Enthalpy of Poly., kJ/mol
TempCorrFac(B)	172	K
Delt Gsub f @T	-90.7	Free Energy Change @T, kJ/mol
TDecompGrp	None	Enter Poly.Grp.for Therm Decomp
TD1/2	583	Temp.ofHalfDecomp, deg., K

Structure

$-(CH_2O)-$

Repeat Structural Groups Selected

$1- CH_2$

$1- O(al)$

194

Table 5.19 Estimated Properties of Poly(vinyl methyl ether)

		StructuralGroup
CALCULATED		Vinyl methyl ether
MOLECULAR		**PolymerClass**
PROPERTIES		Poly(vinyl methyl ether)
Mol.Wt.	58.08	g/mol.
Density Am.	1.008	g/cm3.
Density Gl.	0.981	g/cm3.
Density Cr.	1.139	g/cm3.
VdeWaalsVol	36.18	cm3/mol
Eg	1.63E-02	cm3/mol*K,for glass,therm expan
E1	3.62E-02	cm3/mol*K,forliquid,therm expan
e1	6.23E-04	cm3/g*K,for liq,sp.therm.expan
eg	2.80E-04	cm3/g*k,for glass,sp.therm expan
alpha liq. cubic	6.28E-04	1/K,cubic therm.expan., liq.state
alpha gl. cubic	2.75E-04	1/K, cubic therm.expan., gl.state
alpha liq. linear	2.09E-04	1/K, linear coeff.liq.therm.expan.
alpha gl. linear	9.17E-05	1/K, linear coeff., gl.therm.expan.
Specf Cp(s)	1.526	J/g*K,
Specf Cp(l)	2.132	J/g*K
Hm	12000	J/mol.
Tg	-28	C
Tm	194	C
SolPar(d)	7.382	Based on Vr,Cal1/2cm2/3disper.
SolPar(p)	3.394	Based on Vr,Cal1/2cm2/3polar
SolPar(h)	3.527	Based on Vr.Cal1/2cm2/3 H bond.
SolPrCal/Cm	8.858	Cal1/2cm2/3, based on Vr
SolParJ/Cm	18.123	J1/2/cm3/2
SurTension	33.10	erg/cm2
MChainGrps	2	Enter# of backboneatms/strucunit
C(constant)	15.55	molar lim.visc.,g1/4xcm3/2/mol3/4
K Theta	0.072	lim.visc.#,cm3xmol1/2/g3/2
M(Critical)	32891	critical molecular weight,g
Refr. Index	1.48	refractive index, nD
Dielc Cons	2.50	Dielectric Constant,D.C.
Bulk Mod.(K)	5.16	(K)BulkModulus, GPa
ShearMod.(G)	1.73	(G)Shear or Rigidity Modulus, GPa
Ten. Mod.(E)	4.67	(E)Tensile or Young'sModulus,GPa
PoissonRatio	0.35	Poisson's Ratio
SndVel.Long.	2721.6	Longitudinal SoundVelocity, m/s
SndVel.Shear	1309.5	Transverse Sound Velocity, m/s
SndAbs,Long.	1.97	Long.SoundAbs.Coeff., dB/cm
SndAbs,Shr.	9.87	Shear/Trans Snd.AbsCoeff,dB/cm
ThermalCond	0.21	J/s*m*K
SolPr(d)Cal	7.183	Based on Vg,cal1/2cm2/3
SolPr(p)Cal	3.302	Based on Vg. cal1/2cm2/3
SolPr(h)Cal	3.479	Based on Vg. cal1/2cm2/3
SolPrCal/Cm	8.637	Cal1/2cm2/3, based on Vg
E sub eta Inf.	38.4	E,Act.energy of visc flow,g KJ/mol
M(Actual)	32891	Molecular Wt. @or>M(crit)
T	298	Temp.of interest,degK
A	8.21	Empirical Constant
Log Eta Cr.Inf.	-4.667	Visc.@M(critical),T=inf.,Ns/m2
Log Eta Cr.	2.09	Visc.@T of interest,Ns/m2
LogNewtVisc.	2.089	Newt.Visc.@T,N*s/m2
Gas	N2	Enter Gas of Interest from Table
Log Perm.	-14.5	P,Coeff.of Perm,cm3*cm/cm2*s*Pa
Delta H(A)@298K	-190.7	Free Enthalpy of Poly., kJ/mol
TempCorrFac(B)	387	K
Delt Gsub f @T	-75.4	Free Energy Change @T, kJ/mol
TDecompGrp	None	Enter Poly.Grp.for Therm Decomp
TD1/2	620	Temp.ofHalfDecomp, deg., K

Structure	Repeat Structural Groups Selected	

$$-(CH_2CH)-$$
$$OCH_3$$

1– CH_2	1– CH
1– CH_3	1– O(al)

Table 5.20 Estimated Properties of Poly(phenylene oxide)

		StructuralGroup
CALCULATED		Phenylene oxide
MOLECULAR		PolymerClass
PROPERTIES		Poly(phenylene oxide)
Mol.Wt.	120.14	g/mol.
Density Am.	1.134	g/cm3.
Density Gl.	1.097	g/cm3.
Density Cr.	1.282	g/cm3.
VdeWaalsVol	70.64	cm3/mol
Eg	3.18E-02	cm3/mol*K,for glass,therm expan
E1	7.06E-02	cm3/mol*K,forliquid,therm expan
e1	5.88E-04	cm3/g*K,for liq,sp.therm.expan
eg	2.65E-04	cm3/g*k,for glass,sp.therm expan
alpha liq. cubic	6.67E-04	1/K,cubic therm.expan., liq.state
alpha gl. cubic	2.90E-04	1/K, cubic therm.expan., gl.state
alpha liq. linear	2.22E-04	1/K, linear coeff.liq.therm.expan.
alpha gl. linear	9.68E-05	1/K, linear coeff., gl.therm.expan.
Specf Cp(s)	1.195	J/g*K,
Specf Cp(l)	1.685	J/g*K
Hm	9200	J/mol.
Tg	193	C
Tm	282	C
SolPar(d)	8.723	Based on Vr,Cal1/2cm2/3disper.
SolPar(p)	1.915	Based on Vr,Cal1/2cm2/3polar
SolPar(h)	2.601	Based on Vr.Cal1/2cm2/3 H bond.
SolPrCal/Cm	9.301	Cal1/2cm2/3, based on Vr
SolParJ/Cm	19.031	J1/2/cm3/2
SurTension	25.60	erg/cm2
MChainGrps	5	Enter# of backboneatms/strucunit
C(constant)	28.2	molar lim.visc.,g1/4xcm3/2/mol3/4
K Theta	----	lim.visc.#,cm3xmol1/2/g3/2
M(Critical)	----	critical molecular weight,g
Refr. Index	1.59	refractive index, nD
Dielc Cons	2.89	Dielectric Constant,D.C.
Bulk Mod.(K)	4.79	(K)BulkModulus, GPa
ShearMod.(G)	1.26	(G)Shear or Rigidity Modulus, GPa
Ten. Mod.(E)	3.47	(E)Tensile or Young'sModulus,GPa
PoissonRatio	0.38	Poisson's Ratio
SndVel.Long.	2338.0	Longitudinal SoundVelocity, m/s
SndVel.Shear	1052.5	Transverse Sound Velocity, m/s
SndAbs,Long.	3.18	Long.SoundAbs.Coeff., dB/cm
SndAbs,Shr.	15.89	Shear/Trans Snd.AbsCoeff,dB/cm
ThermalCond	0.16	J/s*m*K
SolPr(d)Cal	8.436	Based on Vg,cal1/2cm2/3
SolPr(p)Cal	1.852	Based on Vg. cal1/2cm2/3
SolPr(h)Cal	2.558	Based on Vg. cal1/2cm2/3
SolPrCal/Cm	9.007	Cal1/2cm2/3, based on Vg
E sub eta Inf.	----	E,Act.energy of visc flow,g KJ/mol
M(Actual)	----	Molecular Wt. @or>M(crit)
T	298	Temp.of interest,degK
A	0.60	Empirical Constant
Log Eta Cr.Inf.	----	Visc.@M(critical),T=inf.,Ns/m2
Log Eta Cr.	----	Visc.@T of interest,Ns/m2
LogNewtVisc.	----	Newt.Visc.@T,N*s/m2
Gas	N2	Enter Gas of Interest from Table
Log Perm.	-12.5	P,Coeff.of Perm,cm3*cm/cm2*s*Pa
Delta H(A)@298K	----	Free Enthalpy of Poly., kJ/mol
TempCorrFac(B)	260	K
Delt Gsub f @T	----	Free Energy Change @T, kJ/mol
TDecompGrp	None	Enter Poly.Grp.for Therm Decomp
TD1/2	753	Temp.ofHalfDecomp, deg., K

Structure

Repeat Structural Groups Selected

$2 - CH_3$

$1 - O(ar)$; $1 - C_6H_2$

196

Table 5.21 Estimated Properties of Poly(hexamethylene adipamide)

CALCULATED MOLECULAR PROPERTIES		StructuralGroup Hexamethylene adipamide PolymerClass Nylon 6,6
Mol.Wt.	226.3	g/mol.
Density Am.	1.096	g/cm3.
Density Gl.	1.086	g/cm3.
Density Cr.	1.238	g/cm3.
VdeWaalsVol	138.3	cm3/mol
Eg	6.22E-02	cm3/mol*K,for glass,therm expan
E1	1.38E-01	cm3/mol*K,forliquid,therm expan
e1	6.11E-04	cm3/g*K,for liq,sp.therm.expan
eg	2.75E-04	cm3/g*k,for glass,sp.therm expan
alpha liq. cubic	6.70E-04	1/K,cubic therm.expan., liq.state
alpha gl. cubic	2.99E-04	1/K, cubic therm.expan., gl.state
alpha liq. linear	2.23E-04	1/K, linear coeff.liq.therm.expan.
alpha gl. linear	9.96E-05	1/K, linear coeff., gl.therm.expan.
Specf Cp(s)	1.527	J/g*K,
Specf Cp(l)	2.140	J/g*K
Hm	44000	J/mol.
Tg	43	C
Tm	226	C
SolPar(d)	8.521	Based on Vr,Cal1/2cm2/3disper.
SolPar(p)	4.639	Based on Vr,Cal1/2cm2/3polar
SolPar(h)	3.435	Based on Vr.Cal1/2cm2/3 H bond.
SolPrCal/Cm	10.292	Cal1/2cm2/3, based on Vr
SolParJ/Cm	21.057	J1/2/cm3/2
SurTension	40.66	erg/cm2
MChainGrps	14	Enter# of backboneatms/strucunit
C(constant)	107.5	molar lim.visc.,g1/4xcm3/2/mol3/4
K Theta	0.226	lim.visc.#,cm3xmol1/2/g3/2
M(Critical)	3319	critical molecular weight,g
Refr. Index	1.50	refractive index, nD
Dielc Cons	4.07	Dielectric Constant,D.C.
Bulk Mod.(K)	6.51	(K)BulkModulus, GPa
ShearMod.(G)	1.46	(G)Shear or Rigidity Modulus, GPa
Ten. Mod.(E)	4.07	(E)Tensile or Young'sModulus,GPa
PoissonRatio	0.40	Poisson's Ratio
SndVel.Long.	2777.6	Longitudinal SoundVelocity, m/s
SndVel.Shear	1152.8	Transverse Sound Velocity, m/s
SndAbs,Long.	3.84	Long.SoundAbs.Coeff., dB/cm
SndAbs,Shr.	19.19	Shear/Trans Snd.AbsCoeff,dB/cm
ThermalCond	0..23	J/s*m*K
SolPr(d)Cal	8.447	Based on Vg,cal1/2cm2/3
SolPr(p)Cal	4.599	Based on Vg. cal1/2cm2/3
SolPr(h)Cal	3.420	Based on Vg. cal1/2cm2/3
SolPrCal/Cm	10.208	Cal1/2cm2/3, based on Vg
E sub eta Inf.	36.4	E,Act.energy of visc flow,g KJ/mol
M(Actual)	3319	Molecular Wt. @or>M(crit)
T	298	Temp.of interest,degK
A	6.04	Empirical Constant
Log Eta Cr.Inf.	-4.494	Visc.@M(critical),T=inf.,Ns/m2
Log Eta Cr.	1.90	Visc.@T of interest,Ns/m2
LogNewtVisc.	1.905	Newt.Visc.@T,N*s/m2
Gas	N2	Enter Gas of Interest from Table
Log Perm.	-15.6	P,Coeff.of Perm,cm3*cm/cm2*s*Pa
Delta H(A)@298K	----	Free Enthalpy of Poly., kJ/mol
TempCorrFac(B)	1020	K
Delt Gsub f @T	----	Free Energy Change @T, kJ/mol
TDecompGrp	None	Enter Poly.Grp.for Therm Decomp
TD1/2	694	Temp.ofHalfDecomp, deg., K

Structure

Repeat Structural Groups Selected

10 – CH$_2$

2 – CONH

$$—(NH(CH_2)_6NH\overset{O}{\overset{\|}{C}}(CH_2)_4\overset{O}{\overset{\|}{C}})—$$

197

Table 5.22 Estimated Properties of Poly(acrylamide)

		StructuralGroup
CALCULATED		Acrylamide
MOLECULAR		PolymerClass
PROPERTIES		Poly(acrylamide)
Mol.Wt.	71.05	g/mol.
Density Am.	1.333	g/cm3.
Density Gl.	1.336	g/cm3.
Density Cr.	1.506	g/cm3.
VdeWaalsVol	33.45	cm3/mol
Eg	1.51E-02	cm3/mol*K,for glass,therm expan
E1	3.35E-02	cm3/mol*K,forliquid,therm expan
e1	4.71E-04	cm3/g*K,for liq,sp.therm.expan
eg	2.12E-04	cm3/g*k,for glass,sp.therm expan
alpha liq. cubic	6.28E-04	1/K,cubic therm.expan., liq.state
alpha gl. cubic	2.83E-04	1/K, cubic therm.expan., gl.state
alpha liq. linear	2.09E-04	1/K, linear coeff.liq.therm.expan.
alpha gl. linear	9.43E-05	1/K, linear coeff., gl.therm.expan.
Specf Cp(s)	1.224	J/g*K,
Specf Cp(l)	1.991	J/g*K
Hm	8300	J/mol.
Tg	163	C
Tm	304	C
SolPar(d)	7.336	Based on Vr,Cal1/2cm2/3disper.
SolPar(p)	8.987	Based on Vr,Cal1/2cm2/3polar
SolPar(h)	4.781	Based on Vr.Cal1/2cm2/3 H bond.
SolPrCal/Cm	12.547	Cal1/2cm2/3, based on Vr
SolParJ/Cm	25.672	J1/2cm3/2
SurTension	38.36	erg/cm2
MChainGrps	2	Enter# of backboneatms/strucunit
C(constant)	34.9	molar lim.visc.,g1/4xcm3/2/mol3/4
K Theta	0.241	lim.visc.#,cm3xmol1/2/g3/2
M(Critical)	2903	critical molecular weight,g
Refr. Index	1.53	refractive index, nD
Dielc Cons	5.67	Dielectric Constant,D.C.
Bulk Mod.(K)	3.35	(K)BulkModulus, GPa
ShearMod.(G)	0.98	(G)Shear or Rigidity Modulus, GPa
Ten. Mod.(E)	2.68	(E)Tensile or Young'sModulus,GPa
PoissonRatio	0.37	Poisson's Ratio
SndVel.Long.	1869.3	Longitudinal SoundVelocity, m/s
SndVel.Shear	857.1	Transverse Sound Velocity, m/s
SndAbs,Long.	2.68	Long.SoundAbs.Coeff., dB/cm
SndAbs,Shr.	13.38	Shear/Trans Snd.AbsCoeff,dB/cm
ThermalCond	0.15	J/s*m*K
SolPr(d)Cal	7.350	Based on Vg,cal1/2cm2/3
SolPr(p)Cal	9.003	Based on Vg. cal1/2cm2/3
SolPr(h)Cal	4.785	Based on Vg. cal1/2cm2/3
SolPrCal/Cm	12.569	Cal1/2cm2/3, based on Vg
E sub eta Inf.	34.8	E,Act.energy of visc flow,g KJ/mol
M(Actual)	2903	Molecular Wt. @or>M(crit)
T	298	Temp.of interest,degK
A	4.18	Empirical Constant
Log Eta Cr.Inf.	-4.359	Visc.@M(critical),T=inf.,Ns/m2
Log Eta Cr.	1.76	Visc.@T of interest,Ns/m2
LogNewtVisc.	1.761	Newt.Visc.@T,N*s/m2
Gas	N2	Enter Gas of Interest from Table
Log Perm.	----	P,Coeff.of Perm,cm3*cm/cm2*s*Pa
Delta H(A)@298K	-145.2	Free Enthalpy of Poly., kJ/mol
TempCorrFac(B)	364.5	K
Delt Gsub f @T	-36.6	Free Energy Change @T, kJ/mol
TDecompGrp	None	Enter Poly.Grp.for Therm Decomp
TD1/2	----	Temp.ofHalfDecomp, deg., K

Structure

$$O=CNH_2$$
$$-(CH_2CH)-$$

Repeat Structural Groups Selected

$1- CH_2$

$1- CH$

$1- CONH_2$

198

Table 5.23 Estimated Properties of Poly(acrylonitrile)

CALCULATED MOLECULAR PROPERTIES		StructuralGroup
		Acrylonitrile
		PolymerClass
		Poly(acrylonitrile)
Mol.Wt.	53.03	g/mol.
Density Am.	1.125	g/cm3.
Density Gl.	1.182	g/cm3.
Density Cr.	1.271	g/cm3.
VdeWaalsVol	31.73	cm3/mol
Eg	1.43E-02	cm3/mol*K,for glass,therm expan
E1	3.17E-02	cm3/mol*K,forliquid,therm expan
e1	5.98E-04	cm3/g*K,for liq,sp.therm.expan
eg	2.69E-04	cm3/g*k,for glass,sp.therm.expan
alpha liq. cubic	6.73E-04	1/K,cubic therm.expan., liq.state
alpha gl. cubic	3.18E-04	1/K, cubic therm.expan., gl.state
alpha liq. linear	2.24E-04	1/K, linear coeff.liq.therm.expan.
alpha gl. linear	1.06E-04	1/K, linear coeff., gl.therm.expan.
Specf Cp(s)	1.244	J/g*K,
Specf Cp(l)	----	J/g*K
Hm	6000	J/mol.
Tg	94	C
Tm	319	C
SolPar(d)	8.086	Based on Vr,Cal1/2cm2/3disper.
SolPar(p)	11.403	Based on Vr,Cal1/2cm2/3polar
SolPar(h)	3.559	Based on Vr.Cal1/2cm2/3 H bond.
SolPrCal/Cm	14.424	Cal1/2cm2/3, based on Vr
SolParJ/Cm	29.512	J1/2/cm3/2
SurTension	49.56	erg/cm2
MChainGrps	2	Enter# of backboneatms/strucunit
C(constant)	26.9	molar lim.visc.,g1/4xcm3/2/mol3/4
K Theta	0.257	lim.visc.#,cm3xmol1/2/g3/2
M(Critical)	2552	critical molecular weight,g
Refr. Index	1.53	refractive index, nD
Dielc Cons	3.15	Dielectric Constant,D.C.
Bulk Mod.(K)	6.54	(K)BulkModulus, GPa
ShearMod.(G)	1.96	(G)Shear or Rigidity Modulus, GPa
Ten. Mod.(E)	5.34	(E)Tensile or Young'sModulus,GPa
PoissonRatio	0.36	Poisson's Ratio
SndVel.Long.	2852.5	Longitudinal SoundVelocity, m/s
SndVel.Shear	1318.8	Transverse Sound Velocity, m/s
SndAbs,Long.	2.56	Long.SoundAbs.Coeff., dB/cm
SndAbs,Shr.	12.81	Shear/Trans Snd.AbsCoeff,dB/cm
ThermalCond	0.20	J/s*m*K
SolPr(d)Cal	8.500	Based on Vg,cal1/2cm2/3
SolPr(p)Cal	11.987	Based on Vg. cal1/2cm2/3
SolPr(h)Cal	3.649	Based on Vg. cal1/2cm2/3
SolPrCal/Cm	15.142	Cal1/2cm2/3, based on Vg
E sub eta Inf.	----	E,Act.energy of visc flow,g KJ/mol
M(Actual)	2552	Molecular Wt. @or>M(crit)
T	298	Temp.of interest,degK
A	----	Empirical Constant
Log Eta Cr.Inf.	----	Visc.@M(critical),T=inf.,Ns/m2
Log Eta Cr.	----	Visc.@T of interest,Ns/m2
LogNewtVisc .	----	Newt.Visc.@T,N*s/m2
Gas	N2	Enter Gas of Interest from Table
Log Perm.	-17.3	P,Coeff.of Perm,cm3*cm/cm2*s*Pa
Delta H(A)@298K	108	Free Enthalpy of Poly., kJ/mol
TempCorrFac(B)	102	K
Delt Gsub f @T	138.4	Free Energy Change @T, kJ/mol
TDecompGrp	None	Enter Poly.Grp.for Therm Decomp
TD1/2	726	Temp.ofHalfDecomp, deg., K

Structure

$$C \equiv N$$
$$-(CH_2CH)-$$

Repeat Structural Groups Selected

1 – CH_2

1 – CHCN

199

Table 5.24 Estimated Properties of Poly(methyl methacrylate)

		StructuralGroup
CALCULATED		methyl methacrylate
MOLECULAR		**PolymerClass**
PROPERTIES		Polyl(methyl methacrylate) PLEXIGLAS
Mol.Wt.	100.1	g/mol.
Density Am.	1.153	g/cm3.
Density Gl.	1.157	g/cm3.
Density Cr.	1.303	g/cm3.
VdeWaalsVol	56.1	cm3/mol
Eg	2.52E-02	cm3/mol*K,for glass,therm expan
E1	5.61E-02	cm3/mol*K,forliquid,therm expan
e1	5.60E-04	cm3/g*K,for liq,sp.therm.expan
eg	2.52E-04	cm3/g*k,for glass,sp.therm expan
alpha liq. cubic	6.46E-04	1/K,cubic therm.expan., liq.state
alpha gl. cubic	2.92E-04	1/K, cubic therm.expan., gl.state
alpha liq. linear	2.15E-04	1/K, linear coeff.liq.therm.expan.
alpha gl. linear	9.72E-05	1/K, linear coeff., gl.therm.expan.
Specf Cp(s)	1.392	J/g*K,
Specf Cp(l)	1.764	J/g*K
Hm	9050	J/mol.
Tg	113	C
Tm	196	C
SolPar(d)	8.052	Based on Vr,Cal1/2cm2/3disper.
SolPar(p)	2.759	Based on Vr,Cal1/2cm2/3polar
SolPar(h)	4.389	Based on Vr.Cal1/2cm2/3 H bond.
SolPrCal/Cm	9.576	Cal1/2cm2/3, based on Vr
SolParJ/Cm	19.594	J1/2/cm3/2
SurTension	42.36	erg/cm2
MChainGrps	2	Enter# of backboneatms/strucunit
C(constant)	26.25	molar lim.visc.,g1/4xcm3/2/mol3/4
K Theta	0.069	lim.visc.#,cm3xmol1/2/g3/2
M(Critical)	35736	critical molecular weight,g
Refr. Index	1.51	refractive index, nD
Dielc Cons	3.14	Dielectric Constant,D.C.
Bulk Mod.(K)	6.47	(K)BulkModulus, GPa
ShearMod.(G)	2.33	(G)Shear or Rigidity Modulus, GPa
Ten. Mod.(E)	6.24	(E)Tensile or Young'sModulus,GPa
PoissonRatio	0.34	Poisson's Ratio
SndVel.Long.	2881.2	Longitudinal SoundVelocity, m/s
SndVel.Shear	1421.5	Transverse Sound Velocity, m/s
SndAbs,Long.	1.57	Long.SoundAbs.Coeff., dB/cm
SndAbs,Shr.	7.83	Shear/Trans Snd.AbsCoeff,dB/cm
ThermalCond	0.31	J/s*m*K
SolPr(d)Cal	8.075	Based on Vg,cal1/2cm2/3
SolPr(p)Cal	2.767	Based on Vg. cal1/2cm2/3
SolPr(h)Cal	4.396	Based on Vg. cal1/2cm2/3
SolPrCal/Cm	9.601	Cal1/2cm2/3, based on Vg
E sub eta Inf.	63.3	E,Act.energy of visc flow,g KJ/mol
M(Actual)	35736	Molecular Wt. @or>M(crit)
T	298	Temp.of interest,degK
A	8.59	Empirical Constant
Log Eta Cr.Inf.	-6.783	Visc.@M(critical),T=inf.,Ns/m2
Log Eta Cr.	4.35	Visc.@T of interest,Ns/m2
LogNewtVisc.	4.35	Newt.Visc.@T,N*s/m2
Gas	N2	Enter Gas of Interest from Table
Log Perm.	-13.8	P,Coeff.of Perm,cm3*cm/cm2*s*Pa
Delta H(A)@298K	-431	Free Enthalpy of Poly., kJ/mol
TempCorrFac(B)	548	K
Delt Gsub f @T	-267.7	Free Energy Change @T, kJ/mol
TDecompGrp	None	Enter Poly.Grp.for Therm Decomp
TD1/2	608	Temp.ofHalfDecomp, deg., K

Structure	Repeat Structural Groups Selected	

$$CH_3$$
$$|$$
$$—(CH_2C)—$$
$$|$$
$$O=C—OCH_3$$

1 – CH_2 2 – CH_3

1 – C(al)   1 – COO(ACRYL)

Table 5.25 Estimated Properties of Poly(cyclohexyl methacrylate)

CALCULATED MOLECULAR PROPERTIES		StructuralGroup Cyclohexyl methacrylate PolymerClass Poly(cyclohexyl methacrylate)
Mol.Wt.	168.24	g/mol.
Density Am.	1.078	g/cm3.
Density Gl.	1.112	g/cm3.
Density Cr.	1.218	g/cm3.
VdeWaalsVol	100.36	cm3/mol
Eg	4.52E-02	cm3/mol*K,for glass,therm expan
E1	1.00E-01	cm3/mol*K,forliquid,therm expan
e1	5.97E-04	cm3/g*K,for liq,sp.therm.expan
eg	2.68E-04	cm3/g*k,for glass,sp.therm expan
alpha liq. cubic	6.43E-04	1/K,cubic therm.expan., liq.state
alpha gl. cubic	2.98E-04	1/K, cubic therm.expan., gl.state
alpha liq. linear	2.14E-04	1/K, linear coeff.liq.therm.expan.
alpha gl. linear	9.95E-05	1/K, linear coeff., gl.therm.expan.
Specf Cp(s)	1.491	J/g*K,
Specf Cp(l)	1.858	J/g*K
Hm	27350	J/mol.
Tg	96	C
Tm	159	C
SolPar(d)	8.235	Based on Vr,Cal1/2cm2/3disper.
SolPar(p)	1.534	Based on Vr,Cal1/2cm2/3polar
SolPar(h)	3.273	Based on Vr.Cal1/2cm2/3 H bond.
SolPrCal/Cm	8.993	Cal1/2cm2/3, based on Vr
SolParJ/Cm	18.400	J1/2/cm3/2
SurTension	40.41	erg/cm2
MChainGrps	2	Enter# of backboneatms/strucunit
C(constant)	35.6	molar lim.visc.,g1/4xcm3/2/mol3/4
K Theta	0.045	lim.visc.#,cm3xmol1/2/g3/2
M(Critical)	84295	critical molecular weight,g
Refr. Index	1.52	refractive index, nD
Dielc Cons	2.90	Dielectric Constant,D.C.
Bulk Mod.(K)	5.46	(K)BulkModulus, GPa
ShearMod.(G)	1.47	(G)Shear or Rigidity Modulus, GPa
Ten. Mod.(E)	4.05	(E)Tensile or Young'sModulus,GPa
PoissonRatio	0.38	Poisson's Ratio
SndVel.Long	2623.5	Longitudinal SoundVelocity, m/s
SndVel.Shear	1167.8	Transverse Sound Velocity, m/s
SndAbs,Long.	3.06	Long.SoundAbs.Coeff., dB/cm
SndAbs,Shr.	15.29	Shear/Trans Snd.AbsCoeff,dB/cm
ThermalCond	0.21	J/s*m*K
SolPr(d)Cal	8.493	Based on Vg,cal1/2cm2/3
SolPr(p)Cal	1.582	Based on Vg. cal1/2cm2/3
SolPr(h)Cal	3.324	Based on Vg. cal1/2cm2/3
SolPrCal/Cm	9.256	Cal1/2cm2/3, based on Vg
E sub eta Inf.	35.5	E,Act.energy of visc flow,g KJ/mol
M(Actual)	84295	Molecular Wt. @or>M(crit)
T	298	Temp.of interest,degK
A	3.80	Empirical Constant
Log Eta Cr.Inf.	-4.419	Visc.@M(critical),T=inf.,Ns/m2
Log Eta Cr.	1.82	Visc.@T of interest,Ns/m2
LogNewtVisc.	1.824	Newt.Visc.@T,N*s/m2
Gas	N2	Enter Gas of Interest from Table
Log Perm.	-15.7	P,Coeff.of Perm,cm3*cm/cm2*s*Pa
Delta H(A)@298K	-497.7	Free Enthalpy of Poly., kJ/mol
TempCorrFac(B)	1083	K
Delt Gsub f @T	-175.0	Free Energy Change @T, kJ/mol
TDecompGrp	None	Enter Poly.Grp.for Therm Decomp
TD1/2	594	Temp.ofHalfDecomp, deg., K

Structure

Repeat Structural Groups Selected

6 – CH$_2$;   1– CH ;   1– CH$_3$

1– C(al) ;   1– COO(ACRYL) and

1– Ring,Saturated

201

5.6   ABBREVIATIONS AND NOTATION

Mol.Wt. (molecular weight)

g/mol.

Density Am. (density, amorphous)

g/cm3 (g/cm^3)

Density Gl. (density, glassy)

g/cm3 (g/cm^3)

Density Cr. (density, crystalline)

g/cm3 (g/cm^3)

VdeWaalsVol(Van der Waals volume)

cm3/mol (cm^3/mol)

Eg (molar glassy thermal expansivity)

cm3/mol*K, for glass,therm expan $(cm^3/mol) \cdot K$

El (molar liquid thermal expansivity)

cm3/mol*K,forliquid,therm expan $(cm^3/mol) \cdot K$

el (specific liquid thermal expansivity)

cm3/g*K,for liq,sp.therm.expan $(cm^3/g \cdot K)$

eg (specific glassy thermal expansivity)

cm3/g*k,for glass,sp.therm expan $(cm^3/mol) \cdot K$

alpha liq. cubic (coefficient of thermal expansion), $\left(\alpha_l\right)$

1/K,cubic therm.expan., liq.state

(liquid state, K^{-1})

alpha gl. cubic (coefficient of thermal expansion), (α_g)

1/K, cubic therm.expan., gl.state,

(glassy state, K^{-1})

alpha liq. linear (linear coefficient of thermal expansion), (α_l^l)

1/K, linear coeff.liq.therm.expan.,

(liquid state, K^{-1})

alpha gl. linear (linear coefficient of thermal expansion), (α_g^l)

1/K, linear coeff., gl.therm.expan.

(glassy state, K^{-1})

Specf Cp(s) (specific solid heat capacity at constant pressure), c_p^s

J/g*K, $(J/g \cdot K)$

Specf Cp(l) (specific liquid heat capacity at constant pressure), c_p^l

J/g*K $(J/g \cdot K)$

Hm (heat of fusion or melting), $\left(H_m\right)$

J/mol. (joules/mol)

Tg (glass transition temperature), $\left(T_g\right)$

C, $(°C)$

Tm (melting temperature), $\left(T_m\right)$

C, $(°C)$

SolPar(d)(dispersive solubility parameter), δ_d

Based on Vr,Cal1/2cm2/3disper.

(rubbery phase), $(cal^{1/2}/cm^{2/3})$

SolPar(p) (polar solubility parameter), δ_p

Based on Vr,Cal1/2cm2/3polar

(rubbery phase), $(cal^{1/2}/cm^{2/3})$

SolPar(h) (hydrogen bonding solubility parameter), δ_h

Based on Vr.Cal1/2cm2/3 H bond.

(rubbery phase), $(cal^{1/2}/cm^{2/3})$

SolPrCal/Cm (total solubility parameter), δ_{dph}

Cal1/2cm2/3, based on Vr

(rubbery phase), $(cal^{1/2}/cm^{2/3})$

SolParJ/Cm (total solubility parameter), δ_{dph}

J1/2/cm3/2

(rubbery phase), $(J^{1/2} / cm^{3/2})$

SurTension, (surface tension), γ

erg/cm2, (erg / cm^2)

MChainGrps (Z)

Enter# of backboneatms/strucunit, (backbone atoms per structural unit)

C (constant)

molar lim.visc.,g1/4xcm3/2/mol3/4

$(cm^{3/4} \cdot g^{1/4} / mol^{3/4})$

K Theta (K_Θ) theta conditions

lim.visc.#,cm3xmol1/2/g3/2, limiting viscosity number,

$(cm^3 \cdot mol^{1/2} / g^{3/2})$

M(Critical), molecular weight

critical molecular weight,g

Refr. Index (refractive index)

refractive index, nD, (n_D)

Dielc Cons (dielectric constant)

Dielectric Constant,D.C., (ε)

Bulk Mod., (K), (bulk modulus)
ShearMod.(G), (shear modulus)
Ten. Mod.(E), (tensile modulus)
PoissonRatio

(K)BulkModulus, GPa
(*G*)Shear or Rigidity Modulus, GPa
(E)Tensile or Young'sModulus,GPa
Poisson's Ratio, (μ)

SndVel.Long.(longitudinal sound velocity), $\left(V_l\right)$

Longitudinal SoundVelocity, m/s, (meters/second)

SndVel.Shear (shear sound velocity), $\left(V_s\right)$

Transverse Sound Velocity, m/s, (meters/second)

SndAbs,Long., (longitudinal sound absorption), $\left(\alpha_L\right)$

Long.SoundAbs.Coeff., dB/cm, coefficient(decibels/cm)

SndAbs,Shr., (shear sound absorption), $\left(\alpha_{sh}\right)$

Shear/Trans Snd.AbsCoeff,dB/cm, coefficient(decibels/cm)

ThermalCond, (thermal conductivity), $\left(\lambda\right)$

J/s*m*K, $(J / s \cdot m \cdot K)$

SolPar(d)(dispersive solubility parameter), δ_d

Based on *Vg*,Cal1/2cm2/3disper.

(glassy phase), $(cal^{1/2} / cm^{2/3})$

SolPar(p) (polar solubility parameter), δ_p

Based on *Vg*,Cal1/2cm2/3polar

(glassy phase), $(cal^{1/2} / cm^{2/3})$

SolPar(h) (hydrogen bonding solubility parameter), δ_h

Based on *Vg*.Cal1/2cm2/3 H bond.

(glassy phase), $(cal^{1/2} / cm^{2/3})$

SolPrCal/Cm (total solubility paramter), δ_{dph}

Cal1/2cm2/3, based on *Vg*

(glassy phase), $(cal^{1/2} / cm^{2/3})$

E sub eta Inf., $\left(E_h(\infty)\right)$

E, Act.energy of visc flow,g

kJ/mol(activation energy for viscous flow), kJ / mol)

M(Actual), molecular weight of interest

Molecular Wt. @or>*M*(crit), molecular weight ≥ critical molecular weight

T

Temp.of interest,degK

A

Empirical Constant

Log Eta Cr.Inf., $\left(\log\eta_{cr}(\infty)\right)$

Visc.@M(critical),T=inf.,Ns/m2,

(viscosity at critical Mol.Wt.)

$(N \cdot s/m^2)$

Log Eta Cr. , viscosity at critical mol. Wt. $\left(\log\eta_{cr}\right)$

Visc.@T of interest,Ns/m2,

$(N \cdot s/m^2)$

LogNewtVisc., η_0

Newt. Visc.@T,N*s/m2, Newtonian

viscosity, $(N \cdot s/m^2)$

Gas
Log Perm. (log permeation)

Enter Gas of Interest from Table
P,Coeff.of Perm,
cm3*cm/cm2*s*Pa,

$(cm^3 \cdot cm/cm^2 \cdot s \cdot Pa)$

Delta H(A)@298K, (ΔH)

Free Enthalpy ofPoly.(polymerization)

kJ/mol

TempCorrFac (B)

K

Delt Gsub f @T $\left(\Delta G_f\right)$

Free Energy Change @T, kJ/mol

TDecompGrp (group for estimating decomposition temperature)

Enter Poly.Grp.for Therm Decomp
(thermal decomposition)

TD1/2 (temperature of half decomposition), $\left(T_{D,1/2}\right)$

Temp.ofHalfDecomp, deg K

5.7   GENERAL REFERENCES

1. Bondi, A. *Physical Properties of Molecular Crystals, Liquids and Glasses*, Wiley, New York, 1968.

2. Exner, O., *Additive Physical Properties*, Collection Czechoslov. Chem. Commun. Vols. 31 and 32.

3. Van Krevelen, D.W. and Hoftyzer, P.J., *Properties of Polymers*, 2nd Ed. Elsevier, New York, N.Y., 1976.

4. Van Krevelen, D.W., *Properties of Polymers*, Elsevier, New York, N.Y., 1990.

5. Bicerano, J., *Prediction of Polymer Properties*, Marcel Dekker, Inc., New York, 1993.

CHAPTER 6

Molecular Forces and Volumetric Properties

6.1 INTRODUCTION

In Chapter 1, fundamental differences in the structures of metals, polymers, and ceramic materials were considered along with the types of bonds responsible for these structures. For many materials, whether or not they be gases, liquids, or solids, we are more concerned with how they interact with themselves and other materials once their particular molecular architecture has been established from their bonding of two or more homo or hetero atoms. This is particularly true in the case of organic liquids and solids and especially in the instance of polymer molecules which consist of many (poly), parts (mers), since we wish to understand how these molecules interact and the forces present that determine the extent of this interaction.

Since the nature of primary forces, i.e., those involved in metallic, ionic, and covalent bonding have been considered previously, we focus our attention on molecular or secondary forces in this chapter which are frequently called van der Waals forces.

6.2 VAN DER WAALS FORCES

The van der Waals forces have their origin in the connection to the ideal gas relationships and are of longer distance in interaction, generally considered significant in the range of $2.5 \overset{\circ}{A}$ - $5.0 \overset{\circ}{A}$. van der Waals law is an extension of the law of

Boyle-Gay Lussac, with corrections for the weak interaction and the proper volume of the gas molecules. For simple gases it is a fair approximation of the P-V-T behavior and can be represented as the following equation of state:

$$(p - \frac{a}{V^2})(V - b) = RT \qquad (6.1)$$

where a and b are the van der Waals constants. It is these van der Waals forces that are responsible for the cohesion of molecules.

In a broad sense, many of these concepts can best be grasped from an analogy to gas molecules. Inert spherical atoms and molecules, such as neon, argon, and methane are examples of van der Waals materials since they are held together solely by dispersion forces. They have low melting points and low latent heats of melting and their bonds are non-directional. As a result, in the crystalline state they surround themselves with the highest possible number of nearest neighbors.

From the point of view of long-term statistical averages, the electron cloud is distributed in these atoms in a spherically symmetrical manner. At times however, greater or lesser deviations from spherical symmetry may occur, the center of the positive and negative charges of the atom will differ and a dipole will be formed. The behavior of this variable dipole may be described by a rough but useful approximation as the motion of a harmonic oscillator. When two such oscillators approach each other the minimum energy corresponds to a state in which the motion of the oscillating dipoles is in phase. Since the energy of two dipoles which interact in this way is less than their energy at infinite distance between them, this causes the oscillators to be influenced by attractive forces which are called "London" or dispersion forces.

6.3   INTERACTION OF FORCES BETWEEN MOLECULES

In principle, the total interaction potential energies between two molecules, V_{ij}, can be represented as:

V_{ij} = overlap repulsion + charge transfer + coulombic interactions

+ charge induced polarization + induced dipole-induced dipole interactions +hydrogen bonding (6.2)

Overlap repulsion occurs when the electron clouds of two closed-shell molecules overlap. The overlap repulsion for two spherical atoms or molecules can be represented as:

$$V(R) = (\sigma / R)^n \qquad (6.3)$$

where n is an integer. When $n = \infty$, an expression for infinitely hard spheres is obtained because for $R > \sigma$, the value of $V(R)$ is effectively zero, while for $R < \sigma$, it is infinity. The value of n, however, is usually between 10 and 15. Taking $n = 12$, the repulsive term used in the Lennard-Jones interaction potential $\psi(r)$ is given as:

$$\psi(r) = 4\varepsilon^* \left[\left(\frac{\sigma}{r} \right)^{12} - \left(\frac{\sigma}{r} \right)^6 \right] \qquad (6.4)$$

The parameter ε^* gives the maximum depth of the potential well and the parameter σ is the point at which the potential is zero, an effective collision diameter for the particle. The twelfth power term is a repulsion term, while the sixth-power term is a dispersive attraction term.

Charge transfer interaction occurs when one molecule donates excess electrons to an acceptor with electron deficiency. This type of interaction occurs when molecules are in close proximity. Acid-base reactions are examples of this. Hydrogen bonding is the most important contributor to this type of interaction force.

Hydrogen bonding is a special type of interaction and it can be considered to be a special case of electrostatic interaction since the low electron density in a hydrogen atom bonded to an electronegative atom, such as oxygen or nitrogen, weakens the Pauli exclusion force and enables the hydrogen atom to come into close proximity to the polar center of the molecule.

Coulombic interactions occur between ions and molecules that possess either net ionic charges or an asymmetrical electron distribution of electrons or nuclei, such as polar molecules.

Induced polarization occurs between a polar molecule with a permanent electrical dipole and another molecule that is polarizable.

Induced dipole-induced dipole interactions are the sources of the London dispersion term which is often the controlling force or component due to its universality.

6.3.1   Coulombic Interactions

Coulomb's law describes attractive and repulsive interactions between fixed charges and depending on the sign of the charges, ion-ion interactions can be attractive or repulsive. The Coulombic force exerted between two charges, Q_1 and Q_2 is given by:

$$F = \frac{Q_1 Q_2}{4\pi\varepsilon_o R^2} \tag{6.5}$$

where:

ε_o = permittivity of a vacuum

Q = ionic charge $Z \cdot e$, and e is the elementary charge of an electron

Z = the ionic valency

The corresponding expression for the potential energy is:

$$V_{ion-ion} = \frac{Q_1 Q_2}{4\pi\varepsilon_o R} = -\frac{Z_1 Z_2 e^2}{4\pi\varepsilon_o R} \tag{6.6}$$

6.3.2   Charge Induced Polarization

Many molecules possess a permanent charge separation which gives rise to a dipole. These molecules are said to be polar molecules, a prime example of which is water. Because dipoles arise from asymmetric displacements of electrons along covalent bonds, they are not observed in singles atoms. The dipole moment of a polar molecule is given by:

$$\mu = qL \tag{6.7}$$

where: L = distance between charges q

μ = dipole moment

Similarly an ion-dipole interaction can be represented as:

$$V_{ion-dipole}(R,\theta) = -\frac{Ze\mu\cos\theta}{4\pi\varepsilon_o R^2} \tag{6.8}$$

where θ is the angle between the ion and the dipoles and the other terms are as before. With a cation, the maximum attraction occurs when the dipole points away from the ion, i.e., $\theta = 0$, while the maximum repulsion occurs when $\theta = 180°$. For the interaction of a sodium ion and a water molecule, the resultant $V_{ion-dipole}/kT$ is $\approx 1/5$ that of $V_{ion-ion}$. Nevertheless, this type of interaction can exert significant effects.

6.3.3 Dipole Interactions

6.3.3.1 Dipole-Dipole

For dipole-dipole interactions of moments m_1 and m_2 at distance R, the orientation of one dipole, relative to the other, plays a key role in the resultant potential energy. Thus the angles between the two must be accounted for as shown in Figure 6.1. The angles between R and the center of the dipoles are defined by q_1 and q_2, and ϕ gives the relative dipole orientations with respect to rotation around the R axis.

$$V(R,\theta_1,\theta_2,\phi) = -\frac{\mu_1\mu_2}{4\pi e_o R^3}\left[2\cos\theta_1\cos\theta_2 - \sin\theta_1\sin\theta_2\cos\phi\right] \tag{6.9}$$

In liquids and solids, where coulombic interactions are

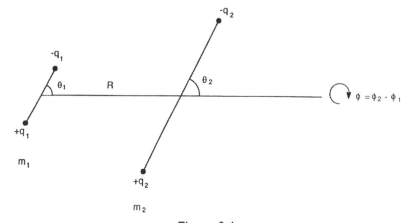

Figure 6.1
Interaction Between Two Dipoles Separated by a Distance R

greatly reduced, dipole-dipole interactions are greatly diminished and are not strong enough to lead to substantial mutual alignment of polar molecules. In the situations in which the dipoles are relatively free to rotate with respect to each other, the potential energy contribution is given by an angle averaged potential from the classical Boltzmann distribution.

$$V(R) = \frac{1}{4\pi} \oint (V(R,\Omega) \exp(-V(R,\Omega)/kT) d\Omega \qquad (6.10)$$

where the integration is performed over the solid angle $d\Omega = \sin\theta d\theta d\phi$. The angle averaged free energy for the dipole-dipole interaction obtained from Equation 6.9 is:

$$V_{dipole-dipole}(R) = -\frac{\mu_1^2 \mu_2^2}{3(4\pi\varepsilon_o)^2 kTR^6}$$

$$\text{for } kT > \frac{\mu_1 \mu_2}{4\pi\varepsilon_o R^3} \qquad (6.11)$$

The $1/R^6$ dependence for the Boltzmann-averaged interaction between two permanent dipoles is known as the Keesom interaction.

For two water molecules, the Keesom force of interaction is about 1/8 that of a sodium ion with water. This allows us to appreciate the differences in the relative magnitudes of the forces between molecules and the small but not insignificant forces resulting from dipole interactions. It should be noted that by using the Boltzmann equation, free energies are obtained.

6.3.4 Induced Polarization Interactions

Attractive interactions occur when electrical fields emanating from a polar molecule induce polarization in nearby molecules. These types of interactions play a major role in polymer-polymer interactions and the resulting interfacial behavior and miscibility.

For weak fields, the magnitude of the induced dipole moment is proportional to the electrical field, E, and the proportionality constant, α, is known as the polarizability of the molecule. The relationship is given by:

$$\mu^{ind} = \alpha E \tag{6.12}$$

and the total energy is given as:

$$V_{ttl} = -\alpha \frac{E^2}{2} \tag{6.13}$$

Polarizability has the dimensionality of $\varepsilon_o \times$ volume and its magnitude typically corresponds to half the molecular volume. For example, in the case of water, $\alpha_o / \varepsilon_o = 1.86 \times 10^{-29}\, m^3$, while the value derived from van der Waals equation of state is $3.0 \times 10^{-29}\, m^3$.

By combining the previous equation with the equation for the electric field associated with an ion of charge Ze:

$$E = \frac{Ze}{4\pi\varepsilon_o R^2} \tag{6.14}$$

which in turn gives:

$$V_{ion-induced\ dipole} = -\frac{\alpha(Ze)^2}{2(4\pi\varepsilon_o)^2 R^4} \qquad (6.15)$$

$$= -\frac{1}{2(4\pi\varepsilon_o)^2 R^4}(Z_1^2 e^2 \alpha_2 + Z_2^2 e^2 \alpha_1) \qquad (6.16)$$

This is the more generalized form.

It turns out that ion-induced dipole interactions with ion-dipole interactions show that induced interactions exhibit only half the energy of direct interactions.

6.3.4.1 Dipole-Induced Dipole Interactions

The dipole-induced dipole interaction is given by:

$$V_{dipole-induced\ dipole} = -1/2\ E^2 = -\frac{\mu^2 \alpha(1 + 3\cos^2\theta)}{2(4\pi\varepsilon_o)^2 R^6} \qquad (6.17)$$

For the general case of two different molecules, each of which possesses a permanent dipole moment μ_1 and μ_2 and polarizabilities α_1 and α_2, the total interaction is given by:

$$V_{dipole-induced\ dipole}(R) = -\frac{1}{2(4\pi\varepsilon_o)^2}\frac{1}{R^6}\left[\begin{array}{c}\mu_1\mu_2(3\cos^2\theta + 1) + \\ \mu_2^2\alpha_1(3\cos^2\theta + 1)\end{array}\right] \qquad (6.18)$$

By using $\langle\cos^2\theta\rangle = 1/3$, the angle-averaged energy is:

$$V_{dipole-induced\ dipole}(R) = -\frac{\mu_1^2\alpha_2 + \mu_2^2\alpha_1}{(4\pi\varepsilon_o)^2 R^6} \qquad (6.19)$$

Equation 6.19 represents the Debye attractive interaction.

6.3.4.2 Induced Dipole-Induced Dipole Interactions

These types of interactions lead to attractive interactions between all molecules and are known as dispersion forces. These forces, as in the previous case, play an important role in interfacial phenomena. Dispersion forces are generally interpreted as emanating from an instantaneous dipole moment determined by the location of the electrons around the nucleus. This dipole generates an electric field, which in turn induces a dipole in nearly neutral atoms. This interaction gives rise to an attractive force between the two atoms.

In 1933 London derived an expression for the attraction between a pair of induced dipoles by solving the Schrödinger equation. Modeling each of the vibration dipoles as a quantified harmonic oscillator with characteristic frequency ν, he obtained:

$$V(R) = -\frac{3}{4}\frac{\alpha_o^2 \hbar \nu}{\left(4\pi\varepsilon_o\right)^2 R^6}$$ (6.20)

and for two similar atoms

$$V(R) = -\frac{3}{2}\frac{\alpha_{o1}\alpha_{o2}}{\left(4\pi\varepsilon_o\right)^2 R^6}\left[\frac{\hbar\nu_1\nu_2}{\left(\nu_1 + \nu_2\right)}\right]$$ (6.21)

6.4 DEFINITION OF VAN DER WAALS FORCES

Building upon the previous abbreviated understanding of various molecular interactions and their resulting forces, we can formally define van der Waals forces for molecules in the vapor state as being comprised of Keesom orientation forces, Debye inductive forces, and London dispersion forces.

For two dissimilar molecules, van der Waals forces can be expressed as:

$$V_{VDW}(R) = \left(\frac{V_{orient.} + V_{ind.} + V_{disp.}}{R^6}\right)$$ (6.22)

Dispersion forces are more important than either orientation or induction forces for all molecules except highly polar molecules like water. In interactions between dissimilar molecules, dispersion forces dominate almost completely if one of the molecules is non-polar.

Expressing Equation 6.22 in terms of the previous relationships presented, we have:

$$V_{VDW}(R) = -\left[\frac{\mu_1^2\mu_1^2}{3kT} + \left(\mu_1^2\alpha_2 + \mu_2^2\alpha_1\right) + \frac{3\alpha_1\alpha_2\hbar\nu_1\nu_2}{2(\nu_1+\nu_2)}\right]\left[(4\pi\varepsilon_o)^2 R^6\right]^{-1}$$

(6.23)

Thus the individual van der Waals forces can be expressed as:

Keesom

$$V_{(dipole-dipole)} \propto -\frac{\mu_1^2\mu_2^2}{kTR^6}$$

(6.24)

Debye

$$V_{(dipole-ind.\,dipole)} \propto -\frac{\mu_1^2\alpha_2 + \mu_2^2\alpha_2}{R^6}$$

(6.25)

London/Dispersion

$$V_{(ind.\,dipole-ind.\,dipole)} \propto -\frac{\alpha^2\hbar\nu}{R^6}$$

(6.26)

6.5   POTENTIAL ENERGY CURVES DESCRIBING MOLECULAR INTERACTIONS

The three forces which contribute to the van der Waals forces account for the attractive interactions between molecules. However, we must remember that the total intermolecular interactions must be considered by adding together the equations for the attractive and repulsive potentials. These concepts are illus-

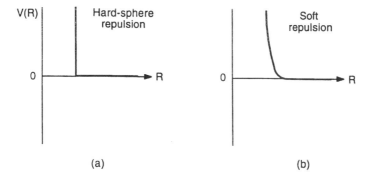

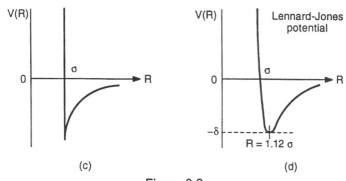

Figure 6.2
Potential Energy Curves

a) hard sphere repulsion $(n = \infty)$

b) soft repulsion $(n = 12)$

c) hard sphere repulsion combined with van der Waals attraction

d) Lennard-Jones 6-12 potential

trated in Figure 6.2 where the hard sphere repulsive potential is:

$$V(R) = \left(\frac{\sigma}{R}\right)^{\infty} \qquad (6.27)$$

and the soft repulsion potential for molecular liquids and solids is:

$$V(R) = \left(\frac{\sigma}{R}\right)^{12} \qquad (6.28)$$

In the case of the hard sphere repulsive potential, the binding energy at contact $(R = \sigma)$ equals the van der Waals energy as illustrated in Figure 6.2.

6.6   MASS AND THE PACKING OF MATTER

We need not be concerned with the estimates of molecular mass since mass is unambiguously defined and measurable. The packing of materials, particularly polymers, is a complicated issue as we have seen in Chapter 1. The packing can vary from 100% amorphous to 100% crystalline as well as be influenced by the morphology of the crystalline phase in semi-crystalline polymers.

6.7   VOLUMETRIC PROPERTIES AND THEIR ESTIMATION

The brief treatment of the fundamentals of molecular forces and interactions have provided the groundwork for a discussion of volumetric properties. The volumetric properties of a material or polymer are important in the measurement of many physical properties which in turn influence processing and applications.

It turns out that the volumetric properties of a polymer are key parameters that are used in estimating many different properties. This means that if the molar volume of a polymer is in error, additional properties that depend on this term will also be in error unless compensating errors offset the volumetric error term. For example, experience has been that slight variations in molecular volume can produce a large change in the estimated solubility parameters.

MPCalc calculates the molecular weight of the structural groups the user has entered into the input column and lists the results in the "Calculated Molecular Properties" column.

As mentioned above, the mass of molecules is obtained simply by summing the mass of each atom comprising the molecule. The volume of the molecule is generally not the simple sum of the atomic masses that constitute the molecule.

Recalling the additivity concepts discussed in Chapter 5, the molecular environment of surrounding functional groups significantly influences the volume occupied by a specific functional group. It would indeed simplify any volumetric computations if the van der Waals radii, which are known quite accurately, could be summed to provide a functional group volume. Unfortunately, this cannot be done since the sum of the van der Waals radii of two atoms of a molecule only determines the closest distance of the atoms and does not take into account the reduced distance from combined atomic volumes as a result of chemical reaction. It can be seen that the van der Waals volume of a molecule is much smaller than the sum of the van der Waals volumes of the atoms constituting the structure.

6.7.1   Rubbery Volumes (V_R)

The rubbery state of amorphous polymers is more closely approximated by the liquid state of organic compounds. These polymers are in the rubbery state and have a glass transition temperature lower than 25 °C. Consequently, we use the molar volumes of organic liquids to approximate the molar volume of each structural unit or functional polymer group as described by Van Krevelen and Hoftyzer [G1].

The volume is estimated using the additivity function for V_R.

$$V_R(298) = \sum_i V_i(298) \qquad (6.29)$$

The value of V_R can also be obtained from the van der Waals volume by using the relationship [G2]:

$$V_R = V_W \times 1.6(\pm 0.035) \qquad (6.30)$$

Although there is good overall correspondence in this relationship, there is sufficient individual variation to justify the use of experimentally determined molar structural volumes. In addition these values have been adjusted, in some cases, based on a significant number of volume estimations and resulting properties for a large number of polymer classes.

This value is then used to estimate the density and is listed as the "Density Am." From the equation

$$\rho_a = \frac{\text{molar mass}}{\text{molar volume}} = \frac{M}{V} = \frac{\text{molecular weight}}{V_R} = \frac{g}{cm^3} \qquad (6.31)$$

6.7.2   Glassy Volumes $\left(V_g\right)$

Amorphous materials or polymers that have a glass transition temperature higher than 25 °C are in the glassy-amorphous state at room temperature.

The volume of V_g is estimated using an approach similar to the previous.

$$V_g(298) = \sum_i V_i(298) \qquad (6.32)$$

The value of V_g can also be obtained from the van der Waals volume from:

$$V_g = V_W \times 1.6(\pm 0.045) \qquad (6.33)$$

It should be noted that the standard deviation has increased thus increasing the potential for greater error in individual estimates if V_g is obtained directly from V_W. It is for this reason that again we use individual V_g structural contributions, many of which have been adjusted, based on polymer results comparisons.

As before, the density is estimated and listed as "Density Gl." from the relationship

$$\rho = \frac{\text{molecular weight}}{V_g} = \frac{g}{cm^3} \qquad (6.34)$$

Although the glassy state will be discussed at length in Chapter 8, it is worth noting here that the glassy state is not one of thermodynamic equilibrium and therefore amorphous glassy polymers exhibit volume relaxation. As a result, V_g and other associated properties represent typical values associated with what would be expected as a result of "normal treatment" procedures as opposed to quick cooling or special treatments.

6.7.3 The van der Waals Volume

The van der Waals volume is approximated by the boundaries of the outer surface of a number of interpenetrating spheres. The radii of the spheres are assumed to be constant atomic radii and the distances between the centers of the spheres are constant bond distances. The van der Waals volume of a molecule then is the molecular space which is impenetrable to other molecules with normal thermal energies.

The van der Waals volume of an atom A of radius R is:

$$V_{W.A} = N_A \left[4/3\pi R^3 - \sum \pi h_i^2 \left(R - \frac{h_i}{3} \right) \right] \tag{6.35}$$

$$= cm^3/mol$$

where:

$$h_i = R - \frac{l_i}{2} - \frac{R^2}{2l_i} + \frac{r_i^2}{2l_i} \tag{6.36}$$

and N_A = Avogadro's number

r_i = radius of atom i

l_i = bond distance between atoms A and i

Again, it should be noted that the van der Waals volume is not constant since its value depends on the nature of the surrounding atoms i. However, the van der Waals volume contribution is approximated by accounting for the volume contribution of the surrounding structural groups. As before, for this reason it is always preferable to choose the largest group(s) which more closely approximate the structural group of the molecule.

6.7.4 Crystalline Volumes

As we have seen in Chapter 2, semi-crystalline polymers contain many crystallites with amorphous regions and are therefore not fully crystalline. The exception is when extraordinary efforts are used to grow single crystals which are fully crystalline. Therefore, calculated densities of polymers can be expected to only approximate actual densities because of the great variation in possible morphologies. This is especially true in the case when estimating the crystalline density of a polymer.

The difference between amorphous, glassy, and crystalline densities, however, can be quite enlightening in allowing us to appreciate the magnitude of the differences involved for a given structure and the total possible variation.

For a number of polymers whose densities for both the "purely" amorphous and the "purely" crystalline states are known, the ratio ρ_c / ρ_a shows a mean value of 1.13. Although there is considerable variation, i.e., 1.13 ± 0.08, the mean value can be used to provide the best estimate of the crystalline density which is estimated by MPCalc according to:

$$\rho_c = \rho_a \times 1.13 \qquad (6.37)$$

6.8   THERMAL EXPANSION

The expansion of a material as a result of the application of heat is an important property that must be taken into account in polymer processing, injection molding, and in application design. The expansion phenomenon depends mostly on internal intermolecular forces. On the application of heat, the atoms which constitute the structure begin vibrating to a greater extent. The general increase in volume with temperature is mainly determined by the increased amplitude of atomic vibrations about a mean position. The repulsion term between atoms changes more rapidly with atomic separation than does the attraction term. Consequently, the minimum energy trough is non-symmetrical (See Figure 6.3).

As thermal energy is added and the lattice energy increases, the increased amplitude of vibration between equivalent energy positions leads to a higher value for the atomic separation corresponding to a lattice expansion. This can be seen from Figure 6.4 where

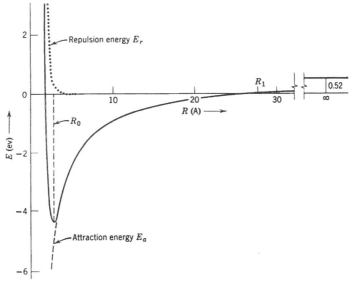

Figure 6.3 [G5]
Lattice Energy as a Function of Atomic Spacing

the dotted line traces the interatomic spacing from the initial average position at *V(T)* to the new mean positions at higher temperatures.

In addition, Figure 6.4 shows that at *T*, the thermal energy results in oscillation between d_a and d_b: the mean position, d_e, is greater than $d_o \left(0° \text{ K}\right)$ because of the asymmetry of the potential energy curve. There is actually a small vibrational energy at $0° \text{K}$.

The potential energy trough is very deep for covalently and ionically bonded solids and some metals of high sublimation temperature, and the corresponding coefficient of thermal expansion is small. The potential energy trough is less deep for many solids with metallic bonds and is relatively shallow for those with molecular bonds, as is the case for polymers. Therefore, such solids expand, melt, sublime, and deform under stress more easily.

This can be seen in the Material Comparison Workbook in MATPROP© where the coefficients of thermal expansion of various materials are listed. The general order for the coefficients of thermal expansion are: ceramics < metals < polymers.

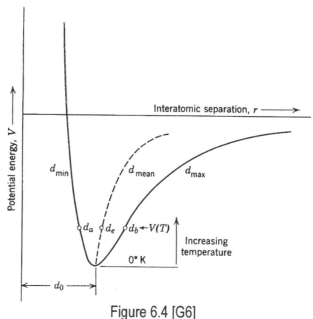

Figure 6.4 [G6]
Variation of Potential Energy as a Function of Interatomic Separation

6.9   GENERAL FACTORS AFFECTING THERMAL EXPANSION

The linear coefficient of thermal expansion is given by:

$$\alpha_l = \frac{\dfrac{dl}{l_o}}{dT} \left(K^{-1} \right) \tag{6.38}$$

where:

l_o = original length
dl = change in length
dT = change in temperature

In general, the following factors are determining influences on α :

1. Composition - α is lower for covalent solids because stretching or expansion is more difficult than for ionic.

This again illustrates the overriding influences of intermolecular forces or lattice effects.

2. Higher melting point materials evidence smaller α's.
3. Lower atomic packing efficiency results in a lower α.
4. Crystalline anisotropy can be a significant factor.
5. Expansion of ceramic (glass, glaze, etc.) must be matched to the substrate to provide a slight compression.

6.10   ESTIMATION OF THERMAL PROPERTIES

6.10.1 Molar Thermal Expansivity

In MPCalc the molar expansivity is estimated for the glassy state E_g or below the glass transition and for the rubbery or liquid state E_l which provides a measure of the expansivity above the glass transition.

A detailed discussion of the nature of the glassy state is given in Chapter 8. At this point it is sufficient to point out that the thermal expansion differs markedly below and above the glass transition and it is useful to have both values even though only the thermal expansion above T_g, i.e., E_l, may be the only useful value.

The molar thermal expansivity, E_g or E_l is given by:

$$\left(\frac{\partial V}{\partial T}\right)_p \equiv E\left(cm^3/mol\,K\right) \tag{6.39}$$

The molar thermal expansivity is estimated using the relationships of E_l and E_g to the van der Waals volume. This is illustrated in Figure 6.5 from which the following relationships have been derived [G2] and are used in MPCalc.

$$E_l\left(E_R\right) = 1.00 \times 10^{-3} V_W \tag{6.40}$$

and:

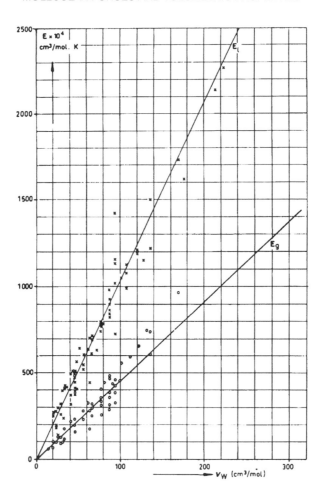

Figure 6.5 [G2]

E_l and E_g As a Function of V_W

$$E_g \left(E_c \right) = 0.45 \times 10^{-3} V_W \qquad (6.41)$$

It will be noted that $E_l \approx E_R$ and $E_g \approx E_c$.

6.10.2 Specific Thermal Expansivity

The specific thermal expansivity is simply the expansivity expressed on a per gram basis.

$$\left(\frac{\partial v}{\partial T}\right)_p \equiv e\left(cm^3 / g \cdot K\right) \tag{6.42}$$

and $\quad e = \dfrac{E_{l\,or\,g}cm^3 / mol.\ K^\circ}{MW\ g / mol.}\left(cm^3 / g \cdot K\right) \tag{6.43}$

6.10.3 Cubic Coefficient of Thermal Expansion

It will be recalled that in Chapter 2 the cubic coefficient of thermal expansion was shown to be :

$$\alpha = \frac{1}{V}\frac{\partial V}{\partial T}\bigg|_P = K^{-1}$$

Thus this value is estimated using the previously estimated values for E_l and E_g with the following:

$$\alpha_2 = \frac{1}{V}\frac{\partial V}{\partial T}\bigg|_p \left(K^{-1}\right) \tag{6.44}$$

$$\alpha_2 = e_l \cdot \rho_l = \frac{cm^3}{g\ K} \times \frac{g}{cm^3} \tag{6.45}$$

6.10.4 Linear Coefficient of Thermal Expansion

When sample uniformity is present, there is no anisotropy and the linear coefficient of expansion can be approximated by:

$$\alpha_l = \frac{\alpha_2}{3}\left(K^{-1}\right) \tag{6.46}$$

6.11 GENERAL EXPRESSIONS FOR MOLAR VOLUMES AS A FUNCTION OF TEMPERATURE

The volume estimations of MPCalc are based on the standard temperature of $298°K$. Volumes can be estimated at temperatures other than $298°K$ but their usefulness of is not as great so the relationships for estimation have not been included in MPCalc.

Using the estimated van der Waals volume, V_R, V_g, and V_c, may be calculated from the following relationships as reported by Van Krevelen.

$$V_R = V_W \left(1.30 + 10^{-3} T \right) \tag{6.47}$$

$$V_g = V_W \left(1.30 + 0.55 \times 10^{-3} T_g + 0.45 \times 10^{-3} T \right) \tag{6.48}$$

$$V_W = V_W \left(1.30 + 0.45 \times 10^{-3} T \right) \tag{6.49}$$

6.12   MPCALC PROPERTY ESTIMATIONS THAT EMPLOY VOLUME ESTIMATIONS AND COMPARISON TO LITERATURE VALUES

Table 6.1 lists MPCalc calculated properties that depend directly on the estimated volumes for a number of polymers. These values are contained in total in the tables of polymer properties listed in Chapter 5.

For purposes of evaluating correspondence with literature values for the cubic coefficient of expansion, the Bueche/Tobolsky [G3,G4] relationship is used which is given as:

$$\alpha_l - \alpha_g \approx 5 \times 10^{-4 \circ} \left(K^{-1} \right) \tag{6.50}$$

This is done because of the wide variation in reported expansion coefficients of polymers and general lack of data for a_l and α_g.

Examination of Table 6.1 shows that in general there is good agreement between the calculated values and the literature values that are available. Literature values have been taken for the most part from the tables compiled by Van Krevelen in "Properties of Polymers" [G2].

It is worth noting that MPCalc estimates values, whether or not there are valid inputs to the selected groups available. As always, the user must check on whether or not inputs for a particular property of interest are available. In addition, MPCalc estimates all values without regard to physical significance. For example, the crystalline density of poly(isobutylene) is calculated even though the 100% crystalline polymer does not exist. There are many other examples here and throughout the book and the user is urged to exercise judgment in using the estimated results.

Table 6.1

Comparison of Estimated Properties Employing Volume or Ancillary Volume Estimations for Various Polymers

	CALCULATED MOLECULAR PROPERTIES	LITERATURE VALUES	
Poly(ethylene)			
Mol.Wt.	28.06	28.06	g/mol.
Density Am.	0.853	0.855	g/cm3.
Density Gl.	0.885	-----	g/cm3.
Density Cr.	0.964	0.96-1.0	g/cm3.
VdeWaalsVol	20.46	20.5	cm3/mol
Eg	9.21E-03	6.7-1.0E-03	cm3/mol*K,for glass,therm expan
E1	2.05E-02	2.1-2.7E-02	cm3/mol*K,forliquid,therm expan
e1	7.29E-04	7.5-9.6E-04	cm3/g*K,for liq,sp.therm.expan
eg	3.28E-04	2.4-3.6E-04	cm3/g*k,for glass,sp.therm expan
alpha liq. cubic	6.22E-04	----	1/K,cubic therm.expan., liq.state
alpha gl. cubic	2.90E-04	----	1/K, cubic therm.expan., gl.state
alpha liq. linear	2.07E-04	----	1/K, linear coeff.liq.therm.expan.
alpha gl. linear	9.68E-05	----	1/K, linear coeff., gl.therm.expan.
$\alpha_2$liq.-$\alpha_2$gl.	3.33E-04	5E-04	1/K
Poly(propylene)			
Mol.Wt.	42.08	42.08	g/mol.
Density Am.	0.857	0.85	g/cm3.
Density Gl.	0.855	-----	g/cm3.
Density Cr.	0.968	0.95	g/cm3.
VdeWaalsVol	30.68	30.7	cm3/mol
Eg	1.38E-02	0.93E-02	cm3/mol*K,for glass,therm expan
E1	3.07E-02	2.3-4.0E-02	cm3/mol*K,forliquid,therm expan
e1	7.29E-04	5.5-9.4E-04	cm3/g*K,for liq,sp.therm.expan
eg	3.28E-04	2.2E-04	cm3/g*k,for glass,sp.therm expan
alpha liq. cubic	6.25E-04	-----	1/K,cubic therm.expan., liq.state
alpha gl. cubic	2.81E-04	-----	1/K, cubic therm.expan., gl.state
alpha liq. linear	2.08E-04	-----	1/K, linear coeff.liq.therm.expan.
alpha gl. linear	9.35E-05	-----	1/K, linear coeff., gl.therm.expan.
$\alpha_2$liq.-$\alpha_2$gl.	3.44E-04	5E-04	1/K
Poly(isobutylene)			
Mol.Wt.	56.1	56.1	g/mol.
Density Am.	0.840	0.84	g/cm3.
Density Gl.	0.822	-----	g/cm3.
Density Cr.	0.949	-----	g/cm3.
VdeWaalsVol	40.9	40.9	cm3/mol
Eg	1.84E-02	0.9-1.1E-02	cm3/mol*K,for glass,therm expan
E1	4.09E-02	3.1-3.9E-02	cm3/mol*K,forliquid,therm expan
e1	7.29E-04	5.6-6.9E-04	cm3/g*K,for liq,sp.therm.expan
eg	3.28E-04	1.6-2.0E-04	cm3/g*k,for glass,sp.therm expan
alpha liq. cubic	6.12E-04	-----	1/K,cubic therm.expan., liq.state
alpha gl. cubic	2.70E-04	-----	1/K, cubic therm.expan., gl.state
alpha liq. linear	2.04E-04	-----	1/K, linear coeff.liq.therm.expan.
alpha gl. linear	8.99E-05	-----	1/K, linear coeff., gl.therm.expan.
$\alpha_2$liq.-$\alpha_2$gl.	3.42E-04	5E-04	1/K

(continued)

Table 6.1 (continued)

	CALCULATED MOLECULAR PROPERTIES	LITERATURE VALUES	

Poly(vinyl chloride)

Mol.Wt.	62.51	62.51	g/mol.
Density Am.	1.398	1.398	g/cm3.
Density Gl.	1.383	1.385	g/cm3.
Density Cr.	1.580	1.52	g/cm3.
VdeWaalsVol	29.23	29.2	cm3/mol
Eg	1.32E-02	8.8-1.3E-02	cm3/mol*K,for glass,therm expan
E1	2.92E-02	2.6-3.3E-02	cm3/mol*K,forliquid,therm expan
e1	4.68E-04	4.2-5.2E-02	cm3/g*K,for liq,sp.therm.expan
eg	2.10E-04	1.4-2.1E-04	cm3/g*k,for glass,sp.therm expan
alpha liq. cubic	6.54E-04	-----	1/K,cubic therm.expan., liq.state
alpha gl. cubic	2.91E-04	-----	1/K, cubic therm.expan., gl.state
alpha liq. linear	2.18E-04	-----	1/K, linear coeff.liq.therm.expan.
alpha gl. linear	9.70E-05	-----	1/K, linear coeff., gl.therm.expan.
$\alpha_2$liq. -$\alpha_2$gl.	3.63E-04	5E-04	1/K

Poly(vinyl fluoride)

Mol.Wt.	46.05	46.05	g/mol.
Density Am.	1.269	-----	g/cm3.
Density Gl.	1.272	-----	g/cm3.
Density Cr.	1.434	1.44	g/cm3.
VdeWaalsVol	23.23	22.8	cm3/mol
Eg	1.05E-02	-----	cm3/mol*K,for glass,therm expan
E1	2.32E-02	-----	cm3/mol*K,forliquid,therm expan
e1	5.04E-04	-----	cm3/g*K,for liq,sp.therm.expan
eg	2.27E-04	-----	cm3/g*k,for glass,sp.therm expan
alpha liq. cubic	6.40E-04	-----	1/K,cubic therm.expan., liq.state
alpha gl. cubic	2.89E-04	-----	1/K, cubic therm.expan., gl.state
alpha liq. linear	2.13E-04	-----	1/K, linear coeff.liq.therm.expan.
alpha gl. linear	9.63E-05	-----	1/K, linear coeff., gl.therm.expan.
$\alpha_2$liq. -$\alpha_2$gl.	3.51E-04	5E-04	1/K

Poly(chlorotrifluoroethylene)

Mol.Wt.	116.48	116.48	g/mol.
Density Am.	2.012	1.9-2.0	g/cm3.
Density Gl.	1.885	-----	g/cm3.
Density Cr.	2.273	2.19	g/cm3.
VdeWaalsVol	35.96	36.9	cm3/mol
Eg	1.62E-02	1.2-1.8E-02	cm3/mol*K,for glass,therm expan
E1	3.60E-02	2.3-4.1E-02	cm3/mol*K,forliquid,therm expan
e1	3.09E-04	2.0-3.5E-04	cm3/g*K,for liq,sp.therm.expan
eg	1.39E-04	1.0-1.5E-04	cm3/g*k,for glass,sp.therm expan
alpha liq. cubic	6.21E-04	-----	1/K,cubic therm.expan., liq.state
alpha gl. cubic	2.62E-04	-----	1/K, cubic therm.expan., gl.state
alpha liq. linear	2.07E-04	-----	1/K, linear coeff.liq.therm.expan.
alpha gl. linear	8.73E-05	-----	1/K, linear coeff., gl.therm.expan.
$\alpha_2$liq. -$\alpha_2$gl.	3.59E-04	5E-04	1/K

(continued)

Table 6.1 (continued)

	CALCULATED MOLECULAR PROPERTIES	LITERATURE VALUES	

Poly(vinyl acetate)

Mol.Wt.	86.09	86.09	g/mol.
Density Am.	1.168	-----	g/cm3.
Density Gl.	1.192	1.19	g/cm3.
Density Cr.	1.320	1.34	g/cm3.
VdeWaalsVol	45.88	45.9	cm3/mol
Eg	2.06E-02	1.6-2.0E-02	cm3/mol*K,for glass,therm expan
E1	4.59E-02	4.1-5.2E-02	cm3/mol*K,forliquid,therm expan
e1	5.33E-04	5.0-6.0E-04	cm3/g*K,for liq,sp.therm.expan
eg	2.40E-04	1.8-2.3E-04	cm3/g*k,for glass,sp.therm expan
alpha liq. cubic	6.23E-04	-----	1/K,cubic therm.expan., liq.state
alpha gl. cubic	2.86E-04	-----	1/K, cubic therm.expan., gl.state
alpha liq. linear	2.08E-04	-----	1/K, linear coeff.liq.therm.expan.
alpha gl. linear	9.53E-05	-----	1/K, linear coeff., gl.therm.expan.
$\alpha_2$liq.-$\alpha_2$gl.	3.37E-04	5E-04	1/K

Poly(butadiene)

Mol.Wt.	54.1	54.1	g/mol.
Density Am.	0.892	0.892	g/cm3.
Density Gl.	0.910	-----	g/cm3.
Density Cr.	1.008	1.01	g/cm3.
VdeWaalsVol	37.4	37.5	cm3/mol
Eg	1.68E-02	1.08E-02	cm3/mol*K,for glass,therm expan
E1	3.74E-02	3.74E-02	cm3/mol*K,forliquid,therm expan
e1	6.91E-04	6.4-7.7E-04	cm3/g*K,for liq,sp.therm.expan
eg	3.11E-04	2.0E-04	cm3/g*k,for glass,sp.therm expan
alpha liq. cubic	6.17E-04	-----	1/K,cubic therm.expan., liq.state
alpha gl. cubic	2.83E-04	-----	1/K, cubic therm.expan., gl.state
alpha liq. linear	2.06E-04	-----	1/K, linear coeff.liq.therm.expan.
alpha gl. linear	9.44E-05	-----	1/K, linear coeff., gl.therm.expan.
$\alpha_2$liq.-$\alpha_2$gl.	3.34E-04	5E-04	1/K

Poly(isoprene)(Poly(cis-1,4 isoprene) Natural Rubber)

Mol.Wt.	68.12	68.12	g/mol.
Density Am.	0.900	≈0.92	g/cm3.
Density Gl.	0.907	-----	g/cm3.
Density Cr.	1.017	1.0	g/cm3.
VdeWaalsVol	47.61	47.5	cm3/mol
Eg	2.14E-02	-----	cm3/mol*K,for glass,therm expan
E1	4.76E-02	4.1-5.7E02	cm3/mol*K,forliquid,therm expan
e1	6.99E-04	6.0-8.3E-04	cm3/g*K,for liq,sp.therm.expan
eg	3.15E-04	-----	cm3/g*k,for glass,sp.therm expan
alpha liq. cubic	6.29E-04	-----	1/K,cubic therm.expan., liq.state
alpha gl. cubic	2.85E-04	-----	1/K, cubic therm.expan., gl.state
alpha liq. linear	2.10E-04	-----	1/K, linear coeff.liq.therm.expan.
alpha gl. linear	9.51E-05	-----	1/K, linear coeff., gl.therm.expan.
$\alpha_2$liq.-$\alpha_2$gl.	3.44E-04	5E-04	1/K

(continued)

230

Table 6.1 (continued)

	CALCULATED MOLECULAR PROPERTIES	LITERATURE VALUES	

Poly(styrene)

Mol.Wt.	104.15	104.15	g/mol.
Density Am.	1.145	-----	g/cm3.
Density Gl.	1.063	1.05	g/cm3.
Density Cr.	1.294	1.13	g/cm3.
VdeWaalsVol	62.85	62.9	cm3/mol
Eg	2.83E-02	1.8-2.8E-02	cm3/mol*K,for glass,therm expan
E1	6.29E-02	4.5-7.1E-02	cm3/mol*K,forliquid,therm expan
e1	6.03E-04	4.3-6.8E-04	cm3/g*K,for liq,sp.therm.expan
eg	2.72E-04	1.7-2.7E-04	cm3/g*k,for glass,sp.therm expan
alpha liq. cubic	6.91E-04	-----	1/K,cubic therm.expan., liq.state
alpha gl. cubic	2.89E-04	-----	1/K, cubic therm.expan., gl.state
alpha liq. linear	2.30E-04	-----	1/K, linear coeff.liq.therm.expan.
alpha gl. linear	9.62E-05	-----	1/K, linear coeff., gl.therm.expan.
$\alpha_2$liq.-$\alpha_2$gl.	4.02E-04	5E-04	1/K

Poly(αmethyl styrene)

Mol.Wt.	118.17	118.17	g/mol.
Density Am.	1.088	-----	g/cm3.
Density Gl.	1.010	1.065	g/cm3.
Density Cr.	1.229	1.16	g/cm3.
VdeWaalsVol	73.07	73.0	cm3/mol
Eg	3.29E-02	-----	cm3/mol*K,for glass,therm expan
E1	7.31E-02	-----	cm3/mol*K,forliquid,therm expan
e1	6.18E-04	-----	cm3/g*K,for liq,sp.therm.expan
eg	2.78E-04	-----	cm3/g*k,for glass,sp.therm expan
alpha liq. cubic	6.73E-04	-----	1/K,cubic therm.expan., liq.state
alpha gl. cubic	2.81E-04	-----	1/K, cubic therm.expan., gl.state
alpha liq. linear	2.24E-04	-----	1/K, linear coeff.liq.therm.expan.
alpha gl. linear	9.36E-05	-----	1/K, linear coeff., gl.therm.expan.
$\alpha_2$liq.-$\alpha_2$gl.	3.92E-04	5E-04	1/K

Poly(carbonate)2,2 propanebis(4-phenyl carbonate)

Mol.Wt.	254.26	254.26	g/mol.
Density Am.	1.197	-----	g/cm3.
Density Gl.	1.208	1.20	g/cm3.
Density Cr.	1.353	1.31	g/cm3.
VdeWaalsVol	136.21	136.2	cm3/mol
Eg	6.13E-02	6.1-7.4E-02	cm3/mol*K,for glass,therm expan
E1	1.36E-01	1.2-1.5E-01	cm3/mol*K,forliquid,therm expan
e1	5.36E-04	4.8-5.9E-04	cm3/g*K,for liq,sp.therm.expan
eg	2.41E-04	2.4-2.9E-04	cm3/g*k,for glass,sp.therm expan
alpha liq. cubic	6.41E-04	-----	1/K,cubic therm.expan., liq.state
alpha gl. cubic	2.91E-04	-----	1/K, cubic therm.expan., gl.state
alpha liq. linear	2.14E-04	-----	1/K, linear coeff.liq.therm.expan.
alpha gl. linear	9.71E-05	-----	1/K, linear coeff., gl.therm.expan.
$\alpha_2$liq.-$\alpha_2$gl.	3.5E-04	5E-04	1/K

(continued)

Table 6.1 (continued)

	CALCULATED MOLECULAR PROPERTIES	LITERATURE VALUES	

Poly(ethylene terephthalate)

Mol.Wt.	192.17	192.17	g/mol.
Density Am.	1.302	-----	g/cm3.
Density Gl.	1.342	1.33	g/cm3.
Density Cr.	1.471	1.48	g/cm3.
VdeWaalsVol	94.18	94.2	cm3/mol
Eg	4.24E-02	4.2-4.6E-02	cm3/mol*K,for glass,therm expan
E1	9.42E-02	1.15-1.42E-02	cm3/mol*K,forliquid,therm expan
e1	4.90E-04	6.0-7.4E-04	cm3/g*K,for liq,sp.therm expan
eg	2.21E-04	2.2-2.4E-04	cm3/g*k,for glass,sp.therm expan
alpha liq. cubic	6.38E-04	-----	1/K,cubic therm.expan., liq.state
alpha gl. cubic	2.96E-04	-----	1/K, cubic therm.expan., gl.state
alpha liq. linear	2.13E-04	-----	1/K, linear coeff.liq.therm.expan.
alpha gl. linear	9.87E-05	-----	1/K, linear coeff., gl.therm.expan.
$\alpha_2$liq.$-\alpha_2$gl.	3.42E-04	5E-04	1/K

Poly(butylene terephthalate)

Mol.Wt.	220.23	220.23	g/mol.
Density Am.	1.220	-----	g/cm3.
Density Gl.	1.259	-----	g/cm3.
Density Cr.	1.379	1.31	g/cm3.
VdeWaalsVol	114.64	-----	cm3/mol
Eg	5.16E-02	-----	cm3/mol*K,for glass,therm expan
E1	1.15E-01	-----	cm3/mol*K,forliquid,therm expan
e1	5.21E-04	-----	cm3/g*K,for liq,sp.therm expan
eg	2.34E-04	-----	cm3/g*k,for glass,sp.therm expan
alpha liq. cubic	6.35E-04	-----	1/K,cubic therm.expan., liq.state
alpha gl. cubic	2.95E-04	-----	1/K, cubic therm.expan., gl.state
alpha liq. linear	2.12E-04	-----	1/K, linear coeff.liq.therm.expan.
alpha gl. linear	9.83E-05	-----	1/K, linear coeff., gl.therm.expan.
$\alpha_2$liq.$-\alpha_2$gl.	3.4E-04	5E-04	1/K

Poly(3,3-bischloromethyl oxacyclobutane)(Penton)

Mol.Wt.	155.05	155.05	g/mol.
Density Am.	1.338	1.386	g/cm3.
Density Gl.	1.316	-----	g/cm3.
Density Cr.	1.512	1.47	g/cm3.
VdeWaalsVol	74.15	71.2	cm3/mol
Eg	3.34E-02	-----	cm3/mol*K,for glass,therm expan
E1	7.42E-02	-----	cm3/mol*K,forliquid,therm expan
e1	4.78E-04	-----	cm3/g*K,for liq,sp.therm expan
eg	2.15E-04	-----	cm3/g*k,for glass,sp.therm expan
alpha liq. cubic	6.40E-04	-----	1/K,cubic therm.expan., liq.state
alpha gl. cubic	2.83E-04	-----	1/K, cubic therm.expan., gl.state
alpha liq. linear	2.13E-04	-----	1/K, linear coeff.liq.therm.expan.
alpha gl. linear	9.44E-05	-----	1/K, linear coeff., gl.therm.expan.
$\alpha_2$liq.$-\alpha_2$gl.	3.57E-04	5E-04	1/K

(continued)

Table 6.1 (continued)

	CALCULATED MOLECULAR PROPERTIES	LITERATURE VALUES	

Poly(methylene oxide)

Mol.Wt.	30.03	30.03	g/mol.
Density Am.	1.204	-----	g/cm3.
Density Gl.	1.162	-----	g/cm3.
Density Cr.	1.360	1.54	g/cm3.
VdeWaalsVol	15.73	13.9	cm3/mol
Eg	7.08E-03	5.4E-03	cm3/mol*K,for glass,therm expan
E1	1.57E-02	-----	cm3/mol*K,forliquid,therm expan
e1	5.24E-04	-----	cm3/g*K,for liq,sp.therm.expan
eg	2.36E-04	1.8E-04	cm3/g*k,for glass,sp.therm expan
alpha liq. cubic	6.30E-04	-----	1/K,cubic therm.expan., liq.state
alpha gl. cubic	2.74E-04	-----	1/K, cubic therm.expan., gl.state
alpha liq. linear	2.10E-04	-----	1/K, linear coeff.liq.therm.expan.
alpha gl. linear	9.13E-05	-----	1/K, linear coeff., gl.therm.expan.
$\alpha_2$liq. -$\alpha_2$gl.	3.56E-04	5E-04	1/K

Poly(vinyl methyl ether)

Mol.Wt.	58.08	58.08	g/mol.
Density Am.	1.008	-----	g/cm3.
Density Gl.	0.981	-----	g/cm3.
Density Cr.	1.139	1.18	g/cm3.
VdeWaalsVol	36.18	35.7	cm3/mol
Eg	1.63E-02	-----	cm3/mol*K,for glass,therm expan
E1	3.62E-02	-----	cm3/mol*K,forliquid,therm expan
e1	6.23E-04	-----	cm3/g*K,for liq,sp.therm.expan
eg	2.80E-04	-----	cm3/g*k,for glass,sp.therm expan
alpha liq. cubic	6.28E-04	-----	1/K,cubic therm.expan., liq.state
alpha gl. cubic	2.75E-04	-----	1/K, cubic therm.expan., gl.state
alpha liq. linear	2.09E-04	-----	1/K, linear coeff.liq.therm.expan.
alpha gl. linear	9.17E-05	-----	1/K, linear coeff., gl.therm.expan.
$\alpha_2$liq. -$\alpha_2$gl.	3.53E-04	5E-04	1/K

Estimated Poly(2,6-dimethyl phenylene oxide)

Mol.Wt.	120.14	120.14	g/mol.
Density Am.	1.134	-----	g/cm3.
Density Gl.	1.097	1.07	g/cm3.
Density Cr.	1.282	1.31	g/cm3.
VdeWaalsVol	70.64	69.3	cm3/mol
Eg	3.18E-02	-----	cm3/mol*K,for glass,therm expan
E1	7.06E-02	-----	cm3/mol*K,forliquid,therm expan
e1	5.88E-04	-----	cm3/g*K,for liq,sp.therm.expan
eg	2.65E-04	-----	cm3/g*k,for glass,sp.therm expan
alpha liq. cubic	6.67E-04	-----	1/K,cubic therm.expan., liq.state
alpha gl. cubic	2.90E-04	-----	1/K, cubic therm.expan., gl.state
alpha liq. linear	2.22E-04	-----	1/K, linear coeff.liq.therm.expan.
alpha gl. linear	9.68E-05	-----	1/K, linear coeff., gl.therm.expan.
$\alpha_2$liq. -$\alpha_2$gl.	3.77E-04	5E-04	1/K

(continued)

233

Table 6.1 (continued)

	CALCULATED MOLECULAR PROPERTIES	LITERATURE VALUES	

Poly(hexamethylene adipamide)

Mol.Wt.	226.3	226.3	g/mol.
Density Am.	1.096	-----	g/cm3.
Density Gl.	1.086	1.07	g/cm3.
Density Cr.	1.238	1.24	g/cm3.
VdeWaalsVol	138.3	128.3	cm3/mol
Eg	6.22E-02	≈5.5E-02	cm3/mol*K,for glass,therm expan
E1	1.38E-01	1≈1.2E-01	cm3/mol*K,forliquid,therm expan
e1	6.11E-04	≈6E-04	cm3/g*K,for liq,sp.therm.expan
eg	2.75E-04	-----	cm3/g*k,for glass,sp.therm expan
alpha liq. cubic	6.70E-04	-----	1/K,cubic therm.expan., liq.state
alpha gl. cubic	2.99E-04	l-----	1/K, cubic therm.expan., gl.state
alpha liq. linear	2.23E-04	-----	1/K, linear coeff.liq.therm.expan.
alpha gl. linear	9.96E-05	-----	1/K, linear coeff., gl.therm.expan.
$\alpha_2$liq. -$\alpha_2$gl.	3.71E-04	5E-04	1/K

Poly(acrylamide)

Mol.Wt.	71.05	71.05	g/mol.
Density Am.	1.333	1.302	g/cm3.
Density Gl.	1.336	-----	g/cm3.
Density Cr.	1.506	-----	g/cm3.
VdeWaalsVol	33.45	----	cm3/mol
Eg	1.51E-02	----	cm3/mol*K,for glass,therm expan
E1	3.35E-02	----	cm3/mol*K,forliquid,therm expan
e1	4.71E-04	----	cm3/g*K,for liq,sp.therm.expan
eg	2.12E-04	----	cm3/g*k,for glass,sp.therm expan
alpha liq. cubic	6.28E-04	----	1/K,cubic therm.expan., liq.state
alpha gl. cubic	2.83E-04	----	1/K, cubic therm.expan., gl.state
alpha liq. linear	2.09E-04	----	1/K, linear coeff.liq.therm.expan.
alpha gl. linear	9.43E-05	----	1/K, linear coeff., gl.therm.expan.
$\alpha_2$liq. -$\alpha_2$gl.	3.45E-04	5E-04	1/K

Poly(acrylonitrile)

Mol.Wt.	53.03	53.03	g/mol.
Density Am.	1.125	1.184	g/cm3.
Density Gl.	1.182	-----	g/cm3.
Density Cr.	1.271	1.28	g/cm3.
VdeWaalsVol	31.73	30.7	cm3/mol
Eg	1.43E-02	-----	cm3/mol*K,for glass,therm expan
E1	3.17E-02	-----	cm3/mol*K,forliquid,therm expan
e1	5.98E-04	3.0E-04	cm3/g*K,for liq,sp.therm.expan
eg	2.69E-04	1.6E-04	cm3/g*k,for glass,sp.therm expan
alpha liq. cubic	6.73E-04	-----	1/K,cubic therm.expan., liq.state
alpha gl. cubic	3.18E-04	-----	1/K, cubic therm.expan., gl.state
alpha liq. linear	2.24E-04	-----	1/K, linear coeff.liq.therm.expan.
alpha gl. linear	1.06E-04	-----	1/K, linear coeff., gl.therm.expan.
$\alpha_2$liq. -$\alpha_2$gl.	3.55E-04	5E-04	1/K

(continued)

234

Table 6.1 (continued)

	CALCULATED MOLECULAR PROPERTIES	LITERATURE VALUES	
Poly(methyl methacrylate) (PLEXIGLAS)			
Mol.Wt.	100.1	100.1	g/mol.
Density Am.	1.153	-----	g/cm3.
Density Gl.	1.157	1.17	g/cm3.
Density Cr.	1.303	-----	g/cm3.
VdeWaalsVol	56.1	56.1	cm3/mol
Eg	2.52E-02	2.3E-02	cm3/mol*K,for glass,therm expan
E1	5.61E-02	5.2-5.5E-02	cm3/mol*K,forliquid,therm expan
e1	5.60E-04	5.2-5.5E-04	cm3/g*K,for liq,sp.therm.expan
eg	2.52E-04	2.3E-04	cm3/g*k,for glass,sp.therm expan
alpha liq. cubic	6.46E-04	-----	1/K,cubic therm.expan., liq.state
alpha gl. cubic	2.92E-04	-----	1/K, cubic therm.expan., gl.state
alpha liq. linear	2.15E-04	-----	1/K, linear coeff.liq.therm.expan.
alpha gl. linear	9.72E-05	-----	1/K, linear coeff., gl.therm.expan.
$\alpha_2$liq. -$\alpha_2$gl.	3.54E-04	5E-04	1/K
Poly(cyclohexyl methacrylate)			
Mol.Wt.	168.24	168.24	g/mol.
Density Am.	1.078	-----	g/cm3.
Density Gl.	1.112	1.10	g/cm3.
Density Cr.	1.218	-----	g/cm3.
VdeWaalsVol	100.36	99.2	cm3/mol
Eg	4.52E-02	4.54E-02	cm3/mol*K,for glass,therm expan
E1	1.00E-01	-----	cm3/mol*K,forliquid,therm expan
e1	5.97E-04	-----	cm3/g*K,for liq,sp.therm.expan
eg	2.68E-04	2.7E-04	cm3/g*k,for glass,sp.therm expan
alpha liq. cubic	6.43E-04	-----	1/K,cubic therm.expan., liq.state
alpha gl. cubic	2.98E-04	-----	1/K, cubic therm.expan., gl.state
alpha liq. linear	2.14E-04	-----	1/K, linear coeff.liq.therm.expan.
alpha gl. linear	9.95E-05	-----	1/K, linear coeff., gl.therm.expan.
$\alpha_2$liq. -$\alpha_2$gl.	3.45E-04	5E-04	1/K

6.13   ABBREVIATIONS AND NOTATION

a	van der Waal constant
a_l	linear coefficient of thermal expansion
b	van der Waal constant
e	elementary charge, (1.602×10^{-19} C)
e	specific thermal expansivity

E	electrical field
E_g	glassy molar thermal espansivity
E_l	liquid molar thermal expansivity
k	Boltzmann constant
l_i	bond distance
N_A	Avagadros' number
p	pressure
r	separation distance
R	separation distance
r_i	atom radius
V	potential energy
V	volume
$V(R)$	overlap repulsion potential
V_R	rubbery volume
V_W	van der Waal volume
V_g	glassy volume
Z	ionic valency
α	molecular polarizatility
α_2	cubical coefficient of thermal expansion
$\psi(r)$	Lennard-Jones interaction potential
ε	Lennard-Jones scaling factor
ε^*	Lennard-Jones scaling factor
ε_o	permittivity of a vacuum
$\hbar\nu$	ionization energy of the molecule
$\hbar$	Planck's constant
μ	dipole moment
ρ	density
ρ_c	density, crystalline state
ρ_a	density, amorphous state
σ	potential length constant

6.14 GENERAL REFERENCES

1. Van Krevelen, D.W., and Hoftyzer, P.J., *J. Appl. Polymer Sci.*, **13**, 871(1969).
2. Van Krevelen, D.W., *Properties of Polymers*, Elsevier, New York, N.Y., 1990.
3. Bueche, F., *Physical Properties of Polymers*, Interscience, New York, 1962.
4. Tobolsky, A.V., *Properties and Structure of Polymers*, J. Wiley, New York, 1960.
5. Kingery, W. D., Bowen, H.K., Uhlmann, D.R., *Introduction to Ceramics*, 2nd Ed., John Wiley & Sons, New York, N.Y., 1976.
6. Hayden, H. W., Moffatt, W. G., and Wulff, J., *The Structure and Properties of Materials*, Vol. III, Mechanical Behavior, John Wiley & Sons, Inc., New York, 1965.

CHAPTER 7

Calorimetric Properties

7.1   INTRODUCTION

Properties such as the thermal expansion coefficient, compressibility, internal energy, specific heat, and the enthalpy of fusion are equilibrium thermodynamic properties and are determined by calorimetric techniques (methods). With the exception of the enthalpy of fusion, these properties are considered to be insensitive to the presence of structural imperfections resulting from processing, heat history, or impurities which affect polymer morphology. The structural imperfections, impurities, and heat history can have a large influence on the enthalpy of fusion of polymers, however. Nevertheless, this property is included in this section since it falls into the calorimetric category.

Thermal expansion was discussed previously in Chapter 6, since the thermal expansion is tied directly to the volume of a material as a function of temperature. The properties of interest in this chapter are heat capacities and the heats of crystallization and fusion and the estimation of specific heats and enthalpies of fusion. Compressibility is discussed later in the chapter on mechanical properties.

We begin by examining a number of fundamental differences between materials that help us in understanding differences in the properties of materials even though many of these concepts can be applied only to simple materials. The basis of estimated polymer properties and comparison with literature values is also included. Specific heat differences between classes of materials are included in the materials properties workbooks contained in MATPROP©.

7.2   INTERNAL ENERGY
The general thermodynamic relationships have been presented previously in Chapter 2.

7.2.1   Crystalline Solids
For a crystalline solid the total lattice energy is given by:

$$\psi_T = \frac{N\left(\psi_{aT}\right)_{r=r_o}}{2(m-n)}\left[m\left(\frac{r_o}{r}\right)^n - n\left(\frac{r_o}{r}\right)^m\right] \qquad (7.1)$$

where:
$$r = \text{nearest neighbor distance}$$
$$r = \text{average separation distance}$$
$$m \text{ and } n = \text{exponents related to the attractive and repulsive force terms}$$
$$N = \text{number of identical atoms}$$
$$\psi_{aT} = \text{total interaction potential for a single particle in the lattice}$$
$$= -z\left(\frac{S_a a}{r^n} - \frac{S_b b}{r^m}\right)$$

and
$$z = \text{number of nearest neighbors}$$
$$S_a \text{ and } S_b = \text{characteristics of the lattice geometry and type of forces}$$

In an ideal crystal, the internal energy may be calculated by assuming that each atom vibrates about its lattice point with a vibration frequency that is determined by the magnitude of the potential energy. The simplest approximation to the real behavior of a material is to assume that a given particle vibrates in an average field that is created by all the other particles at rest in their equilibrium positions. For a particle displaced by an amount y, an average spherically symmetrical potential can be represented as:

$$\left\langle \psi_{aT}\left(y\right)\right\rangle - \psi_{aT}(0) = 2\pi m_r v^2 y^2 \qquad (7.2)$$

where:

m_r = reduced mass

v = vibration frequency

The vibration frequency for the oscillator is:

$$v = \left\{ \frac{\left(\psi_{aT}\right)_{r=r_o} mn}{12 m_r \pi^2 \left(m-n\right) r^2} \left[\left(n-1\right)\left(\frac{r_o}{r}\right)^n - \left(m-1\right)\left(\frac{r_o}{r}\right)^m\right]^{1/2} \right\} \qquad (7.3)$$

A crystal consisting of N particles has $3N-6 \approx 3N$ internal degrees of freedom. In a "monatomic" crystal, all of these are vibrational degrees.

7.2.2 Polymeric Materials

Estimation of thermodynamic properties of polymers from theory is limited due to the difficulty in describing the degrees of freedom of the molecule. Consider a polymer molecule containing n repeating units and z end groups. The end groups have three translational degrees of freedom. Due to the primary carbon-carbon valence bonds that have fixed lengths and angles along the chain, the second atom has two translational degrees of freedom and the third carbon is limited to one rotational degree of freedom. Isomeric conformations may limit positioning of subsequent chain atoms.

For a linear hydrocarbon molecule with n carbons and $2n + 2$ hydrogen atoms, there are $3n + 2$ degrees of freedom which are called external degrees of freedom. The external degrees of freedom are dependent upon the nature of the intermolecular forces and are dependent on the temperature and volume of the material. The details regarding forces have been discussed previously in Chapter 6. For the internal degrees of freedom the one degree of rotational freedom associated with each subsequent carbon atom is actually a fraction (f) of this due to steric hindrance which gives the total degrees of freedom to be equal to $3n + 2 - 3(f)$. The internal degrees depend only upon the nature of the primary bonding.

Since the internal motions depend on the primary bonding and the external motions depend on Van der Waals forces, the characteristic frequencies are of different orders of magnitude. This situation is complicated further by the existence of possible different morphologies based upon heat treatment and the temperature at which the glass transition occurs.

7.3   PARTITION FUNCTIONS

A useful quantity for calculating thermodynamic properties is the partition function Z.

$$Z = \sum_i s_i \exp\left(-\frac{E_i}{kT}\right) \tag{7.4}$$

where:

E_i = total energy of a system of N interacting particles

s_i = number of states

The total internal partition function z is:

$$Z = \left(2 \sinh \frac{\hbar \nu_E}{2kT}\right)^{-3N} \tag{7.5}$$

and the Helmholtz free energy is:

$$A = -kT \ln z \tag{7.6}$$

$$= 3/2 N\hbar\nu_E + 3RT \ln\left[1 - \exp\left(-\frac{\hbar\nu_E}{kT}\right)\right] \tag{7.7}$$

and the internal energy is:

$$U = kT^2 \left(\frac{\partial \ln z}{\partial T}\right)_V \tag{7.8}$$

$$= 3/2 N\hbar\nu_E + \frac{3N\hbar\nu_E}{\left[\exp\left(\hbar\nu_E / kT\right)\right] - 1} \tag{7.9}$$

The first term represents the ground-state vibrational energy (i.e., the energy at absolute zero), while the second term represents the vibrational energy relative to the ground state.

At high temperatures:

$$A = -3RT \ln \frac{kT}{\hbar \nu_E} \tag{7.10}$$

and

$$U = 3RT \tag{7.11}$$

At low temperatures:

$$U = 3N\hbar\nu_E \left[\frac{1}{2} + \exp\left(-\frac{\hbar\nu_E}{kT} \right) \right] \tag{7.12}$$

Equations 7.4 to 7.12 use the Einstein approximation that all $3N$ vibrational degrees of freedom have the same frequency. This is an approximation because each particle does not vibrate independently with all others fixed but is coupled to its neighbors and must move with them. When all the ν_i are identical, Equation 7.5 applies.

If data on the intermolecular-force parameters are available, the thermodynamic properties of the material may be estimated. This occurs generally only for simple materials. When data are not available, these equations can be used to curve fit experimental data and the resulting equations used to predict properties in other ranges of the variables.

7.4   COMPARISON OF PARTITION FUNCTIONS

With the previous section in mind, it is instructive to compare partition functions for well defined crystalline lattices and for polymers. For a crystalline solid we have:

$$Z = \prod_{i=1}^{3N-6} z_i \tag{7.13}$$

$$= \prod_{i=1}^{3N} \left[\exp\left(-\frac{\hbar\nu_i}{2kT}\right) \right] \left[1 - \exp\left(-\left(\frac{\hbar\nu_i}{kT}\right)\right) \right]^{-1} \qquad (7.14)$$

This complex relationship, as above, reduces to Equation 7.5 when all the frequencies are identical. For a polymer, we have:

$$Z = Z_{int}(T) Z_{ext}(T,V) \qquad (7.15)$$

Properties such as internal energy and specific heat can be predicted only from a detailed knowledge of distribution frequencies of both internal and external degrees of freedom. These relationships show how difficult it is to make general statements about the behavior of polymers and that each material must be considered separately and in context with its previous thermal history.

7.5 HEAT CAPACITY

The specific heat of a material is the heat which must be added to raise a unit weight of the substance one degree of temperature at either constant pressure or constant volume. In the case of polymers, the molar heat capacity is the specific heat multiplied by the molar mass of a structural unit of the polymer. The heat added causes a change in the internal energy (U) and in the enthalpy *(H)* or heat content of the material.

We have seen previously (Chapter 2) that the change in internal energy, as a function of temperature is the specific heat capacity of the material and at constant volume is given by:

$$\left(\frac{\partial E}{\partial T}\right)_V = T\frac{\partial S}{\partial T} = \frac{\partial Q}{\partial T} = c_V \left(J / g \cdot K\right) \qquad (7.16)$$

and at constant pressure:

$$\left(\frac{\partial H}{\partial T}\right)_P = T\frac{\partial S}{\partial T} = \frac{\partial Q}{\partial T} = c_p \left(J / g \cdot K\right) \qquad (7.17)$$

The molar heat capacities are simply the specific heat capacities multiplied by the molecular weights:

$$C_V = Mc_V \, (\text{J} / \text{mol.·K}) \qquad (7.18)$$

and

$$C_p = Mc_p \, (\text{J} / \text{mol.·K}) \qquad (7.19)$$

The difference between c_p and c_V is:

$$c_p - c_V = TV \frac{\alpha_V^2}{\kappa_T} \qquad (7.20)$$

where:

α_V = volume coefficient of thermal expansion

κ_T = isothermal compressibility

For most solids, c_V is of the order of 1.4 cal/mole $\times K$, while $c_p - c_V$ is approximately 2.4×10^{-2} cal / mole $\times$ K.

If the vibration frequency of the oscillators are of a single frequency, the specific heat at constant volume is obtained by differentiating Equation 7.9, and the Einstein approximation for c_V becomes:

$$c_V = \frac{3R(\hbar\nu_E / kT)^2 \exp(\hbar\nu_E / kT)}{\left\{\left[\exp(\hbar\nu_E / kT)\right] - 1\right\}^2} \qquad (7.21)$$

We can thus see the fundamental relationship between the vibrational frequencies (energies) of the atoms/molecules of the material constituting the molecular solid and the resulting heat capacity.

At high temperatures, $\hbar\nu_E / kT \ll 1$, the specific heat approaches $3R$ cal/g mole·K, and at low temperatures, $\hbar\nu_E / kT \gg 1$, the specific heat approaches:

$$c_V \cong 3R \left(\frac{\hbar\nu_E}{kT} \right)^2 \exp\left(-\frac{\hbar\nu_E}{kT} \right) \qquad (7.22)$$

The upper limiting value of $3R$ cal./g mole K is in good agreement with the observed values for many crystalline solids at room temperature and above. This value can be compared with specific heat values for many crystalline metallic and non-metallic solids in the materials properties workbooks contained in MATPROP©.

For a more exact expression for the vibration frequencies, Debye [S1] assumes that the sum of the frequencies can be replaced by an integral over a distribution of frequencies:

$$\sum_{i=1}^{3N} \nu_i = \int_0^{\nu_m} \nu df \qquad (7.23)$$

The upper limit ν_m is a maximum cut off frequency which is characteristic of a specific material, assuming a parabolic function of frequencies, since the total number of vibrational degrees is $3N$. For frequencies (df) in the range between ν and $\nu + d\nu$:

$$df = \frac{9N\nu^2 d\nu}{\nu_m^3} \qquad (7.24)$$

and Equation 7.23 becomes

$$\sum_{i=1}^{3N} \nu_i \cong \int_0^{\nu_m} \frac{9N}{\nu_m^3} \nu^3 d\nu = \frac{9}{4} N\nu_m \qquad (7.25)$$

The second term in Equation 7.14 becomes:

$$\sum_{i=1}^{3N} \ln\left[1 - \exp\left(-\frac{\hbar\nu_i}{kT} \right) \right] \qquad (7.26)$$

$$= \frac{9N}{v_m^3} \int_0^{v_m} \ln\left[1 - \exp\left(-\frac{\hbar v_i}{kT}\right)\right] v^2 dv \tag{7.27}$$

and the partition function for a crystalline solid becomes:

$$\ln Z = -\frac{9}{8}N\frac{\theta_D}{T} - 9N\left(\frac{T}{\theta_D}\right)^3 \int_0^{\theta_D/T} x^2 \ln[1 - \exp(-x)] dx \tag{7.28}$$

where

$$x = \frac{\hbar v}{kT}$$

θ_D = Debye characteristic temperature for the material
$\quad = \hbar v_m / k$

At high temperatures, $x \to 0$, Equation 7.28 reduces to:

$$\ln Z = 3N \ln\left(\frac{T}{\theta_D}\right) + N \tag{7.29}$$

which results in:

$$U = 3RT \tag{7.30}$$

and

$$c_V = 3R \tag{7.31}$$

in agreement with the Einstein approximation.

Polymorphic transitions will result in a sudden change in the specific heat because either the latent heat changes (first order transitions) or the more gradual changes in the first derivative properties (second-order transitions). These concepts have been illustrated previously in Chapter 2 (See figure 2.20).

7.6 EQUATIONS OF STATE

7.6.1 Crystalline Materials

Equations of state have been previously presented but we can state formally that an equation of state defines a relationship between pressure, volume, and temperature. It is directly related to the Helmholtz free energy and the partition function by:

$$p = -\left(\frac{\partial A}{\partial V}\right)_T = -kT\left(\frac{\partial \ln Z}{\partial V}\right)_T \qquad (7.32)$$

The usual representation is to express the product, pV/RT, as a function of either pressure, volume, or temperature. Experimental data are often reported in terms of the volumetric thermal expansion coefficient, α, which as we have seen is:

$$\alpha = \left(\frac{1}{V}\frac{\partial V}{\partial T}\right)_p \qquad (7.33)$$

The isothermal compressibility is simply the negative of expansion but at constant temperature.

$$\kappa_T = -\left(\frac{1}{V}\frac{\partial V}{\partial P}\right)_T \qquad (7.34)$$

The equation of state for a crystal can be expressed in terms of the intermolecular forces in the crystal as:

$$\frac{pV}{RT} = -\frac{V}{RT}\left(\frac{\partial \psi_T}{\partial V}\right)_T - \frac{U(v)}{RT}\frac{d\ln v}{d\ln V} \qquad (7.35)$$

where $U_{(v)}$ is the total internal (vibrational) energy of the solid and $\left(\dfrac{d\ln v}{d\ln V}\right)_T$ is the same for all vibration frequencies. The potential energy of interaction, Equation 7.1 and a characteristic vibration frequency, Equation 7.3 are functions of the crystalline structure. For a body centered cubic crystal,

$$V = \frac{Nr^3}{2^{1/2}} \qquad (7.36)$$

7.6.2 Polymeric Materials

The previous relationships for crystals can be used to curve fit experimental data or obtain thermodynamic properties by numerical evaluation of these relationships. This is not the case for polymers as the establishment of an equation of state depends on being able to evaluate the configurational partition function of the molecule (Equation 7.15), the difficulty of which for polymers was pointed out previously.

In regions where there are no phase transitions, the pVT behavior of a polymer can often be represented empirically by a modified form of the Van der Waals equation of state given previously.

$$(V - b)(p + \pi) = \frac{RT}{M} \tag{7.37}$$

where b, π, and M are experimentally determined constants. The net volume, $V - b$, may be interpreted as the free volume of a polymer unit in the solid, and the quantity π is equivalent to the internal pressure of the solid. Thermodynamically, the internal pressure p_i, is:

$$p_i = T\left(\frac{\partial p}{\partial T}\right)_V = \frac{\alpha_V T}{\kappa_T} = \pi \tag{7.38}$$

where α_V is as before the volume coefficient of thermal expansion and κ_T is the thermal compressibility. The internal pressure π is always much larger than a few atmospheres so that the external pressure p can generally be neglected. Differentiating Equation 7.37 with respect to temperature gives the expansion coefficient as:

$$\frac{1}{V_o}\left(\frac{\partial V}{\partial T}\right)_p = \alpha_v = \frac{R}{M\pi V} \tag{7.39}$$

For most polymers, the internal pressure, π, is directly proportional to the cohesive energy density, E_{CD}

$$\pi = 1.1 - 1.6 \times E_{CD} \qquad (7.40)$$

7.7 LATENT HEAT OF CRYSTALLIZATION AND FUSION (MELTING)

The latent heat of fusion or crystallization in terms of the enthalpy difference is given by:

$$H_l\left(T_m\right) - H_c\left(T_m\right) = \Delta H_m\left(T_m\right) \qquad (7.41)$$

The availability of reliable values for ΔH_m is a similar situation to heat capacities availability in that data is available for only a limited number of polymers. For valid experimental values, the degree of crystallinity of the sample has to be taken into account. Van Krevelen [S2] has pointed out that as a general rule, the highest value of ΔH_m for a given polymer is the most probable one.

The ΔH_m estimated values are therefore subject to greater errors from group contribution methods because of the large number of experimental values for ΔH_m which makes it very difficult to derive accurate values for group contributions.

At the melting point, T_m, the entropy of fusion may be calculated as:

$$\Delta S = \frac{\Delta H_m}{T_m} \qquad (7.42)$$

If accurate specific heat measurements can be made, the total enthalpy and entropy of a material are given by:

$$H(ttl) = H(0) + \int_0^T C_p dT + \sum \Delta H_i \qquad (7.43)$$

$$S(ttl) = S(0) + \int_0^T \frac{C_p}{T} dT + \sum \Delta S_i \qquad (7.44)$$

where:

$H(0)$ = enthalpy at $0°\,K$

$S(0)$ = entropy at $0°\,K$

ΔH_i = enthalpy changes at first - order transitions

ΔS_i = entropy changes at first - order transitions

For most polymers the internal pressure is directly proportional to the cohesive energy density,

$$\pi \cong 1.1 - 1.6 \times E_{CD} \tag{7.45}$$

7.8 ESTIMATION OF CALORIMETRIC PROPERTIES

7.8.1 Heat Capacities

Reliable values for molar heat capacities in the liquid and solid state are available for only a limited number of polymers. As before, specific heat values are estimated by Molcalc using the group contribution method.

Two values for the specific heat are calculated both at $298°\,K$. The specific heat of the solid, $c_p(s)$ and the specific heat of the liquid, $c_p(l)$ are estimated to allow heat capacities to be estimated at temperatures above the T_g of the polymer or while the polymer is in the liquid or rubbery state. Expected changes in heat capacities above and below T_g and T_m have been covered previously in Chapter 2 as well as the thermodynamic basis for these heat capacity changes.

Also from Chapter 2, it will be appreciated that because of the possible varying polymer morphologies, a polymer is generally not 100% crystalline or 100% amorphous. Therefore, estimation of the specific heat of a polymer if the % crystallinity is known or can be estimated allows one to better estimate the specific heat capacity of a given polymer.

For example, if a polymer known to be x% crystalline and y% $[(1-x)\%]$ amorphous, the estimated specific heat capacity can be easily estimated by:

$$c_p^{ttl} = x\% \cdot c_p(s) + y\% \cdot c_p(l) \qquad (7.46)$$

The group contribution values used were assembled from Van Krevelen [S2] and were taken from Shaw [S3] and Satoh [S4].

    If experimental data are unavailable, and data are needed to calculate heating requirements for extrusion or other polymer processing, the temperature function of the heat capacity may be approximated by converting the specific heat capacities to molar heat capacities by multiplying the calculated molecular weight of the structural unit and approximating the mean temperature function using the following relationships [S2].

$$c_p^s = c_p^s(298)\left[0.106 + 3 \times 10^{-3} T\right] \qquad (7.47)$$

$$c_p^l = c_p^l(298)\left[0.64 + 1.2 \times 10^{-3} T\right] \qquad (7.48)$$

    These relationships show that, in general, the slopes of the heat capacity lines for solid polymers are approximately constant and the heat capacity slope for polymers in the liquid state $c_p(l)$ is approximately two and one half (2 ½) times greater.

7.8.2   Heat of Fusion

    Molcalc estimates ΔH_m based on additive group contributions, as before. These values are calculated without regard as to whether or not it is possible for the polymer to exhibit a crystalline morphology. Therefore, the user is always urged to exercise proper judgment in interpretation of results since, in reality, all polymers do not crystallize and thus ΔH_m may have no meaning.

    It has been pointed out [S2] that ΔS_m shows a more regular relation with structure than ΔH_m, but ΔS_m values are not used by Molcalc in order to estimate ΔH_m because this would necessitate using the relationship:

$$\Delta S_m = \frac{\Delta H_m}{T_m} \qquad (7.49)$$

The use of this approach would require accurate estimation of both ΔS_m and T_m, and would therefore introduce a second variable whose error would not necessarily be compensating.

7.9   MOLCALC ESTIMATIONS OF CALORIMETRIC PROPERTIES AND COMPARISON TO LITERATURE VALUES FOR A SERIES OF REPRESENTATIVE POLYMERS

With the above concepts in mind, Table 7.1 lists a comparison of estimated calorimetric properties and comparisons to literature values for the set of example polymers.

Table 7.1

Comparison of Estimated Polymer Calorimetric Properties with Literature Values

Polymer	$c_p(s)$; J/g·K		$c_p(l)$; J/g·K		H_m; J/mol	
	Calc.	Lit.	Calc.	Lit.	Calc.	Lit.
Poly(ethylene)	1.81	1.55-1.76	2.17	2.26	8000	8220
Poly(propylene)	1.71	1.63-1.76	2.10	2.14	8500	8700
Poly(isobutylene)	1.66	1.68	1.99	1.97	11350	12000
Poly(vinyl chloride)	1.09	0.96-1.09	1.46	1.22	11000	11000
Poly(vinyl fluoride)	1.35	----	1.57	----	7500	7500
Poly(chloro trifluoro ethylene)	0.90	0.92	1.01	----	7900	5020
Poly(vinyl acetate)	1.37	~1.47	1.78	~1.93	6000	----
Poly(butadiene)	1.63	1.63	1.92	1.89	9000	9000
Poly(isoprene)	1.63	1.59	1.98	1.93	9100	8700
Poly(styrene)	1.22	1.22	1.68	1.72	10400	10000
Poly(αmethyl styrene)	1.25	----	1.68	----	13250	----
Polycarbonate	1.13	1.17	1.61	1.59	31350	33500
Poly(ethylene terephthalate)	1.15	1.13	1.58	~1.55	23000	26900
Poly(butylene terephthalate)	1.24	----	1.66	----	31000	32000
Poly(3,3-bis chloromethyl oxacyclobutane)	----	----	----	----	24650	32000

(continued)

Table 7.1 continued)

Polymer	$c_p(s)$; J/g·K		$c_p(l)$; J/g·K		H_m; J/mol	
	Calc	Lit.	Calc.	Lit.	Calc.	Lit.
Poly(methylene oxide)	1.40	1.42	2.20	~2.10	7500	9790
Poly(vinyl methyl ether)	1.53	----	2.13	----	12000	----
Poly(phenylene oxide)	1.20	1.26	1.69	~1.76	9200	~6000
Poly(hexamethylene adipamide)	1.53	1.47	2.14	----	44000	43000
Poly(acrylamide)	1.22	----	1.99	----	8300	----
Poly(acrylonitrile)	1.24	1.26	----	----	6000	5200
Poly(methyl methacrylate)	1.39	1.45	1.76	----	9050	9600
Poly(cyclohexyl methacrylate)	1.49	----	1.86	----	27350	----

7.10   ABBREVIATIONS AND NOTATION

A	Helmholtz free energy
a	Van der Waals constant
b	Van der Waals constant
C_p	molar heat capacity at constant pressure
$c_p(l)$	specific heat of the liquid
$c_p(s)$	specific heat of the solid
C_V	molar heat capacity at constant volume
E_{CD}	cohesive energy density
E	internal energy
E_i	total energy of N interacting particles
H	enthalpy
H_c	molar enthalpy of the crystalline solid
H_l	molar enthalpy of the liquid
ΔH_m	enthalpy difference or latent heat of fusion
k	Boltzmann constant, 1.38×10^{-23} J·K^{-1}
m	constant/exponent related to attraction/repulsion

M	molar mass
M	experimentally determined constant related to the molar mass
n	exponent
N	number of identical atoms
N	number of particles or atoms
p_i	internal pressure
Q	heat absorbed by the system
r	nearest neighbor distance
r	average separation distance
R	gas constant
S	entropy
S_a and S_b	lattice geometry characteristics
s_i	number of states
T	temperature
U	internal energy
y	displacement
z	number of nearest neighbors
Z	partition function
α_V	volume coefficient of thermal expansion
$\hbar$	Plancks' constant, $1.06 \times 10^{-34}\,\text{J}\cdot\text{s}$
κ_T	isothermal compressibility
m_r	reduced mass
ψ_{ar}	total interaction potential
θ_D	Debye characteristic temperature
π	internal pressure
ν	vibrational frequency

7.11   SPECIFIC REFERENCES

1. Debye, P., *Ann. Physik*, **39**, (1912).
2. Van Krevelen, D.W., *Properties of Polymers,* Elsevier, New York, 1990.
3. Shaw, R., J. *Chem. Eng. Data*, **14**, 461(1969).

4. Satoh, S. *J. Sci. Research Inst.* (Tokyo). **43**, 79(1948).

7.12   GENERAL REFERENCES

1. DiBenedetto, A.T., *The Structure and Properties of Materials*, McGraw-Hill, Inc., New York, N.Y., 1967.
2. Kingery, W.D., Bowen, H.K., Uhlmann, D.R., *Introduction to Ceramics*, 2nd Ed., John Wiley & Sons, New York, N.Y., 1976.
3. Moffatt, W.G., Pearsall, G.W., and Wulff, J., *Structure and Properties of Materials, Vol. 1, Structure*, John Wiley & Sons, New York, 1966.
4. Pauling, L., *Nature of the Chemical Bond*, 3d ed., Cornell University Press, Ithaca, N.Y., 1960.
5. Reed-Hill, R.E., *Physical Metallurgy Principles*, 2nd Ed., D. Van Nostrand Company, New York, N.Y., 1973.
6. Seymour, R.B., Carraher, C.E. Jr., *Polymer Chemistry*, Marcel Dekker, Inc., New York, N.Y., 1988.
7. Van Krevelen, D.W., *Properties of Polymers*, Elsevier, New York, N.Y., 1990.
8. Van Vlack, L.H., *Physical Ceramics for Engineers*, Addison-Wesley Publishing Co., Reading, Massachusetts, 1964.
9. Zachariasen, W.H., "The Atomic Arrangement in Glass", *J. Am. Chem. Soc.*, **54**, 3941(1932).

CHAPTER 8

Structure/Transition Temperature Relationships

8.1 INTRODUCTION

The subject of phase transitions in polymers was briefly introduced in Chapter 2 along with more detailed relationships of thermodynamic quantities. In this chapter the various types of transitions which polymers are capable of undergoing will be discussed along with structural differences responsible for these various transitions. The basis for calculation of transition temperatures and the structural differences that lead to errors in the calculation of the melting (T_m) and glass transition (T_g) temperatures will also be discussed. Finally, comparisons between estimated T_m and T_g transitions and literature values will be compared for the set of example polymer property estimations.

8.2 CLASSIFICATION AND ORIGIN OF MULTIPLE TRANSITIONS IN POLYMERS

Transitions can be classified broadly into the following types [S1]:

1. $T < T_g$; Glassy Phase 1 → Glassy Phase 2

2. $T = T_g$; Glassy Phase → Rubbery Phase

3. $T > T_g$; Amorphous Phase 1 → Amorphous Phase 2

4. $T_g < T < T_m$; Crystalline Phase 1 → Crystalline Phase 2
   Crystalline-Amorphous Interactions;

Amorphous-Amorphous Transitions (Transitions involving motion of the entire polymer chain)

5. $T = T_m$; Crystalline Phase → Amorphous Phase

(Order → Disorder Transitions)

6. $T \rangle T_m$; Order → Disorder Transitions

Isotropic → Chemical/Decomposition Reactions

Designations of the most important transitions which polymers display are illustrated in Table 8.1 as presented by Van Krevelen [S2].

8.3   DETECTION

There are a number of techniques that are used to detect transitions in polymers. The principal techniques of interest are: thermo-mechanical analysis (TMA), dynamic mechanical analysis (DMA), dielectric energy absorption, nuclear magnetic resonance (NMR), and differential scanning calorimetry (DSC). TMA detects discontinuities in the coefficient of thermal expansion while DMA measures the viscoelastic response of a material to periodic forces allowing changes in the modulus and tan δ to be followed and

Table 8.1
Designations of the Most Important Transition Temperatures [S2]

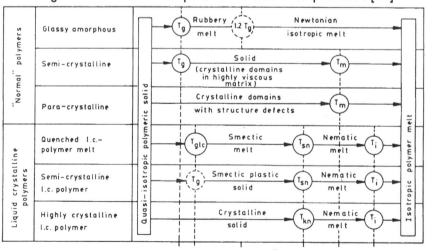

DSC measures changes in heat capacity or enthalpy of fusion, as discussed in Chapter 7. The simplest and therefore the most prevalent techniques used are that of DSC and DMA. These two techniques, or a variation of DMA called DMTA, which is simply dynamic mechanical thermal analysis, allow the viscoelastic behavior of a polymer to be followed as a function of temperature. These techniques often provide definitive results and permit informed interpretation of structure-property relationships of the polymer, based on the transitions occurring.

Idealized examples of the types of results that polymeric

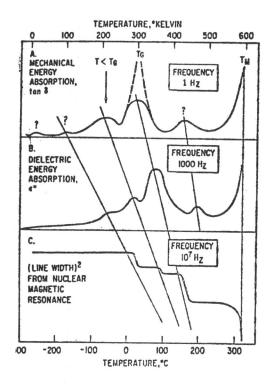

Figure 8.1 [S15]
Schematic Representation of the Absorption Spectra for a Hypothetical, Idealized Partially Crystalline Polymer by Three Common Measuring Techniques
A. Dynamic Mechanical Analysis
B. Dielectric Energy Absorption
C. Nuclear Magnetic Resonance

transitions exhibit along with the frequencies at which the transitions are generally detected using the various techniques are shown in Figure 8.1 [S2]. The DSC scans may also exhibit recrystallization phenomena as illustrated in Figure 8.2.

Recrystallization is shown as an exotherm immediately following the T_g for a rescan of a PET sample after quenching at the conclusion of the initial run. The change in the magnitude of the T_g in the second scan is due to the greater amorphous content of the polymer. It should be noted in both the DMA and DSC scans that transitions where $T \langle T_g$ or at T_g, second order transitions are exhibited, since we have seen in Chapter 2 that the second derivative of G with respect to temperature is:

$$\left(\frac{\partial^2 G}{\partial T^2} \right)_p = -\frac{C_p}{T} \tag{8.1}$$

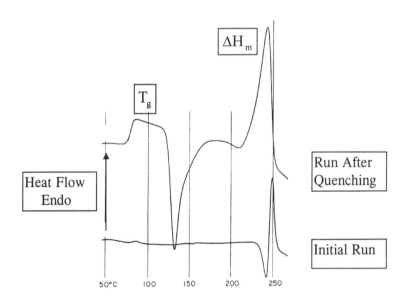

Figure 8.2
Schematic Representation of a DSC Thermogram of Poly(ethylene terephthalate) Illustrating the Changes in Heat Capacity and Enthalpy of Melting

Therefore, these transitions are indicated by changes in heat capacity or baseline shifts. First order transitions are indicated by a step change in volume at the melting point as indicated by the first derivative of free energy with respect to pressure.

8.4 ORIGINS

8.4.1 $T < T_g$

The strong $T < T_g$ relaxation is frequently referred to as the β or γ relaxation and usually lies at a temperature about ¾ of T_g. This relaxation process appears to be related to the toughness of polymers at temperatures below T_g. This relaxation involves motion of a small number (usually 3 to 8) of successive carbon atoms about the chain axis or of 3 or more carbon and/or oxygen groups on the side of the chain. Examples of these in a number of polymers are as follows:

1. Side-chain motion
   a) Ester groups in poly(alkyl acrylates) and methacrylates [S3,S4]
   b) Ether groups in poly(vinyl ethers) [S5]
   c) Methyl group rotation in polypropylene and R group motion in poly alpha-olefins [S6-S10].

2. In-chain motion of sub groups which are smaller in size than the main in-chain groups
   a) Poly(ethylene) terephthalates [S11]
   b) Polycarbonates [S12]
   Polycarbonate is a prime example since this polymer exhibits a T_g about 150°C and would be expected to be brittle at room temperature since this is far below T_g by ≈ 125°C. Yet this amorphous-glassy polymer is known as an "engineering" thermoplastic and is used in applications where toughness is demanded. The explanation for this apparent anomaly is that there is in-

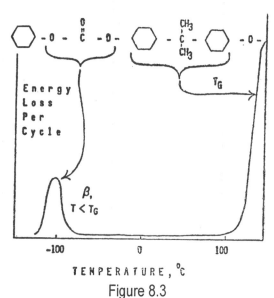

Figure 8.3

Dynamic Mechanical Loss at a Frequency of 1 Hz

chain motion of the carbonate group below T_g [S13, S14]. This is illustrated in the dynamic mechanical loss spectrum shown in Figure 8.3.

3. In-chain motion of the group, $-(CH_2 - CHR)_n -$ about the chain axis. This is referred to as crankshaft motion when n is on the order of 1 to 3. The entire molecule is involved as $n \rightarrow \infty$. The polymer poly(vinyl chloride) is a prime example, since it exhibits a transition at $T < T_g$. However, the β transition is ≈ -30 to $50°C$ which moves to higher temperatures rapidly with increasing frequency of the DMA and which disappears on the addition of plasticizer. The accepted, most probable, explanation is that this transition is related to a crankshaft mechanism, illustrated in Figure 8.4 [S15]. NMR has also been used to study PVC [S16,S17].

4. Interaction of water with polar groups along the polymer chain as in polyamides such as nylon [S18].

A. CRANKSHAFT IN PVC

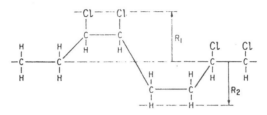

B. CRANKSHAFT IN HEAD TO HEAD PVC

Note: $R = R_1 > R_2$

Figure 8.4 [S15]
Schematic Two-Dimensional Representation of the Schatzke
Crankshaft Motion
A. Normal (head to tail) PVC
B. Head to head PVC

8.4.2 $T = T_g$

The glassy state is not confined to high polymers but is general to substances which can be cooled below their melting points without the occurrence of crystallization. Polymers are especially prone to glass formation because their high molecular weight nature facilitates super cooling and prevents or greatly inhibits complete crystallization.

In order to better understand the origin of the glass transition and its nature, we list the following characteristics of this transition.

1. The glass transition is confined to the amorphous areas and hence does not involve a phase change. All physical properties of a material change at T_g, the magnitude of which are a function of the amorphous content.

2. The change from a glassy state to a leathery and then rubbery state is characterized by an abrupt change in the thermodynamic state (cf. Chapter 2) occurring in a narrow temperature range and is accompanied by characteristic features of relaxation involving net cooperative molecular movement. This is generally thought to be coordinated crankshaft rotation involving 30-40 carbon atoms alone the chain.

3. T_g is affected by crystallinity in a manner that cannot be predicted "a priori" as Figure 8.5 illustrates [S1]. In general, crystalline moieties prevent the free motion of the amorphous portions of the polymer which results in an apparently higher T_g value for the polymer.

4. T_g increases with number-average molecular weight $\overline{M}_N$, and finally reaches an asymptotic value, as given by:

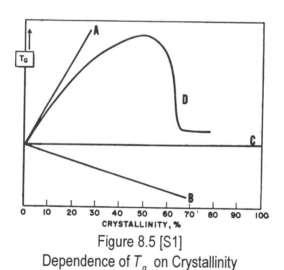

Figure 8.5 [S1]
Dependence of T_g on Crystallinity

A. Represents expected restraining influence of crystallites
B. Represents concentration of impurities in the amorphous phase
C. C indicates cancellation of effect A and B
D. D is actual experimental data

$$T_g = T_g(\infty) - Q / \overline{M}_N \qquad (8.2)$$

where:

$T_g(\infty)$ = limiting value of T_g

Q = constant depending upon the particular polymer or series

T_g generally reaches $T_g(\infty) \approx 10000 - 30000 \ \overline{M}_w$

5. T_g determined by DMA increases with frequency according to $\ln f \propto -\dfrac{H_a}{RT}$ where H_a is the activation energy.

6. T_g is an iso-free volume state. This is indicated by:

$$\Delta \alpha T_g = K_1 \qquad (8.3)$$

or $$\alpha_l T_g = K_2 \qquad (8.4)$$

where K_1 has a value of 0.113 and K_2 has a value of 0.164 as measures of the free volume at T_g. The free volume is often referred to as "empty lattice sites".

The implication of T_g as an iso-free volume state (Equations 8.3 and 8.4) is supported by the data in Figure 8.6 [S1]. Figure 8.7 [S1] allows us to better understand the effects of Equations 8.3 and 8.4.

In Figure 8.7 polymers closest to the ordinate have flexible chains, small substituents, and low molecular force while those farthest away from the ordinate have stiff chains and/or bulky substituents and/or high intermolecular forces. The effect on specific volume is immediately apparent.

As can be seen, there are various ways of defining T_g as an iso-free volume state. These definitions are represented schematically in Figure 8.8 [S1].

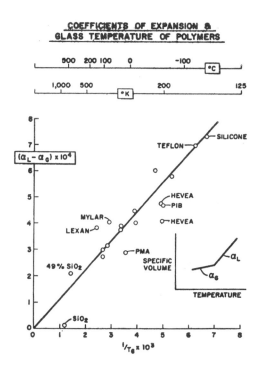

Figure 8.6 [S1]
Difference in Coefficients of Cubical Expansion Above and
Below T_g vs. $1/T_g$ in K

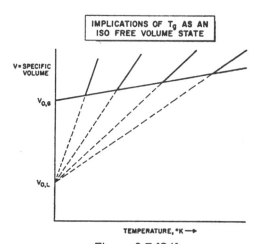

Figure 8.7 [S1]
Specific Volume vs. Temperature

266

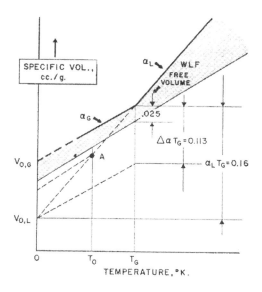

Figure 8.8 [S1]
Specific Volume vs. Temperature Illustrating Definitions
of the Free Volume at T_g

0.025=WLF
0.113=Equation 8.3
0.16 =Equation 8.4

Four structural factors affect the value of T_g, viz., cohesive energy density, chain stiffness, geometrical factors, and chain symmetry. Their influence is described as follows:

1. Cohesive Energy Density (CED)

The T_g is related to the CED as follows:

$$T_g = a\left[\frac{k_1 + k_2(\text{CED})}{\Delta S_m}\right] + b \qquad (8.3)$$

where a, b, k_1, and k_2 are constants and ΔS_m = entropy change which is primarily affected by chain stiffness

2. Backbone Chain Stiffness

A good example of the effect of chain stiffness can be seen

by comparing PMMA $\left(T_g \approx 105°C\right)$ and PMA $\left(T_g \approx 7\right)$. The difference in the T_g's between poly(methyl methacrylate) (PMMA) and poly(methyl acrylate) (PMA) is attributed primarily to the difference in chain stiffness of the two polymers.

3. Geometrical Factors Including Bulkiness of Substituent Groups

There are two main types which influence T_g. The first effect emanates from structural differences where the chemical compositions are identical as in *cis* and *trans* polymers. An example of this occurs in the *cis* and *trans* polyisoprenes where rotation about the single bonds adjacent to the double bonds is equally easy in the *cis* form but hindered in the *trans* form because of geometrical interference between methyl groups. The second effect comes from the size, shape, and stiffness of the side groups on the polymer chain.

4. Chain Symmetry

Chain symmetry also plays a role in influencing the T_g as can be seen by the effects of symmetry on the T_g of poly(propylene) and poly(isobutylene).

8.4.3 $T > T_g$

This type of transition involves coordinated rotation involving the entire polymer chain.

8.4.4 $T_g < T < T_m$

This type of transition usually involves a transition from one type of crystalline phase to a different type.

8.4.5 $T = T_m$ (Melting Point Temperature)

For polymers, the dominant factors that affect T_m are similar to those that affect T_g :

- intermolecular forces

- chain stiffness
- geometry

The T_m increases as the intermolecular forces and chain stiffness increase. From Chapter 2 we have seen that

$$T_m = \frac{\Delta H_f}{\Delta S_f} \qquad (8.4)$$

where:

ΔH_f = heat of fusion

ΔS_f = entropy of fusion

T_m can be expressed as:

$$T_m = \frac{k_1 + k_2(\text{CED})}{\Delta S_m} \qquad (8.5)$$

where:

k_1 and k_2 = constants

CED = cohesive energy density

ΔS_m = entropy change which is primarily affected by the chain stiffness

8.4.6 $T > T_m$

These types of transitions occur in materials such as liquid crystalline polymers where mesophases or multi-phase states may exist and in other polymers where short-range order has been preserved even though the material has the characteristics of a liquid. All liquid crystal polymers contain stiff mesogenic groups, often inserted in flexible chain systems. The flexible chain groups are called spacers and are connected to the mesogenic groups via functional groups. The mesogenic unit is inserted either in the main chain or in side chains. In addition to these types of transitions, many polymers begin undergoing decomposition reactions immediately after melting which may show up as relatively sharp spikes on a DSC thermogram and may be misinterpreted. In other cases, broad endotherms or exotherms may be exhibited.

Table 8.1 shows the transitions possible in liquid crystalline polymers and compares them with conventional polymers. In Table

8.1, the following new designations have been used: T_{sn} is the transition of smectic phase into nematic phase; T_{kn} is the transition of crystalline phase into nematic phase; T_i is the isotropic or clearing temperature where the transition from liquid crystalline state to the isotropic melt takes place.

8.5   RELATIONSHIP OF T_g TO T_m

Since T_g and T_m have been shown previously to be related to the CED from Equations 8.3 and 8.5, it is not surprising that there would be a general relationship [S19] between these two transition temperatures as follows:

$$T_g = aT_m + b \qquad (8.6)$$

where: a and b are constants. Expression of this general relationship in numerical quantities gives:

$$\frac{T_g}{T_m} \approx 1/2 \text{ for symmetrical polymers} \qquad (8.7)$$

$$\frac{T_g}{T_m} \approx 2/3 \text{ for unsymmetrical polymers} \qquad (8.8)$$

Unsymmetrical polymers are defined as those containing a main-chain atom which does not have two identical substituents. Other polymers are regarded as symmetrical.

8.6   OTHER FACTORS INFLUENCING T_g

Although some of these factors have been discussed previously, we summarize important effects and relationships which provide insight into the magnitude of their effect.

8.6.1   Pressure Dependence of T_g

The effect of pressure on T_g is a linear function [S20] and can be related to the pressure of interest p, from a reference pressure p_o by applying a weighted correction factor $f_{wt.}$:

$$T_g(p) = T_g(p_o) + f_{wt}(p) \qquad (8.9)$$

For semi-rigid aromatic polymers $f_{wt.} \approx 0.55 \text{K} / \text{MPa}$. This is consistent with the expected effects of bulky side groups and rigid main chain groups.

8.6.2   Molecular Mass

This aspect has been discussed previously, the influence of which is given by:

$$T_g = T_g(\infty) - \frac{Q}{\overline{M}_N} \qquad (8.10)$$

where Q is a constant and $T_g(\infty)$ is the limiting value of T_g.

8.6.3 Tacticity

The T_g of vinyl polymers of the general form $\left(CH_2 - \underset{\underset{Q}{|}}{\overset{\overset{P}{|}}{C}} \right)_X$
is affected only if $P \neq Q$ and neither P nor Q is hydrogen [S21]. For polyalkyl methacrylates, the relationship is:

$$T_g(\text{syndiotactic}) - T_g(\text{isotactic}) \approx 112° \qquad (8.11)$$

8.6.4   Crosslinking

The qualitative effects of crosslinking have been presented in an earlier section. Empirical formulae for the effects of crosslinks are given as follows [S22]:

$$T_{g,crl} - T_{g,o} = \frac{3.9 \times 10^4}{\overline{M}_{crl}} \qquad (8.12)$$

$$T_{g,crl} - T_{g,o} = \frac{788}{n_{crl}} \tag{8.13}$$

where:

$T_{g,crl}$ = glass transition of the crosslinked polymer

$T_{g,o}$ = glass transition of the polymer with out

crosslinks

$\overline{M}_{crl}$ = average mass of polymer between

crosslinks

n_{crl} = average number of backbone atoms

between two crosslinks

8.7   METALS AND CERAMICS

Many of the concepts applicable to phase transitions in polymers are also applicable to metals and ceramics. In the case of ceramics, at the glass transition temperature, changes in the specific volume of the material is displayed as for polymers at the temperature of intersection between the curve for the glassy state and that for the supercooled liquid as illustrated in Figure 8.9 [G3]. In addition, the ceramic materials also display different relaxation times that are dependent upon the rate of cooling.

As the liquid is cooled from a high temperature without crystallizing, a region of temperature is reached in which a bend appears in the specific volume-temperature relationship. In this region, the viscosity of the material increases and generally reaches a limiting value of about 10^{12} to 10^{3} Pascals.

The term glassy is only infrequently applied to metal. In the case of metals the equivalent to the glass transition in other materials is referred to as a ductile brittle transition. For metals, in place of specific volume-temperature plots an impact vs. temperature test is used. This test is know as a "Charpy" impact test and uses a notched impact specimen and is carried out as illustrated in Figure 8.10 [G6].

This test is valuable because of its simplicity, universal acceptance, and reproduces the ductile-brittle transformation of steel in about the same temperature range as it is actually observed in engineering structures. In an ordinary tension test, the transition

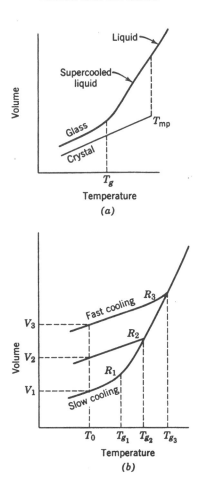

Figure 8.9 [G3]
Specific Volume-Temperature Relationships
a) Relationship for liquid, glass, and crystal
b) Glasses formed at different cooling rates; $R_1 < R_2 < R_3$

of iron or steel occurs at a much lower temperature.

A representative Charpy impact ductile to brittle fracture transition curve is shown in Figure 8.11. Completely ductile specimens exhibit surfaces that are rough or fibrous, while those of brittle specimens contain an irregular array of small bright facets, each corresponding to the surface of a cleaved crystal. In specimens where the fracture is part ductile and part brittle, the bri-

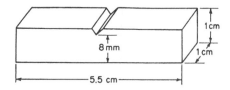

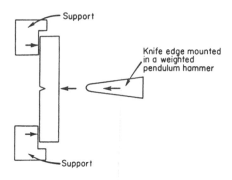

Figure 8.10 [G6]
V-Notch Charpy Impact Test Specimen and Method of Applying a Load

ittle or bright area is found at the center of the cross-section.

The Charpy impact test has been widely used to measure the effect of composition and microstructure. We can understand the importance of the heat treatment of steels and the effects on

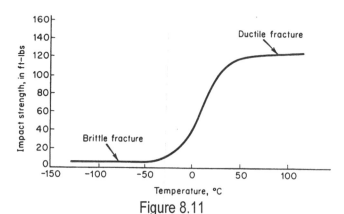

Figure 8.11
Example of a Charpy Impact Ductile to Brittle Fracture Transition Curve

microstructure from the Charpy impact curves in Figure 8.12 [G6]. Both heat treatments produce the same hardness in the steel, but widely different transition temperatures. One steel was quenched and tempered to produce spheroidized cementite (See Figure 2.3), and the other the steel was normalized or air-cooled and then tempered to produce pearlite. The difference between the two structures is remarkable and the ductile-brittle transition temperature for the spheroidized cementite structure is about 300°F lower than the pearlite structure.

Composition also has a large effect on the transition temperature. In general, other variables being held constant, the higher the carbon content of a commercial steel, the greater likelihood that it will be brittle at or near room temperature. Phosphorus has an even stronger effect and is known to segregate at grain boundaries and decrease the mechanical properties.

8.8   ESTIMATION OF THE T_g AND T_m OF POLYMERS

8.8.1   The Glass Transition $\left(T_g\right)$

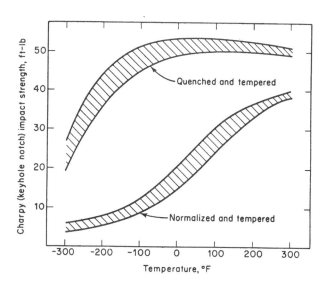

Figure 8.12 [G6]
Effect of Two Different Heat Treatments on the Low-Temperature Impact
Strength of a Medium Carbon Steel

A number of investigators have proposed correlations between the chemical structure and the glass transition temperature of polymers. These methods are based on the assumption that the structural groups in the repeating units provide weighted additive contributions to T_g.

The product $T_g \cdot M$ has been found to behave as an additive function which is termed the Molar Glass Transition Function [S23]. The contribution to this function is given by:

$$Y_g = \sum_i Y_{gi} = T_g \cdot M \qquad (8.14)$$

where:

$\quad Y_g$ = contribution to T_g of group i

and

$$T_g = \frac{Y_g}{M} = \frac{\sum_i Y_{gi}}{M} \qquad (8.15)$$

8.8.2 The Crystalline Melting Point (T_m)

In the previous section, Equations 8.7 and 8.8 showed that the T_g is related to the T_m for many polymers. However, these relationships are only empirical and caution needs to be exercised especially when estimating a T_m based upon an estimated T_g. Therefore, we prefer using the following relationships:

$$Y_m = \sum_i Y_{mi} = T_m \cdot M \qquad (8.16)$$

where:

$\quad Y_{mi}$ = contribution to T_m of group i

and

$$T_m = \frac{Y_m}{M} = \frac{\sum_i Y_{mi}}{M} \qquad (8.17)$$

8.9 EXCEPTIONS

The group contributions of Y_{gi} to T_g and Y_{mi} to T_m have been selected from the literature and Van Krevelen. Many of these values have been modified to approximate, as closely as possible, the $T_g s$ and $T_m s$ of polymers. This allows changes in the structure of a polymer to be evaluated for its effect on T_g and T_m. It also provides a rational procedure for estimating transition temperatures of T_g and T_m for polymers for which there are no literature values.

There are special types of polymers that have been identified as being exceptions to the more general listing of group contributions or their modified forms as employed here, and for which additional detailed treatment must be used to provide results that compare favorably to literature values. We have not included them in the internal databases of the software since they represent highly specialized cases and add complexity to the calculations that become cumbersome to the user.

Major exceptions to the general group additive contributions for Y_{gi} and Y_{mi} fall into the following categories.

8.9.1 Aliphatic

These substances consist of carbon main chains with long paraffinic side chains ("comb polymers") of the following type.

$$-(CH_2R_1)_x^{-}$$
$$\overset{|}{\underset{R_2}{(CH_2)}}_n$$

This results in major changes in estimated transition temperatures as the number of methylene groups in the side chain increases. Short lengths cause a disordering effect which becomes ordered as more methylene groups are added to the side chain. This is probably due to side-chain crystallization. Thus different effects on both T_g and T_m could be expected based on the length of methylene groups which results in corrections that are non-linear where T_g reaches its minimum value at $N = 9$ while T_m reaches its minimum value at $N = 5$. This results in the following set of equations.

For T_g:

$$N = 9; \; Y_g \approx 0.2\,M_9 = Y_{g9} \tag{8.18}$$

$$N < 9; \; Y_g \approx Y_{g0} + \frac{N}{9}\left(Y_{g9} - Y_{g0}\right) \tag{8.19}$$

$$N > 9; \; Y_g \approx Y_{g9} + 7.5(N - 9) \tag{8.20}$$

Table 8.2

Data For Vinyl Polymers with Long Paraffinic Side Chains and Methyl End Groups Affecting T_g

Series	Basic Polymer	Trivalent Group	Y_{g0}	Y_{g9}	M_0	M_9
Polyolefins	Polypropylene	$-\mathrm{CH}-$	10.7	33.6	42.0	168.3
Polyalkylstyrenes	Poly(p-methyl styrene)	$-\mathrm{CH}-$ (phenyl)	48.8	48.8	118.2	244.4
Polyvinyl ethers	Poly(vinyl methyl ether)	$-\mathrm{CH}-$ O	14.6	36.8	58.1	184.3
Polyvinyl esters	Poly(vinyl acetate)	$-\mathrm{CH}-$ O $\mathrm{C}=\mathrm{O}$	26.0	42.4	86.1	212.3
Polyacrylates	Poly(methyl acrylate)	$-\mathrm{CH}-$ O $\mathrm{C}=\mathrm{O}$	26.0	42.4	86.1	212.3
Polymethacrylates	Poly(methyl methacrylate)	$-\mathrm{C(CH_3)}-$ $\mathrm{C}=\mathrm{O}$ O	37.8	45.2	100.1	226.3

The values for Y_{g0}, Y_{g9}, and M_9 are tabulated in Van Krevelen [S23] and are listed in Table 8.2 for convenience.

For T_m:

$$N = 5;\ Y_m \approx 0.235 \cdot M_5 = Y_5 \tag{8.21}$$

$$N < 5;\ Y_m \approx Y_{m0} + \frac{N}{5}\left(Y_{m5} - Y_{m0}\right) \tag{8.22}$$

$$N > 5;\ Y_m \approx Y_{m5} + 5.7\left(N - 5\right) \tag{8.23}$$

The values for, Y_{m0} Y_{m5}, and M_5 are tabulated in Van Krevelen [S2] and are listed in Table 8.3 for convenience.

8.10   MOLCALC TRANSITION TEMPERATURE ESTIMATES OF T_g AND T_m AND COMPARISION TO LITERATURE VALUES FOR A SERIES OF REPRESENTATIVE POLYMERS

Molcalc estimates the T_g and T_m of a polymer employing the group additive function and tabulated values are contained in the internal database. These values are calculated without regard as to whether or not the particular polymer is capable of displaying a melting point, as of course, amorphous polymers are not.

The effects of geometry of the polymer structure have been discussed previously. Since Molcalc uses additive group contributions, there is no way of distinguishing whether or not a CH_2, CH_3, or other group is in the *cis* or *trans* position relative to an asymmetric carbon. Thus the user is cautioned to exercise judgment in interpreting the calculated T_g and T_m for polymers capable of existing in one or more stereoregular forms.

Comparisons of T_g and T_m values for the series of representative polymers, used previously to illustrate calculated properties, are listed in Table 8.4.

In the case of "comb-polymers", there is considerable dis-

Table 8.3
Data for Vinyl Polymers With Long Paraffinic Side Chains with Methyl End Groups Affecting T_m

Series	Basic Polymer	Trivalent Group	Y_{m0}	Y_{m5}	M_0	M_5
Polyolefins	Polypropylene	$-CH-$	18.7	26.3	42.0	112.2
Polyvinyl ethers	Poly(vinyl methyl ether)	$-CH-$ O	24.4	30.1	58.1	128.2
Polyvinyl esters	Poly(vinyl acetate)	$-CH-$ O $C=O$	44	36.7	86.1	156.2
Polyacrylates	Poly(methyl acrylate)	$-CH-$ O $C=O$	44	36.7	86.1	156.2
Polymethacrylates	Poly(methyl methacrylate)	$-C(CH_3)-$ $C=O$ O	47.3	40.0	100.1	170.2

Table 8.4
Estimated T_g and T_m Values of Various Polymers and Comparison with Literature Values

Polymer	T_g, °C		T_m, °C	
	Calculated	Literature	Calculated	Literature
Poly(ethylene)*	-120 ±5	-120 ±5	133	137.5
Poly(propylene)	-18	-17	181	187.7
Poly(isobutylene)*	-83	-73	117	128
Poly(vinyl chloride)	81	81	278	273
Poly(vinyl fluoride)	-18	----	192	230
Poly(chloro trifluoro ethylene)	70	52	253	220
Poly(vinyl acetate)	14	31	239	Melt Unstable
Poly(cis butadiene)**	-116	-102	30	11
Poly(cis 1,4-isoprene)**	-67	-72	37	35.5

(continued)

Table 8.4   (continued)

Polymer	T_g, °C		T_m, °C	
Poly(styrene)	99	100	197	243
Poly(αmethyl styrene)	168	168	216	Melt Unstable
Polycarbonate*	148	145	231	220
Poly(ethylene terephthalate)	64	69	249	280
Poly(butylene terephthalate)	41	22-80	235	232
Poly(3,3-bis chloromethyl oxacyclobutane)	2	18	193	186
Poly(methylene oxide)	-85	-83	183	184
Poly(vinyl methyl ether)	-28	-31	194	147
Poly(phenylene oxide)	193	210	282	307
Poly(hexamethylene adipamide)	43	50	226	267
Poly(acrylamide)	163	165	304	----
Poly(acrylonitrile)	94	105	319	318
Poly(methyl methacrylate)*	113	105	196	200
Poly(cyclohexyl methacrylate)*	96	104	159	----

*Polymers for which the literature lists multiple and conflicting values for either T_g or T_m.

**T_g and T_m values correspond to the *cis* forms.

agreement between the calculated T_g from Molcalc vs. literature values. As an example of the effect of increasing "carbons" (methylene groups) in the side-chain, Table 8.5 is included. This table lists values for a series of poly(methacrylates).

The result of increasing length of side-chain methylene groups is as expected in that short lengths cause a disordering effect which becomes ordered as more methylene groups are added to the side chain. Molcalc shows a systematic lowering of T_g since it cannot account for the effects of the changing polymer morphologies. This is illustrated in Figure 8.13. It is interesting to note that the minimum value is reached at a value of $(CH_2)_9$.

Table 8.5

Comparison of Calculated vs. Literature Values For A Series of Poly(methacrylates)

Polymer	T_g,°C	
	Calculated	Literature
Poly(methyl methacrylate)	113	105
Poly(ethyl methacrylate)	84	65
Poly(hexyl methacrylate)	17	-5
Poly(decyl methacrylate)	-17	-55
Poly(dodecyl methacrylate)	-45	15

For this and other applications, an interpolating polynomial can easily be calculated which will correlate calculated values with literature values. From this polynomial any intermediate values can be estimated with good accuracy.

Figure 8.14 shows the relationship of the literature T_g's to

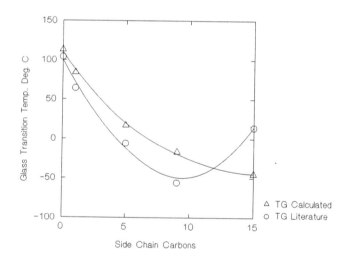

Figure 8.13

The Effect of Carbon Side Chain Length on T_g for a Series of Poly(methacrylates)

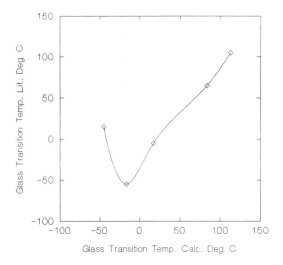

Figure 8.14

Relationship of Literature T_g's to MPCalc Estimated T_g's for a Series of Poly(methacrylates) with Long Side Chains

the calculated T_g's. This curve can be closely approximated by the following polynomial.

$$T_g = -37.559 + 1.665 \cdot T_{g,\,calc.} + 2.508 \times 10^{-2} \cdot T_{g,\,calc.}^2$$
$$- 6.744 \times 10^{-4} \cdot T_{g,\,calc.}^3 + 3.724 \times 10^{-6} \cdot T_{g,\,calc.}^4 \qquad (8.24)$$

Since the above and many similar polymers are amorphous and/or only a few of a specific series exhibit melting phenomena, there were not sufficient literature data to explore similar relationships in the case of melting phenomena.

8.11 ABBREVIATIONS AND NOTATION

a constant

b	constant
CED	cohesive energy density
C_p	molar heat capacity at constant pressure
H_a	activation energy
K_1	constant related to iso-free volume state
K_2	constant related to iso-free volume state
$\overline{M}_{crl}$	average mass of a polymer between crosslinks
$\overline{M}_N$	number average molecular weight
n_{crl}	average number of backbone atoms between two crosslinks
Q	constant
T_g	glass transition temperature
$T_{g,crl}$	glass transition temperature of a crosslinked polymer
$T_{g,o}$	glass transition temperature of a polymer without crosslinks
T_i	transition from an intermediate phase into an isotropic melt
T_m	melting temperature
T_{kn}	transition of a crystalline phase to a nematic phase
T_{sn}	transition of a smectic phase to a nematic melt
$T_g(\infty)$	glass transition temperature limiting value
Y_g	contribution to T_g of group i
Y_m	contribution to T_m of group i
α	coefficient of cubical expansion
ΔS_m	entropy change on melting

8.12 SPECIFIC REFERENCES

1. Boyer, R.F., *Rubber Chem. Tech.*, **36**, 1303-1421(1963).
2. Van Krevelen, D.W., *Properties of Polymers*, Elsevier, New York, N.Y., 1990.
3. Deutsch, K., Hoff, E.A.W., and Reddish, W., *J. Polymer Sci.*, **13**, 565(1954).
4. "Reports on Progress in Polymer Physics in Japan", **6**, pg. 113-154, Kobayasi Institute of Physical Research. Kokubunji, Tokyo, Japan, 1963.
5. Schmieder, K. and Wolf, K., *Kolloid-Z.* **134**, 149(1953).
6. Slichter, W.P. and Mandell, E.R., *J. Appl. Phys.*, **29**, 1438(1958).
7. Muns, T., McCrum, N.G., and McGrew, F.C., APE J., 15, No.5, 368(1959).
8. Woodward, A.E., Sauer, J.A., and Wall, R.A., *J. Chem. Phys.*, **30**, 854(1959).
9. Sauer, J.A., Woodward, A.E., and Fuschello, N., *J. Appl. Phys.*, **30**, 1488(1959).
10. Slichter, W.P., *Makromol. Chem.*, **34**, 67(1959).
11. Bateman, J., Richards, R.E., Farrow, G., and Ward, I.M., *Polymer*, **1**, 63(1960).
12. Reding, F.P., Faucher, J.A., and Whitman, R.D., *J. Polymer Sci.*, **54**, 5-56(1961).
13. Nielsen, L.E., *Mechanical Properties of Polymers*, Reinhold Publishing Co., New York, 1962.
14. Illers, K.H., Kilian, H.G., and Kosfeld, R., *Annual Reviews of Physical Chemistry*, **12**, published by Annual Reviews, Palo Alto, Calif., 1961.
15. Boyer, R.F., *Dependence of Mechanical Properties In Molecular Motion in Polymers, SPE International Award in Plastics and Science Engineering*, New York, May 8, 1968.
16. McBrierty, V .J., *Faraday Discuss. Chem. Soc.*, **68**, 78(1979).
17. Douglass, D. C., *ACS Sym. Ser.*, **142**, 147(1980).
18. Woodward, A.E., Crusman. J.M., and Sauer, J.A., *J. Polymer Sci.*, **44**, 23(1960).

19. Boyer, R.F., "Changements de Phases", p. 383, published by *Soc. De Chimie Physique*, Paris, 1952.
20. Zoller, P., Starkweather, H., and Jones, G., *J. Polymer Sci.*, Phys. Ed., **20**, 1453(1982).
21. Karasz, F.E., and MacKnight, W.J., *Macromolecules*, **1**, 537(1968).
22. Nielsen, L.E., *J. Macromol. Sci.*, Part C3, 69(1969).
23. Van Krevelen, D.W. and Hoftyzer, P.J., unpublished 1975.

8.13   GENERAL REFERENCES

1. Brandup, J., and Immergut, E.H., *Polymer Handbook*, Interscience Publishers, New York, N.Y., 1966.
2. DiBenedetto, A.T., *The Structure and Properties of Materials*, McGraw-Hill, Inc., New York, N.Y., 1967.
3. Kingery, W. D., Bowen, H.K., Uhlmann, D.R., *Introduction to Ceramics*, 2nd Ed., John Wiley & Sons, New York, N.Y., 1976.
4. Moffatt, W.G., Pearsall, G.W., and Wulff, J., *Structure and Properties of Materials*, Vol. 1, Structure, John Wiley & Sons, New York, 1966.
5. Pauling, L., *Nature of the Chemical Bond*, 3d ed., Cornell University Press, Ithaca, N.Y., 1960.
6. Reed-Hill, R.E., *Physical Metallurgy Principles*, 2nd Ed., D. Van Nostrand Company, New York, N.Y., 1973.
7. Seymour, R.B., Carraher, C.E. Jr., *Polymer Chemistry*, Marcel Dekker, Inc., New York, N.Y., 1988.
8. Van Krevelen, D.W., *Properties of Polymers*, Elsevier, New York, N.Y., 1990.
9. McBrierty, V.J., and Packer, K.J., *Nuclear Magnetic Resonance In Solid Polymers*, Cambridge University Press, Cambridge, Great Britain, 1993.

CHAPTER 9

Cohesive Properties and Solubility Parameter Concepts

9.1   INTRODUCTION AND BACKGROUND

From the earliest times people have been fascinated with the reasons why some materials and substances mix while others do not. There are many examples in a wide variety of applications that illustrate the importance of mixing and its influence on advancements which hold the potential to improve living conditions.

Applications are as diverse as the mixing of metals to form alloys which improve properties to the mixing of oils and water for the preparation and breaking of emulsions. Scientists in almost every field have been attempting to understand the underlying fundamental science and formulate general laws that allow one to estimate the tendency of materials to mix or blend without the necessity of experimental "trial" and "error" procedures.

In order to appreciate why some materials mix and form a uniform solution and others do not, we must have some understanding of the forces which hold materials or substances together and how they may interact. Although much effort has been devoted to this understanding on an atomic scale, our interests and goals are to attempt to understand molecular interactions and the theory that provides useful approaches.

An approach that has proven to be very useful for polymers over the years, is an approach pioneered by Hildebrand [S1] and improved by Scatchard [S2]. A major step forward was subsequently outlined by Hildebrand and Scott [S3] which resulted in the concept of "cohesive energy density" or "solubility parameter" concept of estimating the behavior of solvents. Although many other investigators have subsequently refined this approach, this early pioneering work provided the basis.

9.2   COHESIVE ENERGY DENSITY AND THE SOLUBILITY PARAMETER

The cohesive properties of a substance or material are expressed quantitatively in the cohesive energy. This is related directly to a material's solubility in a solvent. Hildebrand [S1] pointed out that the order of a given solute in a series of solvents is determined by the internal pressure of the solvents. Scatchard [S2] built upon this idea and introduced the concept of "cohesive energy density".

The symbols U or E (cf. Chapter 2) are usually used to define the molar internal energy of a material relative to the ideal vapor at the same temperature. In this context, U has a negative value and the molar cohesive energy or the energy associated with net attractive interactions $(-U)$, has a positive value.

The molar cohesive energy can be divided into two parts:

1. The molar vaporization energy, $\Delta_l^g U$ required to vaporize a mole of liquid to its saturated vapor

2. The energy $\Delta_g^\infty U$ required to completely separate the molecules

The complete expression for the internal energy is:

$$-U = \Delta H_l^g + \Delta_g^\infty H - RT + p_s V \qquad (9.1)$$

where:

$\Delta_l^g H$	=	molar enthalpy of vaporization
$\Delta_g^\infty H$	=	increase in enthalpy on isothermal expansion of one mole of saturated vapor to zero pressure
R	=	gas constant

$$
\begin{aligned}
T &= \text{temperature, K} \\
p_s &= \text{saturated vapor pressure at } T \\
V &= \text{molar volume}
\end{aligned}
$$

At pressures below the normal boiling point, U is:

$$-U = \Delta_l^g H - RT \tag{9.2}$$

In condensed phases, the cohesive energy is expressed as the cohesive energy per unit volume, or the "cohesive energy density", i.e., *C.E.D.* where the cohesive energy is given by:

$$E_{coh} = \Delta U = \Delta E \text{ (J / mol)} \tag{9.3}$$

and the cohesive energy density is:

$$C.E.D. = \frac{E_{coh}}{V} \text{ (J / cm}^3\text{)} \tag{9.4}$$

Cohesive energy density is the basis of the original definition by Hildebrand and Scott and is now generally referred to as the solubility parameter and is given the symbol, δ, and is equal to the square root of the *C.E.D.*

$$\delta = \left(\frac{E_{coh}}{V} \right)^{1/2} \approx \left(\frac{\Delta_l^g U}{V} \right)^{1/2} \left(@\,298^\circ \text{K} \right) \tag{9.5}$$

and, in general

$$\delta_i = \left(\frac{\Delta E_i}{V_i} \right)^{1/2} \tag{9.6}$$

where:

δ = solubility parameter of component i which describes the attractive strength between molecules of component i

ΔE_i = energy of vaporization per cm^3

This parameter was intended to describe the enthalpy of the mixing of simple liquids (nonpolar, non-associating solvents) but has been extended to polar solvents and systems as well as polymers. The form of Equations 9.5 and 9.6 suggest a close relationship or link between solubility or miscibility and that of cohesion or vaporization since in a mixing process like molecules of each component in a mixture are separated from one another by an infinite distance, similar in some respects to what occurs in the vaporization process.

9.3   SOLUBILITY PARAMETERS IN MIXING
In Chapter 2 we saw that the free energy of mixing is:

$$\Delta G_m = \Delta H_m - T\Delta S_m \tag{9.7}$$

where:

ΔG_m = change in Gibbs free energy of mixing

ΔH_m = the enthalpy change in mixing

T = absolute temperature

ΔS_m = entropy change on mixing

A negative ΔG_m predicts that a mixing process will occur spontaneously. When polymers mix or dissolve, there is always a large increase in entropy and the magnitude and sign of the enthalpy term is the deciding factor. This is more fully discussed in Chapter 22.

Hildebrand and Scott [S3] and Scatchard [S4] proposed that the enthalpy term is described by:

$$\Delta H_m = V\left[\left(\frac{\Delta E_1}{V_1}\right)^{1/2} - \left(\frac{\Delta E_2}{V_2}\right)^{1/2}\right]^2 \phi_1\phi_2 \tag{9.8}$$

where:

$$V \qquad = \text{volume of the mixture}$$

ΔE_1 and ΔE_2 = energy of vaporization of components 1 and 2

V_1 and V_2 = molar volumes of components 1 and 2

ϕ = volume fraction of components 1 and 2

The solubility parameter can be considered to represent the internal pressure of the solvent. For a binary mixture Equation 9.8 can be expressed as:

$$\frac{\Delta H_m}{V \phi_1 \phi_2} = \left(\delta_1 - \delta_2 \right)^2 \qquad (9.9)$$

This equation gives the heat of mixing of regular solutions in which the components mix with:

1. No volume change on mixing at constant pressure
2. No reaction between the components
3. No complex formation

9.4 CALCULATION OF SOLUBILITY PARAMETERS FROM PHYSICAL CONSTANTS

For low molecular weight materials, ΔE can be calculated from $\Delta E = \Delta H - RT$ since ΔH values or latent heat of vaporization at a given temperature, T, may be found in the literature. For many solvents where literature values are not available, the Hildebrand equation can be used for ΔH estimations.

$$\Delta H_{25°C} = 23.7 T_b + 0.020 T_b^2 - 2950 \qquad (9.10)$$

where T_b is the boiling point in K. Figure 9.1 [S5] illustrates this expression graphically in which RT has already been subtracted from ΔH and is applicable to those liquids which are not hydrogen bonded. It can be seen that the solubility parameter of a solvent can be estimated from its energy of vaporization.

Estimation of solubility parameters of polymers is a more

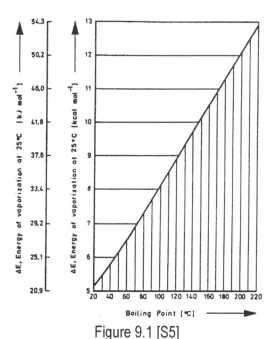

Figure 9.1 [S5]
The Relationship of the Solubility Parameter (δ) of a Solvent to Its
Energy of Vaporization

complex problem. In this case, polymer solubility parameters can only be determined indirectly and may be affected by variations in their chemical constitution, i.e., the number of crosslinks and uniform chain branches or substitutive groups along the chain, the compositional heterogeneity of which cannot be accurately represented by uniform repeating structures.

Methods for determining the solubility parameters of polymers are based upon indirect measurements since polymers cannot be evaporated. Their solubility parameters or cohesive energies must be determined by comparative swelling or dissolving experiments in liquids of known cohesive energy density. This type of measurement is illustrated in Figure 9.2.

9.5 SOLUBILITY PARAMETERS BASED UPON GROUP CONTRIBUTIONS

Many investigators use solubility parameters based upon group contribution methods because of the difficulties in perform-

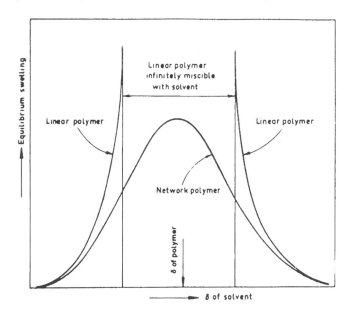

Figure 9.2 [S9]
Equilibrium Swelling as a Function of the Solubility Parameter of the
Solvent for Linear and Cross-Linked Polystyrene

ing accurate measurements as discussed above. These methods are based on the use of group contributions to a quantity termed the molar attraction constant, F, which is given by:

$$F = \left(E_{coh} V_{(298)} \right)^{1/2} \qquad (9.11)$$

The most prevalent values used are those of Small [S6], Hoy [S7], and Van Krevelen [S8]. These can be found in many texts but have been tabulated and extensively discussed by Van Krevelen [S9].

9.6   APPLICATION OF THE SOLUBILITY PARAMETER CONCEPT TO A WIDE RANGE OF MATERIALS

The solubility parameter gives a good measure of the enthalpy change on the mixing of non-polar solvents but does not provide results that are nearly as reliable when extended to polar systems. The reason for this lies with the fact that the solubility parameter, as such, is a gross measure of the forces involved. These different types of forces do not interact effectively which results in a similarity of solubility parameters that do not always ensure miscibility.

It should be noted also that the solubility parameter is based on the concept of applicability to regular solutions, despite the existence of interactions which lead to a non-ideal enthalpy of formation. This effectively restricts regular mixtures to those systems in which only dispersion forces are important, because the orientation effects of polar molecules cause nonrandom molecular distributions.

In Chapter 6, we have seen that the origin of forces between molecules, precluding covalent bond formation during chemical reactions, arise from van der Waals forces. These forces consist of Keesom orientation forces, Debye inductive forces, London dispersion forces, and hydrogen bonding which is the most important contributor to acid-base or electrostatic forces.

Two materials having the same solubility parameter does not ensure that there is an equal balance of London, Keesom, and hydrogen bonding forces. It is this that accounts for the fact that a substance will not dissolve in a solvent in spite of the fact that the solute and solvent have similar solubility parameters. Another factor that contributes to this apparent discrepancy is that some materials containing hydroxyls will self associate.

For these reasons, a number of investigators have decomposed the classical Hildebrand solubility parameter into several terms, as Hildebrand [S4] himself did using dispersive and polar. In 1966, Crowley et al. [S10, S11] showed that solubility parameter exceptions could be accounted for by using a group of solvents with the same δ^H (Hildebrand solubility parameter) and dipole moment but with different hydrogen bonding affinities.

Hansen [S12-S16], and Hansen and Skarup [S17] assumed that the cohesive energy arises from dispersive, permanent dipole-dipole interactions, and hydrogen bonding forces. This allowed the total solubility parameter to be resolved and is expressed as:

$$\delta^2 = \delta_d^2 + \delta_p^2 + \delta_h^2 \tag{9.12}$$

The term δ_h may very well account for a variety of association bonds such as electrostatic interactions. This treatment gives improved agreement with data but is still not completely accurate in predicting solution thermodynamics for every system.

Nevertheless, the use of the individual solubility parameter contributions, δ_d, δ_p, and δ_h to the total allows one to evaluate various combinations, viz., δ_{dp}, δ_{dh}, and δ_{ph}, to examine the forces playing a role in molecular interactions and, in some cases at least, driving solubility. For these reasons we employ that approach here.

Each of the individual solubility parameters can be represented as one axis in the three dimensional Cartesian coordinate axis system as Figure 9.3 shows. For two materials, say polymers P_1 and P_2, the resulting solubility parameter for each component is:

$$\left(\delta_1^P\right)^2 = \delta_d^2 + \delta_p^2 + \delta_h^2 \tag{9.13}$$

$$\left(\delta_2^P\right)^2 = \delta_d^2 + \delta_p^2 + \delta_h^2 \tag{9.14}$$

From Figure 9.3, it can be seen that each total solubility parameter can be represented as a vector quantity of magnitude $\left|\vec{\delta}_1^P\right|$ and $\left|\vec{\delta}_2^P\right|$. In this approach δ_1^P and δ_2^P are:

$$\left|\vec{\delta}_1^P\right| = \left[\delta_d^2 + \delta_p^2 + \delta_h^2\right]^{1/2} \tag{9.15}$$

$$\left|\vec{\delta}_2^P\right| = \left[\delta_d^2 + \delta_p^2 + \delta_h^2\right]^{1/2} \tag{9.16}$$

In the case of the individual contributions, the vector quantities lie in the planes δ_{dp}, δ_{dh}, or δ_{ph} and can be expressed as:

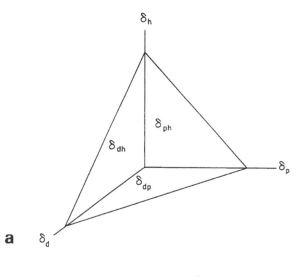

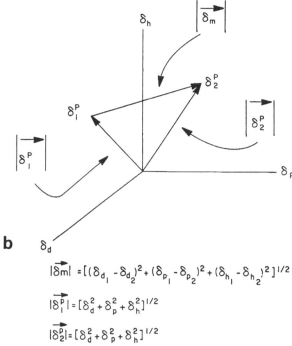

$$|\overrightarrow{\delta_m}| = [(\delta_{d_1} - \delta_{d_2})^2 + (\delta_{p_1} - \delta_{p_2})^2 + (\delta_{h_1} - \delta_{h_2})^2]^{1/2}$$

$$|\overrightarrow{\delta_1^p}| = [\delta_d^2 + \delta_p^2 + \delta_h^2]^{1/2}$$

$$|\overrightarrow{\delta_2^p}| = [\delta_d^2 + \delta_p^2 + \delta_h^2]^{1/2}$$

Figure 9.3 [S18]
a. Solubility Parameter Coordinate Axis System
b. Illustration of Solubility Parameter Vector Quantities

$$\left|\vec{\delta}_{dh}\right| = \left[\delta_d^2 + \delta_h^2\right]^{1/2} \tag{9.17}$$

$$\left|\vec{\delta}_{dp}\right| = \left[\delta_d^2 + \delta_p^2\right]^{1/2} \tag{9.18}$$

$$\left|\vec{\delta}_{ph}\right| = \left[\delta_p^2 + \delta_h^2\right]^{1/2} \tag{9.19}$$

This treatment using the individual components will allow examination of group interactions and is illustrated for polymer miscibility estimations in Chapter 21.

9.7 ESTIMATION OF SOLUBILITY PARAMETERS

The individual solubility parameter components may be estimated from group contributions, using the equations of Hoftyzer and Van Krevelen [S19].

$$\delta_d = \frac{\sum F_{di}}{V} \tag{9.20}$$

$$\delta_p = \frac{\sqrt{\sum F_{pi}^2}}{V} \tag{9.21}$$

$$\delta_h = \sqrt{\frac{\sum E_{hi}}{V}} \tag{9.22}$$

One serious problem, using literature values of solubility parameters, based on either individual contributions or the total solubility parameter is the fact that there are generally different values listed for the same material depending upon the investigator and the method of measurement. This, by itself, can cause very serious errors in determining whether or not $\Delta\delta_{1-2}$ is zero or very close to it.

Utilizing the group contribution technique ensures that the same database is used for comparing and using solubility parameters of different materials to provide consistency in making comparisons. In this way accurate relative differences between solubility parameters and/or various combinations can be quite meaningful even if the absolute values may not be highly accurate.

9.8   MOLCALC ESTIMATIONS OF SOLUBILITY
        PARAMETERS AND COMPARISON TO
        LITERATURE VALUES FOR A SERIES OF
        REPRESENTATIVE POLYMERS

Molcalc uses the relationships, illustrated above, to calculate the individual solubility parameter contributions as well as the total solubility parameter. These relationships show that the contributions from the specific polymer groups and the total volume of the repeating groups are important in achieving accurate solubility parameters. Thus, the solubility parameters are simultaneously subject to two different errors.

If the errors from group solubility parameter contributions and group volume contributions are random errors, then there is a high probability that these errors will be self compensating. There will undoubtedly be situations where this is not the case. The extent of error is especially important in the case of solubility parameter estimations since the individual solubility parameters are used in polymer miscibility calculations, as discussed in Chapter 21.

Table 9.1 provides a listing of comparisons between calculated and literature values for the selected list of polymers. Table 9.1 shows considerable variation in the literature values for the total solubility parameters. This reflects differences between investigators and the method of measurement. However, it will be noted that the calculated solubility parameters generally fall in the mid range of the literature values and show good correspondence with reported values.

An obvious exception is in the case of polyamide(nylon 6,6) where the calculated value falls on the low end. Literature values [S20] show the largest discrepancy in the polar contribution of nylon 6,6. A large variation in this component will obviously affect the results of any miscibility/compatibility estimations.

Table 9.1
Comparison of Estimated Polymer Solubility Parameters with Literature
Values

Polymer	Solubility Parameter, $\delta(J^{1/2}cm^{3/2})$: Based on V_R	
	Calculated	Literature
Poly(ethylene)*	16.41	15.8-17.1
Poly(propylene)	15.68	16.8-18.8
Poly(isobutylene)*	15.57	16.0-16.6
Poly(vinyl chloride)	21.92	19.2-22.1
Poly(vinyl fluoride)	13.36	----
Poly(chloro trifluoro ethylene)	15.80	14.7-16.2
Poly(vinyl acetate)	19.67	19.1-22.6
Poly(cis butadiene)**	16.39	16.6-17.6
Poly(cis 1,4-isoprene)**	16.20	16.2-20.5
Poly(styrene)	19.61	17.4-19.0
Poly(αmethyl styrene)	18.90	----
Polycarbonate*	19.65	~20.3
Poly(ethylene terephthalate)	21.15	19.9-21.9
Poly(butylene terephthalate)	20.20	----
Poly(3,3-bis chloromethyl oxacyclobutane)	20.88	----
Poly(methylene oxide)	24.44	20.9-22.5
Poly(vinyl methyl ether)	18.12	----
Poly(phenylene oxide)	19.03	~19.0
Poly(hexamethylene adipamide)	21.06	22.8-27.8
Poly(acrylamide)	25.67	----
Poly(acrylonitrile)	29.51	25.61-31.54
Poly(methyl methacrylate)*	19.59	18.6-26.2
Poly(cyclohexyl methacrylate)*	18.40	----

* Polymers for which the literature lists multiple and conflicting values
** δ values correspond to cis forms

On the positive side, the use of a single database for calculating properties will provide consistency and hopefully, good, relative comparisons. Again, this is very important for estimating polymer miscibilities.

The group contributions contained in Molcalc reflect values for polymers and not for simple solvents and therefore Molcalc should not be used to estimate the solubility parameters of organic solvents. It is instructive to note that when the solubility parameter of an organic solvent is calculated it is generally higher than the literature values obtained from experimental measurements.

The reasons for this are due in large part to the large differences calculated for the molar volumes between simple organic solvents and polymers containing similar repeat groups. For example, the calculated molar volume from Molcalc is generally less than 50% of the actual molar volume which is reflected in density, solubility parameter, and other such estimations.

In the case of non-halogenated solvents, the estimated solubility parameters ($J^{1/2} cm^{3/2}$) are generally about $2\times$ greater than literature values. For halogenated solvents, this difference can range from 3 to 19, the difference widening as the number of halogens is increased.

9.9   SOLUBILITY PARAMETERS OF METALS

The concept of solubility parameters has been extended to metals by Hildebrand and Scott [S3]. Although the solubility parameters of metals have not proven to be as useful, compared with polymers, they are included here in Table 9.2 for general reference and for drawing distinctions between metals and polymers in Chapter 21. In addition, the applicability and use of metallic solubility parameters for indicating the miscibility of metals is discussed briefly in Chapter 21, section 21.11.

Table 9.2
Volumes and Solubility Parameters of Metals At 298 °K

Element	V, cm^3	$\delta, \left(cal/cm^3\right)^{1/2}$
Re	8.9	146
W	9.6	145
B	(4.3)	145
Os	8.4	144
Ir	8.6	139
Ru	8.3	139
Ta	10.8	136
Rh	8.3	129
Be	4.9	129
Mo	9.4	128
Cb	10.8	127
αCo	6.6	126
Ni	6.6	124
Pt	9.1	121
V	8.5	119
αFe	7.1	117
αHf	13.5	112
Cr	7.3	108
Cu	7.2	107
Pd	8.9	102
U	12.7	100
αMn	7.6	95
αZr	14.0	94
αTi	10.7	94
Au	10.3	93
Si	11.7	88
Al	10.0	86
Ag	10.3	82
Sc	14.5	80
Ge	13.7	76
Ga	11.8	74
Y	(19)	72
As	13.1	66
βSn	16.3	65
αLa	22.4	63
In	15.7	60
Sb	18.2	59

(continued)

Table 9.2 (continued)

Element	V, cm^3	$\delta, (cal/cm^3)^{1/2}$
Zn	9.2	58
Li	13.0	54
Pb	18.3	51
Mg	14.0	50
αTl	17.3	49
Bi	21.3	48
Cd	13.0	45
αCa	26.0	40
α	32.9	34
Ba	37	33
Na	23.7	33
Hg	14.8	31
K	45.4	21
Rb	56	19
Cs	70	16

9.10   ABBREVIATIONS AND NOTATION

C.E.D.	cohesive energy density per unit volume
E_{coh}	cohesive energy density
F	molar attraction constant of one mole of saturated vapor to zero pressure
p_s	saturated vapor pressure at T
R	gas constant
T	temperature
U	internal energy
V	molar volume
$\Delta_l^g H$	molar enthalpy of vaporization
$\Delta_g^\infty H$	increase in enthalpy on isothermal expansion
δ_i	solubility parameter of component i
δ	solubility parameter
ΔE_i	energy of vaporization per cm^3
ΔG_m	change in the Gibbs free energy of mixing

ΔH_m	change in enthalpy of mixing
ΔS_m	change in entropy of mixing
δ_i^P	solubility parameter of a polymer i
δ_d	dispersive solubility parameter
δ_p	polar solubility parameter
δ_h	hydrogen bonding solubility parameter
ϕ	volume fraction of component

9.11 SPECIFIC REFERENCES

1. Hildebrand, J.H., *J. Am. Chem. Soc.*, **38**, 1452(1916).
2. Scatchard, G., *Chem. Revs.*, 8, 321(1931).
3. Hildebrand, J.H. and Scott, R.L., *The Solubility of Non-Electrolytes*, 3rd ed., Reinhold Publishing Corp., New York, 1959.
4. Scatchard, G., *Chem. Rev.*, **44**, 7(1949).
5. Burrell, H., "Solubility Parameter Values", in *Polymer Handbook*, 2nd ed., J. Brandrup and E.H. Immergut, Eds., J. Wiley & Sons, New York, 1975.
6. Small, P.A., *J. Appl. Chem.*, **3**, 71(1953).
7. Hoy, K.L., *J. Paint Technol.*, **42**, 76(1970). "Tables of Solubility Parameters", Solvent and Coatings Materials Research and Development Department, Union Carbide Corp. (1985). *J. Coated Fabrics*, **19**, 53(1989).
8. Van Krevelen, D.W., *Fuel*, **44**, 236(1965).
9. Van Krevelen, D.W., *Properties of Polymers*, Elsevier, New York, 1990.
10. Crowley, J.D., Teague, G.S., Lowe, J.W., *J. Paint Technol.*, **38**, 296, 496(1966).
11. Crowley, J.D., Teague, G.S., Lowe, J.W., *J. Paint Technol.*, **39**, 19, 504(1967).
12. Hansen, C.M., *Ind. Eng. Chem. Prod. Res. Dev.*, **8**, 2(1969).
13. Hansen, C.M., Doctoral Dissertation, Technical University of Denmark, Danish Technical Press, Copenhagen, 1967.

14. Hansen, C.M., *Farg Lack*, **14**(1), 18; (2) 23(1968).
15. Hansen, C.M., *J. Paint Technol.*, **39**(505), 104(1967).
16. Hansen, C.M., *J. Paint Technol.* **39**(511), 505(1967).
17. Hansen, C.M., Skaarup, K., *J. Paint Technol.*, **39**(511), 511(1967)
18. Hofytzer, P.J. and Van Krevelen, D.W., *Properties of Polymers*, 2nd ed., Elsevier, New York, 1976.
19. David, D. J., and Sincock, T. F., *Polymer*, **33**, 4505(1992).
20. Barton, A.F.M., *CRC Handbook of Solubility Parameters and Other Cohesion Parameters*, CRC Press, Inc., Boca Raton, Florida(1988).

CHAPTER 10

Interfacial Properties

10.1   INTRODUCTION

It is difficult to overemphasize the importance of this topic since materials communicate with each other through their surfaces. In order for surfaces to interact at the molecular level, the surfaces must first come into contact with each other and the initial contact and interaction will first be determined by the roughness and hardness of each surface along with the temperatures and pressures at contact. Further communication and interaction at the points or areas of contact will be determined by forces at molecular levels. It may at first seem odd to talk or think about the interaction of surfaces down to the molecular level but molecular interactions occur which are favorable or unfavorable. The degree to which this occurs will influence the range of interaction from chemical bonding to miscibility.

Interfacial properties influence the behavior of materials whether they be metals or metal alloys, ceramic materials, or polymers. Many applications, such as, film formation, wetting behavior, coating and adhesion, among others, depend strictly on the surface properties of two or more materials.

10.2   SURFACE CONCEPTS

10.2.1 Definition

As a prelude to a detailed discussion of surface and interfacial properties, we need to define a surface.

Definition: A surface is the outermost atomic layer, including foreign atoms absorbed into it and those absorbed onto it.

The importance of this concept will be immediately apparent if one wishes to measure the surface energy of a solid, adhere two surfaces together with an adhesive, or coat a surface. In each instance the cleanliness or pristine nature of the surfaces will dramatically affect the results.

Surfaces that are formed by machining, molding, pressing, etc. are normally thought of as being flat. While this may be true on a macro or bulk scale, on a molecular or atomic scale, the surface will be comparable to the relative roughness of the Andes mountains as we perceive them. If one wishes to have surfaces that are smooth on an atomic scale, it is necessary to cleave the surface along an appropriate plane.

On a molecular or atomic scale, atoms in a surface are in an environment vastly different from the environment of atoms in the bulk of the solid since they are surrounded by fewer neighbors than bulk atoms and there is an anisotropic distribution of these neighbors. This is characteristic only of a surface.

10.2.2 Surface Thermodynamic Functions

In a homogeneous crystalline body that contains N atoms and is surrounded by plane surfaces, the energy and the entropy of the solid per atom can be denoted by E^o and S^o. The specific surface energy, E^s or energy per unit area is given by:

$$E = NE^o + AE^s \qquad (10.1)$$

where:

E = total energy of the body

A = the surface area

Thus E^s is the excess of the total energy E that the solid has. Of course NE^o is the energy of the solid itself in a state of thermodynamic equilibrium. Similarly, the total entropy of the solid is:

$$S = NS^o + AS^s \qquad (10.2)$$

where:

S^s = specific surface entropy (entropy per unit area of surface)

The surface work content, A^s (energy per unit area) is:

$$A^s = E^s - TS^s \qquad (10.3)$$

The surface free energy (energy per unit area) is:

$$G^s = H^s - TS^s \qquad (10.4)$$

where:

H^s = specific surface enthalpy or the heat absorbed per unit surface area created

and finally, the total free energy of a system is:

$$G = NG^o + AG^s \qquad (10.5)$$

10.2.3 Surface Work and Energy of a One Component System

To create a surface we have to do work on the system that often involves breaking bonds and removing neighboring atoms. Under conditions of equilibrium, the reversible surface work, δW^s required to increase the surface area A by an amount dA is:

$$\left(\delta W^s\right)_{T,P} = \gamma dA \qquad (10.6)$$

Equation 10.6 is comparable with the reversible work to increase the volume of a one component system at constant pressure, PdV. It should be noted that γ is the two dimensional analog of the pressure and is the surface tension while the volume change is replaced by the change in surface area. The surface tension, γ, is the pressure along the surface plane that opposes the creation of more surface. The surface tension or pressure, γ is given by the force per unit length (dynes/cm) or $ergs/cm^2$.

The total surface free energy is thus equal to the specific free energy times the surface area and the reversible surface work can be expressed as:

$$\left(\delta W^s\right)_{T,P} = d\left(G^s A\right) \tag{10.7}$$

An important point to appreciate is that the creation of a stable interface always has a positive free energy of formation. This reluctance of a solid or liquid to form a surface defines many of the interfacial properties of condensed phases. This explains why liquids tend to minimize their surface area by assuming a spherical shape (minimum surface) while solids, at or near equilibrium, will form surfaces that have the lowest free energy at the expense of other surfaces.

There are two ways of forming a new surface:

1. by increasing the surface area
2. by stretching an already existing surface

This latter condition alters the state of strain assuming the number of atoms per unit are fixed.

Therefore, Equation 10.7 becomes:

$$\left(\delta W^s\right)_{T,P} = \left[\frac{\partial\left(G^s A\right)}{\partial A}\right]_{T,P} dA = \left[G^s + A\left(\frac{\partial G^s}{\partial A}\right)_{T,P}\right] dA \tag{10.8}$$

When a new surface is created by increasing the surface area.

$$\left(\frac{\partial G^s}{\partial A}\right)_{T,P} = 0 \tag{10.9}$$

and the surface work is:

$$\delta W^s\big|_{T,P} = G^s dA \tag{10.10}$$

When new surfaces are created by cleavage or strained under conditions such that atoms/molecules diffuse to the surface from the bulk and become part of a new, unstrained surface then Equations 10.6 and 10.10 can be combined to give:

$$\gamma = G^s = \left(\frac{\partial G}{\partial A}\right)_{T,P} \tag{10.11}$$

i.e.:

Surface Tension $\equiv$ Specific Surface Free Energy

At constant temperature and pressure, the change in the Gibbs free energy (ΔG) equals the reversible non-pressure - volume work. Therefore, we relate Equation 10.10 to thermodynamic properties and write:

$$dG = \gamma dA \tag{10.12}$$

For a system at constant temperature, we have seen previously, cf. Chapter 2,:

$$dG = -PdV + \sum_i \mu_i^\sigma dn_i^\sigma \tag{10.13}$$

where μ_i^σ and n_i^σ are the surface chemical potentials and concentrations, respectively and from our discussion above, dG can be expressed as:

$$dG = \gamma dA + \sum_i \mu_i^\sigma dn_i^\sigma \tag{10.14}$$

10.3 CONTACT ANGLE EQUILIBRIUM: THE YOUNG EQUATION

A liquid in contact with a solid will exhibit a contact angle dependent upon the nature of the liquid and the surface with which it is in contact. If the system is at rest a static contact angle is obtained. A liquid drop in contact with a smooth, homogeneous, planar, rigid surface, under equilibrium conditions, is represented in Figure 10.1.

A balance of forces provides the change in the Gibbs free energy as the meniscus slides reversibly on the solid surface.

$$dG = \gamma_{LV} dA_{LV} + \gamma_{SV} dA_{SV} + \gamma_{SL} dA_{SL} \equiv 0 \qquad (10.15)$$

But
$$dA_{SL} = -dA_{SV} \qquad (10.16)$$

and
$$dA_{LV} = \cos\theta dA_{SL} \qquad (10.17)$$

$$\therefore 0 = -\gamma_{LV} dA_{LV} + \gamma_{SV} dA_{SL} - \gamma_{SL} dA_{SL} \qquad (10.18)$$

$$= -\gamma_{LV} \cos\theta dA_{SL} + \gamma_{SV} dA_{SL} - \gamma_{SL} dA_{SL} \qquad (10.19)$$

and $\gamma_{LV} \cos\theta = \gamma_{SV} - \gamma_{SL} =$ Young's Equation [S1]
$$\qquad (10.20)$$

where:
   γ_{LV} = surface tension of the liquid in equilibrium
        with its saturated vapor
   γ_{SV} = surface tension of the solid in equilibrium
        with saturated vapor of the liquid
   γ_{SL} = interfacial tension between the solid and liquid

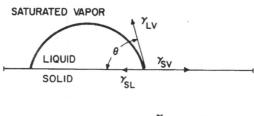

Figure 10.1
Illustration of Contact Angle Equilibrium on a Smooth, Homogeneous, Planar, and Rigid Surface

Adsorption of a vapor or liquid will occur if the free energy of the system is reduced thus changing the surface tension of the substrate. This occurs when the condensed vapor has a surface tension similar to or lower than that of the substrate.

10.3.1 Spreading Pressure

A term that is often used to describe the behavior of a liquid at a solid surface is "spreading pressure". This refers to movement of a phase front whereby the interfacial area is increased. This is represented in Figure 10.1.

The equilibrium spreading pressure of the vapor on the substrate, illustrated in Figure 10.1, is:

$$\pi_e = \gamma_S - \gamma_{SV} \tag{10.21}$$

From Young's equation, we have:

$$\pi_e = \gamma_S - \gamma_{LV}\cos\theta - \gamma_{SL} \tag{10.22}$$

This illustrates that spreading pressure is simply a balance of forces in opposite directions. If we do work on a "dirty" glass surface, for example, and clean it, a drop of water placed on that surface will spread since it will have a small contact angle. This clean surface will begin to immediately lower its surface energy by absorbing materials onto its surface.

10.3.2 Work of Adhesion

The work of adhesion is also represented by a balance of forces:

$$W_a = \gamma_S + \gamma_{LV} - \gamma_{SL} \tag{10.23}$$

$$= \gamma_{LV}(1 + \cos\theta) + \pi_e \tag{10.24}$$

For solids we have a similar situation where the work required to separate reversibly the interface between two bulk phases, α and β, from their equilibrium separation to infinity, is the work of adhesion.

$$W_a = \gamma_\alpha + \gamma_\beta - \gamma_{\alpha\beta} \qquad (10.25)$$

where W_a is as before
and:

γ_α = surface tension of phase α

γ_β = surface tension of phase β

$\gamma_{\alpha\beta}$ = interfacial tension between phases α and β

This is shown graphically in Figure 10.2 and was proposed by Dupré [S2].When the two phases are identical, the reversible work is the work of cohesion.

$$W_{cp} = 2\gamma_p \qquad (10.26)$$

where:

W_{cp} = work of cohesion for phase p

γ_p = surface tension of phase p

Equation 10.25 shows that the greater the interfacial attraction, the smaller the interfacial tension will be.

10.4   SURFACE TENSIONS OF MULTI-COMPONENT MIXTURES

For a multi-component system, the chemical potential of the ith component, μ_i, of a multi-component system is given by:

$$\mu_i = \overline{G}_i = \left(\frac{\partial G}{\partial n_i}\right)_{n_j,P,T} \qquad (10.27)$$

where $\overline{G}_i$ is the partial molal free energy and $\dfrac{\partial G}{\partial n_i}$ is the change of

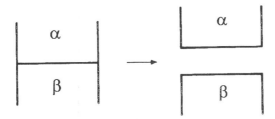

Figure 10.2
Conceptual Representation of the Work of Adhesion

the total free energy of the system with respect to changes in the number of moles of the ith component. The total free energy change of a multi-component system can be expressed as:

$$dG = \left(\frac{\partial G}{\partial T}\right)dT + \left(\frac{\partial G}{\partial P}\right)dP + \left(\frac{\partial G}{\partial A}\right)dA + \sum_i \left(\frac{\partial G}{\partial n_i}\right)dn_i \qquad (10.28)$$

$$= -SdT + VdP + \gamma dA + \sum_i \mu_i dn_i \qquad (10.29)$$

When dn_i moles are transferred from the bulk to the surface at constant temperature and pressure, the change in the Gibbs free energy is given by:

$$dG_{T,P} = \gamma dA - \sum_i \mu_i dn_i \qquad (10.30)$$

If the bulk concentration decreases by transferring dn_i moles to the freshly created surface, there is a simultaneous increase in the number of moles of the ith component of the surface by the same amount.

$$-dn_i = C_i^s dA \qquad (10.31)$$

where C_i^s is the surface concentration in moles / cm^2.

Substituting Equations 10.31 and 10.11 into 10.30:

$$\left(G^{s}dA\right)_{T,P} = \gamma dA + \sum_{i}\mu_{i}C_{i}^{s}dA \qquad (10.32)$$

$$\text{or } \gamma = G^{s} - \sum_{i}\mu_{i}C_{i}^{s} \qquad (10.33)$$

and differentiating:

$$d\gamma = dG^{s} - \sum_{i}\mu_{i}dC_{i}^{s} - \sum_{i}C_{i}^{s}d\mu_{i} \qquad (10.34)$$

The change in the excess free energy for a constant unit surface area is similar to Equation 10.28 and is given by:

$$dG^{s} = -S + \sum_{i}\mu_{i}dC_{i}^{s} \qquad (10.35)$$

Substitution of this into the previous equation gives the Gibbs equation [S3]:

$$d\gamma = -S^{s}\sum_{i}C_{i}^{s}d\mu_{i} \qquad (10.36)$$

10.5   SURFACE TENSIONS OF BINARY MIXTURES

We now consider ideal binary systems as examples that allow appreciation of how or why a surface may become enriched with a particular component over its equilibrium bulk concentration. In the case of metals, there are numerous examples of significant surface enrichment over that of the bulk compositions. Two such examples occur for Cu-Ni [S4] and Ag-Au alloys [S5].

Controlled experiments were carried out using annealed standard alloys and cleaning the alloy surfaces by "ion sputtering". In each instance, substantial surface enrichment occurred for the component having the lower surface tension. Thus Cu enriched the surface of the Cu-Ni alloy and Ag enriched the surface of the Ag-Au alloy, consistent with the surface tensions determined at their melting points.

The Gibbs equation may be expressed as:

$$d\gamma = -C_1^s d\mu_1 - C_2^s d\mu_2 \tag{10.37}$$

For a one component system, the surface tension, γ, takes the form:

$$g = g_1 + \frac{RT}{A_1} \ln \frac{X_1^s}{X_1^b} \tag{10.38}$$

where:

γ_1 = surface tension of component 1

X_1^s = surface molar concentration of component 1

X_1^b = bulk molar concentration of component 1

A_1 = area(cm^2 / mole) of component

For a second component, we have:

$$\gamma = \gamma_2 + \frac{RT}{A_2} \ln \frac{X_2^s}{X_2^b} \tag{10.39}$$

where the nomenclature is as above but for a second component.

In the case of a two component system, assuming ideal behavior, i.e., $A_1 = A_2 = A$, we can combine Equations 10.38 and 10.39:

$$\gamma_1 + \frac{RT}{A} \ln \frac{X_1^s}{X_1^b} = \gamma_2 + \frac{RT}{A} \ln \frac{X_2^s}{X_2^b} \tag{10.40}$$

to obtain [S6]:

$$\frac{X_1^s}{X_2^s} = \frac{X_1^b}{X_2^b} \exp(\gamma_2 - \gamma_1) \frac{A}{RT} \tag{10.41}$$

Equation 10.40 shows that the surface composition depends on the bulk composition and on the surface tension of the pure components. The surface will be richer in the component that has a smaller surface tension.

10.6   EFFECT OF TEMPERATURE

From the Gibbs surface free energy, $G^s = H^s - TS^s$,

(Equation 10.4), we have:

$$\left(\frac{\partial G^s}{\partial T}\right)_P = \left(\frac{\partial H^s}{\partial T}\right)_p - S\left(\frac{\partial T}{\partial T}\right)_p - T\left(\frac{\partial S^s}{\partial T}\right)_p \qquad (10.42)$$

$$= C_p^s - S^s - C_p^s \qquad (10.43)$$

$$= -S^s = \left(\frac{\partial \gamma}{\partial T}\right)_p \qquad (10.44)$$

but from Equation 10.4, $H^s = G^s + TS^s$,

$$H^s = \gamma - T\left(\frac{\partial \gamma}{\partial T}\right)_p \qquad (10.45)$$

$$\left(\frac{\partial H^s}{\partial T}\right)_p = \left(\frac{\partial G^s}{\partial T}\right)_p + T\left(\frac{\partial S^s}{\partial T}\right)_p + S^s\left(\frac{\partial T}{\partial T}\right)_p \qquad (10.46)$$

$$= -S^s - T\left(\frac{\partial^2 \gamma}{\partial T^2}\right)_p + S^s \qquad (10.47)$$

At constant pressure, P,

$$C_p^s = -T\left(\frac{\partial^2 \gamma}{\partial T^2}\right)_p \qquad (10.48)$$

$$H^s = E_{p,v}^s = \gamma - T\left(\frac{\partial \gamma}{\partial T}\right)_p \qquad (10.49)$$

Using a harmonic model [S6]:
$$E \approx 3Nk_B T \qquad (10.50)$$

where:

N = number of atoms
E = internal energy
k_B = Boltzmann's constant

$$\therefore 3Nk_BT = \gamma - T\left(\frac{\partial \gamma}{\partial T}\right)_p \tag{10.51}$$

and

$$N = \frac{\gamma}{T3k_B} - \left(\frac{\partial \gamma}{\partial T}\right)_p \Big/ 3k_B \tag{10.52}$$

A plot of $1/T$ vs. N gives a slope of $\gamma/3k_B$ and an intercept of:

$$-\left(\frac{\partial \gamma}{\partial T}\right)_p \Big/ 3k_B \tag{10.53}$$

Thus it is possible to estimate the surface tension of a component of a binary mixture that enriches the surface as a function of temperature [S6].

It should be noted from Equation 10.44 that a decrease in surface tension would increase the entropy of the surface. This is consistent with the Guggenheim equation [S7], developed for small molecule liquids but which has been found to apply to polymers.

$$\gamma = \gamma_o\left(1 - \frac{T}{T_c}\right)^{11/9} \tag{10.54}$$

where:

γ_o = surface tension at $T = 0°$ K
T_c = critical temperature

and

$$-\frac{d\gamma}{dT} = \frac{11}{9}\frac{\gamma_o}{T_c}\left(1 - \frac{T}{T_c}\right)^{2/9} \tag{10.55}$$

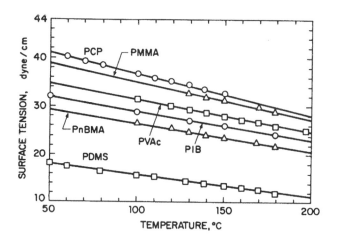

Figure 10.3 [S8]
Variation of Surface Tension of Example Polymer Melts and Liquids as a
Function of Temperature

The effect of temperature is illustrated in Figure 10.3 [S8].

Surface tensions of small molecule liquids vary linearly with temperature and $-d\gamma/dT$ exhibits a change of about 0.1 dyne/cm at temperatures below the critical temperature. Similarly, surface tensions of polymers also vary linearly with temperature, with $-d\gamma/dT$ being about 0.05 dynes/cm. As we have seen above, the $-d\gamma/dT$ for a polymer is the surface entropy, the smaller change for polymers is attributed to conformational restrictions of long chain molecules.

10.7   EFFECT OF CHEMICAL COMPOSITION

For small molecule liquids, Sugden [S9] showed that γ is determined by chemical constitution alone and is given by:

$$\gamma = \left(\frac{P}{M}\rho\right)^n \qquad (10.56)$$

where:

γ = surface tension

P = the parachor

M = molecular weight of a repeat unit
ρ = density

The parachor was introduce by Sugden [S9] who gave a list of atomic and group contributions. Later the atomic and group contributions were slightly modified and improved by Mumford and Phillips [S10] and Quayle [S11]. This relationship can predict the surface tension of polymers within 1-2 dynes/cm over wide temperature ranges in most cases.

10.8 EFFECT OF MOLECULAR WEIGHT

There are no general relationships that hold for a wide variety of polymers. Plots of γ versus M^{-1} show various degrees of curvature [S12-S15] although this is the only general relationship. This is illustrated in Figure 10.4.

For a homologous series, the surface tension varies linearly with $M_n^{-2/3}$ [S16, S17]. The surface tension is given by Equation 10.57.

$$\gamma = \gamma_\infty - \frac{k_e}{M_n^{2/3}} \qquad (10.57)$$

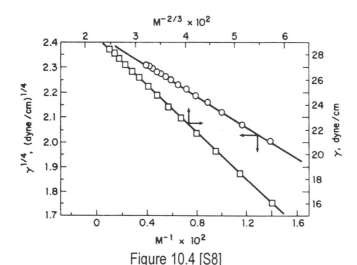

Figure 10.4 [S8]
Variation of Surface Tension as a Function of Molecular Weight
for n-Alkanes [S15]

where:

γ_∞ = surface tension at infinite molecular weight

k_e = constant

M_n = number average molecular weight

10.9   EFFECTS OF PHASE TRANSITIONS

The temperature coefficient of surface tension in the glassy region is related to that in the rubbery region by:

$$\left(\frac{d\gamma}{dT}\right)_g = \frac{\alpha_g}{\rho_r}\left(\frac{d\gamma}{dT}\right)_r \qquad (10.58)$$

where:

α = isobaric thermal expansion coefficient

$$\alpha = \left(\frac{1}{V}\right)\left(\frac{dV}{dT}\right)_p$$

Thus a plot of γ versus T is continuous but changes slope at T_g as exhibited by α, as we have seen earlier (cf. Chapters 2 and 8).

At the crystal-melt transition, the surface tension of the crystalline phase γ_c is related to that of the amorphous phase, γ_a by:

$$\gamma_c = \left(\frac{\rho_c}{\rho_a}\right)n_\gamma a \qquad (10.59)$$

where:

ρ_c = crystalline density

ρ_a = amorphous density

n_γ = constant

a = constant

Thus at the crystal melt transition, the surface tension changes discontinuously since the density is discontinuous. As ρ_c

is greater than ρ_a, the crystalline phase will have a higher surface tension than the amorphous phase.

10.10 ESTIMATION OF SURFACE ENERGY OF LIQUIDS AND POLYMERS

There are a number of experimental methods for determining the surface tension of liquids. The most popular of the direct methods is the measurement of the force required to pull a metal disk or ring out of a liquid. This is done while controlling the rate at which the ring is being pulled from the liquid so as to control reproducibility and validity of results comparison.

For polymers, other techniques must be used. One such technique is the "pendant drop" method whereby the shape characteristics of a molten drop of polymer are measured which allows calculation of the surface tension of the material.

Surface tensions of solid polymers may also be obtained by extrapolation of melt data (γ vs. *T*) or by extrapolation of liquid homolog data (γ vs. *M* relation). When extrapolating from melt data, the effects of primary and secondary transitions should be considered.

The surface tensions of solid polymers may also be determined from wettability data. This method uses the contact angles of a series of testing liquids to obtain a spectrum of critical surface tensions, $\gamma_{c,\phi}$ by using a proposed equation of state.

$$\cos\theta = 2\left(\frac{\gamma_{c,\phi}}{\gamma_{LV}}\right)^{1/2} - 1 \qquad (10.60)$$

The $\gamma_{c,\phi}$ is then plotted against γ_{LV} to obtain a curve (the equation of state plot), whose maximum value $\gamma_{c,\phi}^{max}$ equals the surface tension of the solid, γ_s. This is illustrated in Figure 10.5.

The equation of state defines a spectrum of critical surface tensions for a given solid when a series of testing liquids are used. Rearranging Equation 10.60 gives:

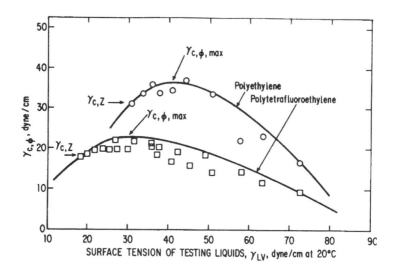

Figure 10.5 [S8]
Equation of State Plots for Poly(ethylene) and Poly(tetrafluoroethylene)

$$\gamma_{c,\phi} = 1/4 (1 + \cos\theta)^2 \gamma_{LV} \qquad (10.61)$$

This relationship can be used to calculate $\gamma_{c,\phi}$ using the θ of any testing liquid.

A similar approach uses this concept first proposed by Fox and Zisman [S18-S20]. When homologous liquids are used as the testing liquids, a straight line is often obtained. It can be seen from Equation 10.61 that at 90°:

$$\gamma_{c,\phi} = \gamma_{LV} \qquad (10.62)$$

Therefore, the extrapolated intercept at 90° is the critical surface tension γ_c. This is illustrated in Figure 10.6.

10.11 ESTIMATION OF SURFACE TENSION OF LIQUIDS FROM RELATED PROPERTIES

Since surface tension is a result of intermolecular forces, it may be expected to alter properties resulting from these forces.

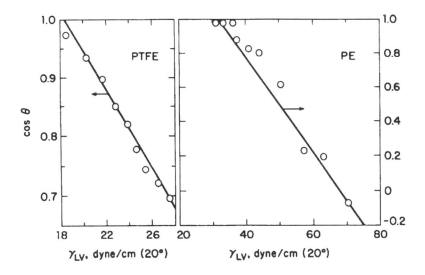

Figure 10.6 [S8]
Zisman Plots for Poly(tetrafluorethylene) using n-Alkanes as the Testing
Liquids, and for Poly(ethylene) Using Various Testing Liquids

This is indeed the case and internal pressure, compressibility, and cohesion energy density are correlated with surface tension.

Similarly, McGowan [S21] showed the correlation with compressibility κ, to be:

$$\gamma^{3/2} = \frac{1.33 \times 10^{-8} \, (\text{cgs units})}{\kappa} \quad (10.63)$$

Hildebrand and Scott [21] showed there was a relationship between solubility parameter and surface tension.

$$\gamma = \left(\frac{\delta}{4.1}\right)^{3.326} V^{1/3} \quad (10.64)$$

where:

δ = solubility parameter
V = molar volume /structural unit

10.12 MOLCALC ESTIMATIONS OF SOLUBILITY PARAMETERS AND COMPARISON TO LITERATURE VALUES FOR A SERIES OF REPRESENTATIVE POLYMERS

The quantity Molcalc uses for estimating surface tensions is known as the molar parachor. It is based on the additive quantity:

$$P_s = \gamma^{1/4} \frac{M}{\rho} = \gamma^{1/4} V \tag{10.65}$$

where:

P_s = molar parachor
M = molecular weight of repeat structural unit
V = volume of the repeat structural unit

and the surface tension is expressed as:

$$\gamma = \left(\frac{P_s}{V} \right)^4 \tag{10.66}$$

Values for group parachors are available generally only for atomic groups, double and triple bonds, and three to seven membered rings. Surface tensions obtained from the group contributions in Molcalc have been constructed using the additivity of the parachor values of the individual substituents. The final estimation is, of course, based on Equation 10.66 and is by:

$$\gamma = \sum_{i}^{\infty} \left(\frac{P_s^i}{V_i} + \frac{P_s^j}{V_j} + ... \right)^4 \tag{10.67}$$

We prefer to use the contributions of the individual parachor values to estimate surface tension as opposed to the estimation of surface tensions based on solubility parameters since this would provide an estimate based upon an estimated value rather than an experimentally determined value.

Table 10.1 provides a listing of comparisons between calculated and literature values for the selected list of polymers.

Table 10.1
Comparison of Estimated Polymer Surface Tensions with Literature Values

Polymer	Surface Tension, γ (ergs/cm^2) on V_g	
	Calculated	Literature
Poly(ethylene)	36.66	35.7(1),31(3)
Poly(propylene)	31.98	29.61(1), 32(3)
Poly(isobutylene)	27.30	33.6(1), 27(3)
Poly(vinyl chloride)	42.19	41.5(2)
Poly(vinyl fluoride)	32.75	36.7(2),28(3)
Poly(chloro trifluoro ethylene)	27.10	30.8(1), 31(3)
Poly(vinyl acetate)	40.20	36.5(1), 37(3)
Poly(*cis* butadiene)	35.39	32(3)
Poly(*cis* 1,4-isoprene)	36.03	31(3)
Poly(styrene)	43.72	40.7(1) 42(2), 34.5(3)
Poly(αmethyl styrene)	35.10	----
Polycarbonate	44.36	45.00(3)
Poly(ethylene terephthalate)	42.49	44.6(1), 43(2), 41.5(3)
Poly(butylene terephthalate)	41.38	----
Poly(3,3-bis chloromethyl oxacyclobutane)	39.01	----
Poly(methylene oxide)	37.57	45(1), 38(3)
Poly(vinyl methyl ether)	33.10	29(1)
Poly(phenylene oxide)	23.15	----
Poly(hexamethylene adipamide)	40.66	46.5(1), 39.3(2), 44(3)
Poly(acrylamide)	38.36	35/40
Poly(acrylonitrile)	49.56	50(3)
Poly(methyl methacrylate)	42.36	41.1(1), 40.2(2), 39(3)
Poly(cyclohexyl methacrylate)	43.06	----

(1)-extrapolation of γ_{melt}

(2)-from contact angle measurement

(3)-from γ_{cr} measurement

10.13 ABBREVIATIONS AND NOTATION

A	surface area
A^s	surface work or energy per unit area
a	constant
C_i^s	surface concentration
C_p^s	surface heat capacity
E	total energy of a body (internal energy)
E^o	energy per atom of a material
E^s	specific surface energy (energy per unit area)
G	total Gibbs free energy of a system
G^o	Gibbs free energy per atom of a material
G^s	surface free energy per unit area
$\overline{G}_i$	partial molal free energy of component i
H^s	specific surface enthalpy
k_B	Boltzmann's constant
k_e	constant
M	molecular weight of a polymer repeat unit
M_n	number average molecular weight
N	number of atoms contained in the material
n_γ	constant
n_i^σ	number of surface atoms of specie i
P	Parachor
P_s	molar Parachor
S^o	entropy per atom of a material
S^s	specific surface entropy
T_c	critical temperature
W_a	work of adhesion
W_{cp}	work of cohesion for phase p
$X_{1\,or\,2}^s$	surface molar concentration of component 1 or 2

$X^b_{1 \text{ or } 2}$	bulk molar concentration of component 1 or 2
α	isobaric thermal expansion coefficient
δW^s	reversible surface work, not independent of the reaction path (not a total differential)
γ_α	surface tension of phase α
γ_β	surface tension of phase β
$\gamma_{\alpha\beta}$	interfacial tension between phases α and β
$\gamma_{c,\phi}$	critical surface tension of a solid
γ_σ	surface tension
γ_{LV}	surface tension of liquid in equilibrium with its saturated vapor
γ_{SV}	surface tension of the solid in equilibrium with saturated vapor of the liquid
γ_{SL}	interfacial tension between solid and liquid
π_e	equilibrium spreading pressure
γ_s	surface tension of a solid material
γ_p	surface tension of phase p
$\gamma_{1 \text{ or } 2}$	surface tension of component 1 or 2
μ_i	surface chemical potential
γ_0	surface tension at T=0°K
ρ	density
γ_∞	surface tension at infinite molecular weight
ρ_c	crystalline density
ρ_a	amorphous density

10.14 SPECIFIC REFERENCES

1. Young, T., *Proc. Roy. Soc.(Lond.)*, December 1804, Collected Works, D. Peacock, ed.: *Royal Society of London*, London; also, *Phil. Trans.*, **95**, **65**, 84(1805).

2. Dupré, A., *Théorie mécanique de la chaleur*, Paris, 1869, p.368.

3. Gibbs, J.W., *The Collected Works of J.W. Gibbs*, Longmans, Green and Co. Ltd., London, 1931.

4. Helms, C.R., *J. of Catalysis*, **36**, 114(1975).

5. Nelson, G.C., *Surface Science*, **59**, 310(1976).

6. David, D.J., *Doctoral Dissertation*, "High Temperature Thermodynamics and Kinetics of The Reaction of Oxygen With T-11(Ta-8W-2Hf)", University of Dayton, 1981.

7. Guggenheim, E.A., *J. Chem. Phys.*, **13**, 253(1945).

8. Wu, S., *Polymer Interface and Adhesion*, Marcel Dekker, Inc., New York, 1982.

9. Sugden, S.J., *J. Chem. Soc.*, **125**, 32(1924).

10. Mumford, S.A., and Phillips, J.W.C., *J. Chem. Soc.*, **130**, 2112(1929).

11. Quayle, O.R., *Chem. Revs.*, **53**, 439(1953).

12. Phillips, M.C., and Riddiford, A.C., *J. Colloid Interface Sci.*, **22**, 149(1966).

13. Dettre, R.H. and Johnson, R.E., *J. Phys. Chem.*, **71**, 1529(1967).

14. Edwards, H., *J. Appl. Polym. Sci.*, **12**, 2213(1968)

15. Wu, S., in *Polymer Blends*, Vol. 1, D.R. Paul and S. Newman, eds., Academic Press, New York, 1978.

16. LeGrand., D.G., and Gaines, Jr., G.L., *J. Colloid. Interface Sci.*, **31**, 162(1969).

17. LeGrand., D.G., and Gaines, Jr., G.L., *J. Colloid. Interface Sci.*, **42**, 181(1973).

18. Fox, H.W. and Zisman, W.A., *J. Colloid Sci.*, **5**, 514(1950).

19. Fox, H.W. and Zisman, W.A., *J. Colloid Sci.*, **7**, 109(1952).

20. Fox, H.W. and Zisman, W.A., *J. Colloid Sci.*, **7**, 428(1952).

21. McGowan, J.C., *Polymer*, **8**, 57(1967).

22. Hildebrand, J.H. and Scott, R.L. "The Solubility of Non-Electrolytes", Van Nostrand, Princeton, N. J., 1950.

10.15 GENERAL REFERENCES

1. Van Krevelen, D.W. and Hoftyzer, P.J., *Properties of Polymers*, 2nd Ed. Elsevier, New York, N.Y., 1976.
2. Van Krevelen, D.W., *Properties of Polymers*, Elsevier, New York, N.Y., 1990.
3. Wu, S., Polymer Interface and Adhesion, Marcel Dekker, Inc., New York, N.Y., 1982.

CHAPTER 11

Solution Property Relationships

11.1 INTRODUCTION

Solutions can be prepared from metals, metal alloys, and polymers using appropriate solvents. Ceramic materials generally cannot be solubilized using any of the common solvents and must first be fused at high temperatures with alkali salts in order to form solutions.

Soluble salts, metals, metal alloys, and polymers, form dilute solutions on the molecular level with appropriate solvents but the distinguishing difference between polymers and the other materials is that the size of the polymer molecule is exceedingly large compared to simple atomic species. As such, the properties of very dilute polymer solutions are determined by the conformational state of the separate polymer molecules. However, the same fundamental thermodynamic considerations that apply to polymer solutions also apply to metal solid solutions (alloys) and will be treated in greater detail in Chapter 21.

11.2 TYPES OF SOLUTIONS

11.2.1 Basic Considerations

Before proceeding to the details of a dilute polymer solution, we consider briefly the conditions governing the formation of a solution in binary systems whose components are of comparable molecular size.

331

The partial molal free energies $\overline{G}_i$, are customarily referred to as the chemical potentials. The activity of species i, denoted by a_i, is defined by:

$$\overline{\Delta G_i} \equiv \overline{G}_i - \overline{G}_i^o = RT \ln a_i \tag{11.1}$$

The above equation is consistent with the previous thermodynamic considerations (cf. Chapters 2 and 10):

$$\overline{\Delta G_i} = \overline{\Delta H_i} - T\overline{\Delta S_i} \tag{11.2}$$

where:

$$\overline{\Delta H_i} \equiv \overline{H}_i - \overline{H}_i^o$$
$$\overline{\Delta S_i} \equiv \overline{S}_i - \overline{S}_i^o$$
$$R \quad = \text{gas constant}$$

The superscript " o " refers to a standard state.

The free energy of mixing of a binary system is related to the chemical potential of the components and is given by:

$$\Delta G_m = n_1 \overline{\Delta G_1} + n_2 \overline{\Delta G_2} \tag{11.3}$$

This is represented graphically in Figure 11.1.

The free energy of mixing for a two component miscible system is shown in curve (a) in Figure 11.1. This curve has no inflection points. For systems consisting of two coexisting phases, a higher free energy of mixing will result when a single phase forms as a result of the two phases merging. This is illustrated in curve (b) of Figure 11.1 which shows two points that have a common tangent displayed for the change in the free energy of mixing curve. In this instance the two phases, x_1' and x_1'', have identical values for $\overline{\Delta G_1}$ and $\overline{\Delta G_2}$ so that they are in equilibrium with each other. Any other phase state between x_1' and x_1'', corresponds to a state of higher free energy than the corresponding point representing a two phase system of the same composition and will therefore be thermodynamically unstable.

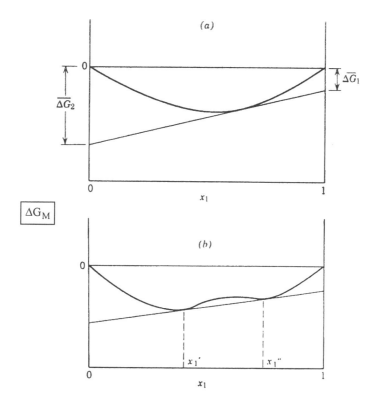

Figure 11.1
Representation of the Free Energy of Mixing

a) System of two components miscible in all proportions
b) System of two components with limited miscibility

11.2.2 Ideal Solutions

Ideal solutions are solutions in which the activity is equal to the mole fraction for all components and all compositions of the system. This is given by:

$$a_i = x_i \tag{11.4}$$

The mixing of molecules of different sizes may lead to deviations from ideal solution behavior even if no heat accompanies dissolution.

The entropy of the pure components is the sum of contributions due to molecular translation, rotation, and vibration. When components are mixed to form an ideal solution, these entropy contributions may be considered to remain unchanged.

For a binary system, for the ideal entropy of mixing:

$$\Delta S_m^i = -R\left[n_1 \ln x_1 + n_2 \ln x_2\right] \qquad (11.5)$$

where n_1 and n_2 are the moles of n_1 and n_2 respectively, and x_1 and x_2 are the respective molar concentrations.

Since the solution is athermal $\left(\overline{\Delta H_1} = \overline{\Delta H_2} = 0\right)$, the ideal free energy of mixing is given by:

$$\overline{\Delta G_m^i} = -T\Delta S_m^i = RT\left(n_1 \ln x_1 + n_2 \ln x_2\right) \qquad (11.6)$$

11.2.3 Regular Solutions

Hildebrand [S1] introduced the concept of a regular solution as one in which the partial molal entropies of the components are those to be expected from the ideal solution law. Therefore, deviations from ideal behavior must be accounted for by the heat of mixing as a result of differences in the forces between similar and dissimilar molecules in a solution. This will cause the mixing process to be accompanied by the absorption or evolution of heat and will cause a deviation from the random distribution of interacting molecules. In other words, ΔS_m^i will be lower than those predicted for an ideal solution.

In the case of polymers, their molecular size difference from that of a solvent may lead to large deviations from solution ideality even if no heat accompanies formation of the solution. Therefore, Hildebrand [S2] later defined a regular solution as "one in which thermal agitation is sufficient to give practically complete randomness". This means the entropy of mixing of a regular solution need not be ideal. This allows the entropy of mixing and heat of solution to be considered separately.

In regular solutions, the free energy of mixing may be treated as being the additive result arising from contributions due to configurational probability and a free energy arising from nearest neighbor interactions. The latter are characterized by the "Flory-Huggins interaction parameter", χ, which specifies in units of RT, the excess free energy for the transfer of a mole of solvent molecules from the pure solvent to the pure polymer phase. This leads to:

$$\Delta G_m = RT(n_1 \ln \phi_1 + n_2 \ln \phi_2 + n_1 \chi \phi_2) \qquad (11.7)$$

The activity of the solvent is given by:

$$\ln a_1 = \frac{\overline{\Delta G_1}}{RT} = \ln \phi_1 + \left(1 - \frac{V_1}{V_2}\right)\phi_2 + \chi \phi_2^2 \qquad (11.8)$$

which yields on expansion of the $\ln \phi_1 = \ln(1 - \phi_2)$ term,

$$\ln a_1 = -\frac{\phi_2 V_1}{V_2} - \left(\frac{1}{2} - \chi\right)\phi_2^2 - \frac{1}{3}\phi_2^3.... \qquad (11.9)$$

11.3   THE CONFIGURATION OF POLYMER MOLECULES

A number of important interacting variables influence the viscosity of polymer solutions, among which are: molecular conformations, solvent viscosity, temperature, and shear rate.

There are two key factors which influence the conformation polymer molecules take in solution. These key factors are the intramolecular interactions of chain units and the interactions with solvent molecules, assuming that intermolecular interactions are absent. Further, these interactions are divided into two groups, short-range and long-range.

The short-range interactions depend on the bond structure and local interactions between atoms and groups which are separated by only a few bonds along the chain. They determine the degree of flexibility of the chain, and are responsible for the stability of helical conformations of a number of natural and synthetic polymers.

Long range interactions involve pairs of chain units which are separated by many bonds along the chain. The regions of space occupied by these macromolecules are not clearly defined and it is possible for the molecular coils to interpenetrate one another. Such interpenetration will be resisted because of the restriction it imposes on the number of possible chain conformations. These chains have a finite volume and as they approach one another, they cannot occupy the same volume element.

This "volume element" that cannot be simultaneously occupied is referred to as the "excluded volume". This effect will be influenced by mutual attraction of chain segments in poor solvent media, or magnified by a mutual repulsion of the polymer segments if polymer-solvent interactions are favored. A positive value of the excluded volume will lead to an expansion of the molecular coil, since the fraction of conformations which has to be excluded will decrease with an increasing separation of chain ends.

There may be cases, however, in which the use of a poor solvent causes chain segments to attract one another just enough to compensate for the effect of the physically occupied volume of the chain. In this instance, the excluded volume effect will vanish, and the chain should behave according to predictions based on mathematical volume chains of zero volume [S3], provided that the short-range restraints to chain flexibility are taken properly into account.

The elimination of binary interaction effects between segments of a single chain implies that binary interactions between solute molecules make no contribution to deviation from ideality, so that $\frac{1}{2} - \chi$ in Equation 11.9 must be zero. Solvents satisfying this condition are called Flory Θ solvents, and chain dimensions are referred to as "unperturbed". For a given polymer-solvent system, if the temperature of the system is adjusted for the attractive forces between chain units to compensate for their finite volumes so that the effective volume of the chain unit is zero, the temperature is known as the Flory Θ temperature.

11.4 AVERAGE DIMENSIONS OF MACROMOLECULES

The average size of linear polymer molecules can be characterized by the mean-square end-to-end distance, $\langle R^2 \rangle$ or $\langle h^2 \rangle$, which is equal to the square of the magnitude of the end-to-end vector averaged over all configurations.

Another measure of the average dimensions is based on the analysis of the distribution of chain units about the center of mass. The mean-square distance of chain units from the center of mass averaged over all configurations is the mean square radius of gyration $\langle S^2 \rangle$ [S3]. While the mean square end-to-end distance is defined only for linear chains, the mean square radius of gyration is also meaningful with branched and ring-shaped molecules.

11.5 DEFINITIONS OF END-TO-END DISTANCE

In the absence of any type of interaction, except for covalent bonding which fixes the length of the chain links, and assuming completely free internal rotation, a long chain molecule obeys Gaussian random flight statistics. For a polymer chain having n bonds each of fixed length, l, the mean square value of the end-to-end distance of the chain is given by:

$$\langle h^2 \rangle = nl^2 \tag{11.10}$$

and the mean-square radius of gyration by:

$$\langle S^2 \rangle = \frac{1}{6} nl^2 \tag{11.11}$$

which is applicable for high n values.

Short-range interactions resulting from interactions between groups separated by only a small number of valence bonds, result in angles of rotation about each bond that are not uniformly distributed owing to steric and energetic hindrances caused by interactions between atomic groups along the backbone chain.

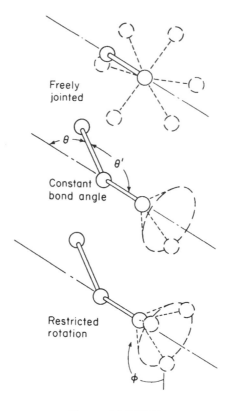

Figure 11.2 [G1]
Polymer Chain Conformations

A chain in which the bond angles are fixed but rotations around a single bond are unrestricted is the freely rotating chain model [S3-S6]. In this instance the mean square end-to-end distance is given by:

$$\left\langle h^2 \right\rangle_o = nl^2 \frac{1+\cos\theta}{1-\cos\theta} \tag{11.12}$$

$$\approx nl^2 \text{ for carbon} \tag{11.13}$$

where:

θ' = polymer backbone chain bond angles

θ = $180° - \theta'$

$$\phi \quad = \text{angle of restricted group rotation}$$

These concepts are illustrated in Figure 11.2.

For fixed bond angles and rotations restricted by short range interactions the polymer is in an "unperturbed" state and the effective end-to-end distance becomes:

$$\langle h^2 \rangle_o = \sigma_c \langle h^2 \rangle \qquad (11.14)$$

where σ_c = conformation factor. The factor σ_c represents the effect of steric or energetic hindrances on rotations around the main chain bonds. It depends on temperature, and sometimes on the solvent, but is independent of n.

In the unperturbed state, i.e., where the excluded volume is not a factor, the dimensions of the polymer molecules are $\langle R^2 \rangle_o$ and $\langle S^2 \rangle_o$. If the excluded volume is not zero, the chain dimensions are changed by expansion factors, α_R^2 and α_S^2, as given by:

$$\langle R^2 \rangle = \alpha_R^2 \langle R^2 \rangle_o \qquad (11.15)$$

and

$$\langle S^2 \rangle = \alpha_S^2 \langle S^2 \rangle_o \qquad (11.16)$$

11.6   CONFORMATIONAL MODELS

The above equations attempt to relate the size of macromolecules inherent in these concepts and definitions of sizes statistically to the dimensions of real chain coiled polymer molecules and the mutual displaceability of the chain atoms. The proportionality of $\langle R^2 \rangle_o$ and n, postulated by the random flight model, is valid also with a real chain of sufficient length l. The real chain can be represented by a freely jointed chain (equivalent random flight chain) which is composed of n' equivalent bonds (Kuhn statistically independent segments) of length l' [S7].

$$nl = n'l' = l^* \qquad (11.17)$$

where l^* is the contour length of the chain. This model defines as a statistical element not a single chain atom but a short section of the polymer chain containing several chain atoms. The length of the section is chosen so that both of its ends may be regarded as completely free joints in the chain.

Another model proposed by Porod and Kartky [S8, S9] is the "wormlike chain model". This model will not be pursued further here.

New theories have recently emerged that have provided new insights into traditional models of polymers in solution. These new theories have been proposed by Wilson and Kogut [S10] and DeGennes' [S11] "scaling concepts".

DeGennes' model allows us to better visualize how a polymer chain moves in a melt and solution and much experimental evidence bears out the concepts he proposed. DeGennes considered an ideal single, flexible, polymer single chain that is characterized by: 1) Gaussian statistics, 2) a size proportional to $p^{1/2}$ (p = number of structural units per chain), 3) a large domain of linear relation between force and elongation, and 4) a scattering law of the $1/q^2$ type.

Real chains in good solvents have the same universal features as self avoiding walks on a lattice. These features are described by exponents, γ, which is related to chain entropy and to, ν, which is related to chain size. However, a real chain has a size much larger than that of the ideal chain $\left(p^\nu (\nu \approx 3/5)\right)$ (instead of ½) and in good solvents is "swollen".

A single chain behaves like a coil whose relaxation mode is described by an elastic spring constant and a friction constant. When a coiled chain is under traction it will break up into a series of subunits or blobs of size ξ, under sufficiently large tension. The blob size or ξ can be considered equivalent to a correlation distance. Viewed at larger scales $(r)\xi)$ the conformation of the chain may be considered as a string of "blobs". This is illustrated graphically in Figure 11.3. In some respects, this is similar to previous bead models.

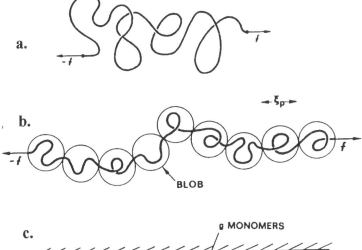

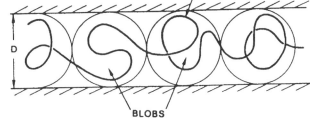

Figure 11.3 [G2]
Coiled Chains and Blobs

a) Coiled chain under traction.
b) At large tension the chain breaks up into a series of blobs.
   Successive blobs act as hard spheres.
c) A real chain is squeezed in a tube.

If a polymer chain, such as one in a melt, is confined in a tube, and the tube is squeezed, it is assumed to behave as a sequence of "blobs" of diameter equal to the tube. The surrounding polymer chains form the wall of the tube and the movements of the chain or string of "blobs" is characterized by the "reptation" model with snakelike motions back and forth characterizing its movement.

Swollen polymer gels can be visualized similarly with each "blob" associated with one chain and having properties similar to those of a single chain.

The scaling concepts are characterized by three types of spatial scales:

1. the region between the size of the structural unit and the minimum "ideal blob" size behavior
2. the region between the minimum and maximum "blob" size or non-ideal behavior which is the excluded volume type
3. the region between the maximum "blob" size and the size of the coils which characterizes ideal behavior and the region in which the "blobs" behave as a flexible chain

11.7   FACTORS INFLUENCING THE FLOW OF SOLUTIONS

When an external stress, σ, is exerted on a fluid, deformation occurs until the stress is removed. A steady state can be reached in which the rate of deformation is constant because the deformation process is retarded by internal frictional forces.

For a real fluid between two infinite parallel plates separated by a distance h' where only one is moving with a constant velocity in the x direction, a transverse velocity gradient or shear rate, $\dot{\gamma}$, exists in the fluid. The velocity, v, of the flow, as illustrated in Figure 11.4, is given by:

$$v = \left(\dot{\gamma}y;0;0\right) \qquad (11.18)$$

The higher the internal friction in the fluid, the higher is the stress which must be applied to maintain the velocity gradient. The shear stress is related to the shear rate as illustrated by the following relationship.

$$\sigma = \sigma_{xy} = \eta\left(\frac{dv_x}{dy}\right) = \eta\dot{\gamma} \qquad (11.19)$$

where:

σ = shear stress
η = viscosity

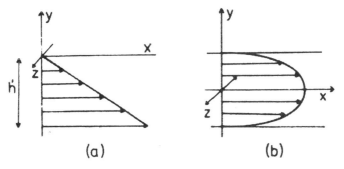

Figure 11.4
The Velocity Profile of Laminar Flow
(a) Between Parallel Plates
(b) In Capillary

For Newtonian flow, a plot of σ vs. $\dot\gamma$ results in straight line passing through the origin which means that the viscosity is directly proportional to shear rate and that the viscosity coefficient is a constant.

For flow other than Newtonian, such as pseudoplastic and dilatent, the dependence of $\dot\gamma$ on σ is linear at very low and high shear rates while it is non-linear in the intermediate portion. If the ratio $\sigma/\dot\gamma$, in the intermediate part, is a decreasing function of γ, so that the viscosity coefficient decreases with increasing $\dot\gamma$ (i.e. $d\eta/d\dot\gamma \langle 0$), pseudoplastic flow is indicated. If $d\eta/d\dot\gamma \rangle 0$, the flow is dilatent. Dilatent behavior is exhibited mostly by particulate dispersions such as latexes, slurries, and concentrated suspensions. In slurries and concentrated suspensions there may be almost no flow, $\dot\gamma = 0$, until σ reaches some yield value. This is known as "Bingham plastic" behavior of which ordinary toothpaste is a prime example. This and other types of flow behavior are illustrated in Figure 11.5.

Polymer solutions often exhibit time-dependent rheological properties. The two main categories exhibited are thixotopic and rheopectic behavior. In the former case, the viscosity diminishes with time, in the latter, it increases.

The frictional force acting on the moving plate of area A in Figure 11.4 (a) is:

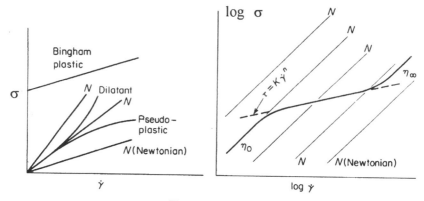

Figure 11.5 [G1]
Stress vs. Rate of Shear for Newtonian and Non-Newtonian Fluids

$$F \equiv \sigma A = \eta \dot{\gamma} A \qquad (11.20)$$

and the velocity of the displacement of the plate is:

$$v = \dot{\gamma} h' \qquad (11.21)$$

The work dissipated per unit time and unit volume is:

$$W \equiv Fv/h'A = \eta \dot{\gamma}^2 \qquad (11.22)$$

This shows that the increase in viscosity of polymer solutions is due to a greater amount of work to overcome the frictional forces exerted by the dissolved molecules in solution.

A spherical particle of radius r_a, in a flowing fluid, will accommodate the forces exerted by the flowing solution. The center of the particle mass will assume the translational speed of the surrounding fluid while other parts of particle exposed to different layers of the fluid which have different velocities will cause the particle to attempt to follow these velocities. These forces give rise to a torque which rotates the particle. The angular velocity at which the total torque is zero becomes:

$$\Omega = 1/2 \cdot \dot{\gamma} \qquad (11.23)$$

The velocity at different points on the surface, i.e., A, B, C, and D in Figure 11.6 differ in direction but not in absolute value.

$$u = \Omega r_a = 1/2 \cdot \dot{\gamma} r_a \qquad (11.24)$$

11.8 MACROMOLECULE RELATIONSHIPS IN DILUTE SOLUTION

11.8.1 Extended Linear Macromolecules

For a fully extended polymer chain, the end-to-end distance is:

$$h_{max} = \frac{n}{\cos\theta} \qquad (11.25)$$

For a polyethylene chain, $\theta = 70°$ and the end-to-end distance becomes:

$$h_{max} \approx 0.83nl = 0.83l \left(\frac{M_w}{M_{su}} \right) Z \qquad (11.26)$$

where:

M_w = molecular mass
M_{su} = molecular mass/structural unit
Z = number of backbone atoms/structural unit
l = bond length

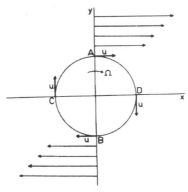

Figure 11.6 [G4]
A Spherical Particle in Laminar Flow

11.8.2 Unperturbed Random Coil Macromolecules

For unperturbed macromolecules, the skeletal factor s or conformation factor, σ_c, has a value between 4 and 16 for most polymers. Equation 11.26 becomes:

$$\langle h^2 \rangle_o^{1/2} = s^{1/2} l \left(\frac{M_w}{M_{su}} \right)^{1/2} Z^{1/2} \tag{11.27}$$

For polymethylene, $s \approx 6.5$. The maximum extension ratio, Λ_{max} of an isolated unperturbed polymethylene molecule becomes:

$$\Lambda_{max} = \frac{h_{max}}{\langle h^2 \rangle_o^{1/2}} \approx \frac{0.83}{6.5^{1/2}} \left(\frac{M_w}{M_{su}} \right)^{1/2} Z^{1/2} \tag{11.28}$$

$$\approx 0.33 \left(\frac{M_w}{M_{su}} \right)^{1/2} Z^{1/2} \tag{11.29}$$

Thus one can obtain an appreciation of the extent to which a chain can be extended. For example, a polymer coil with $Z = 2$ and a $D.P. \left(\dfrac{M_w}{M_{su}} \right)$ of 10^4, can be extended 50-fold.

From Equation 11.27, the real polymer coil dimensions, in solution, become:

$$\langle R^2 \rangle^{1/2} = \alpha_R \langle R^2 \rangle_o^{1/2} = \alpha_R \langle h^2 \rangle_o^{1/2} \tag{11.30}$$

$$= \alpha_R s^{1/2} l \left(\frac{M_w}{M_{su}} \right)^{1/2} Z^{1/2} \tag{11.31}$$

The expansion factor α varies generally from about 1 to 2 depending on the $D.P.$ An empirical approximation is:

$$\langle h^2 \rangle^{1/2} = K_h M_w^b \qquad (11.32)$$

11.9 STEADY STATE VISCOSITY OF POLYMER SOLUTIONS

Interpretation of the flow properties of polymer solutions is based on:

- The solvent is incompressible and its viscosity is Newtonian.
- The macromolecules behave in a manner similar to corresponding models.
- The polymer chains are the centers of the resistance to flow with intramolecular and intermolecular hydrodynamic interactions.
- In an external force field, rigid molecules undergo translational and rotational motions whereas flexible molecules also undergo deformation.

11.10 EXPRESSSIONS FOR THE VISCOSITY OF POLYMER SOLUTIONS

From the previous discussions, we can expect that the viscosity of a dilute polymer solution depends on the nature of the polymer, the concentration of the polymer, the nature of the solvent, the molecular mass of the polymer and its distribution (M_w / M_n), the temperature, and the rate of deformation. For dilute solution viscosities and viscosity measurements, the rate of deformation is low and this factor can therefore be neglected.

There are a number of ways in which to define the viscosity of a polymer in dilute solution. The various terms defining solution viscosities will be found in the literature and are widely used in academia and industry for control as well as characterization of solutions. These solution viscosity terms have important inter-relationships and are defined as follows:

1. Relative Viscosity (Poise)

$$\eta_{rel} = \frac{\eta}{\eta_s} \approx \frac{t}{t_s} \tag{11.33}$$

Relative viscosity is simply a dimensionless ratio of the solution viscosity of a polymer in a solvent to that of the solvent. In an actual measurement these viscosity values are approximated by the flow times of each fluid, i.e., t and t_s, respectively.

2. Specific Viscosity

$$\eta_{sp} = \frac{\eta - \eta_s}{\eta_s} = \eta_{rel} - 1 \approx \frac{t - t_s}{t_s} \tag{11.34}$$

3. Reduced Viscosity (Deciliters/g)

$$\eta_{red} = \frac{\eta_{sp}}{c} \tag{11.35}$$

The reduced viscosity accounts for the effects of concentration.

4. Inherent Viscosity (Deciliters/g)

$$\eta_{inh} \frac{\ln \eta_{rel}}{c} \tag{11.36}$$

5. Intrinsic Viscosity (Deciliters/g or in S.I. Units, cm^3/g

$$[\eta] = \left(\frac{\eta_{sp}}{c}\right)_{c \longrightarrow 0} \tag{11.37}$$

This is the most important viscosity term since it gives a basic viscosity characteristic of the polymer. It is obtained by making measurements of η_{red} at a number of dilute concentrations and extrapolating these values to obtain the viscosity, $[\eta]$ at infinite dilution, i.e., at zero concentration. The dimension of $[\eta]$ is volume per unit mass so that this quantity is referred to as the "effective hydrodynamic specific volume" of the polymer in solution.

11.10.1 Inter-relationships of Intrinsic Viscosity

The dependence of the intrinsic viscosity $[\eta]$ to the molar-mass was first proposed by Mark [S12] and Houwink [S13].

$$[\eta] = KM_w^\nu \tag{11.38}$$

where:

K and ν = empirical parameters and they depend upon the polymer-solvent pair and on temperature.

The parameters K and ν are evaluated from the intercept and slope of the log plot of $[\eta]$ vs. M_w determined for a series of samples that differ only in their molar masses. Very low polydispersity and a broad range of molecular masses are required for reliable values of K and ν to be obtained [S14, S15].

For polydispersive polymers, M_w is replaced by $\overline{M}_v$, the so-called viscosity average molecular mass so that we obtain the relationship known as the Mark-Houwink equation:

$$[\eta] = K\overline{M}_v^a \tag{11.39}$$

Under theta conditions (unperturbed random coil):

$$[\eta]_\theta = K_\theta \overline{M}_v^{1/2} \tag{11.40}$$

A theoretical approach relating the hydrodynamic expansion factor, α_h, to viscosity leads to:

$$[\eta] = \alpha_h^3 [\eta]_\theta = \alpha_h^3 K_\theta \overline{M}_v^{1/2} \tag{11.41}$$

For rod-like molecules "a" in Equation 11.39 approaches the value of 2.

With flexible chain polymers, the exponent v, in Equation 11.38, assumes values between 0.5 and 0.8, higher values corresponding to good solvents. With stiff-chain polymers, the exponent is higher and less dependent on the solvent.

A simple relationship between α_h and M has been proposed by Stockmayer and Fixman (16)

$$\alpha_h^3 = \frac{1 + BM^{1/2}}{K_\theta} \tag{11.42}$$

in which B is an interaction parameter. Combining Equations 11.41 and 11.42 gives:

$$[\eta] = K_\theta M^{1/2} + BM \tag{11.43}$$

which is the Stockmayer-Fixman equation which is often used for the determination of K_θ from viscosity measurements on an arbitrary polymer-solvent system.

11.11 MOLCALC ESTIMATIONS OF THE UNPERTURBED VISCOSITY COEFFICIENT, K_θ, AND COMPARISON TO LITERATURE VALUES FOR A SERIES OF REPRESENTATIVE POLYMERS

The quantity Molcalc uses for estimating the unperturbed viscosity coefficient, K_θ, is the molar intrinsic viscosity function based on the estimation method of Van Krevelen and Hoftyzer and is given by:

$$J = K_\theta^{1/2} M - 4.2Z \tag{11.44}$$

where:

$$J = \sum_i n_i J_i = g^{1/4} cm^{3/2} / mol^{3/4}$$

Z = number of backbone atoms per structural unit
M = molar mass per structural unit

$$K_\theta \quad = cm^3 mol^{1/2}/g^{3/2}$$

The critical molecular weight, M_{cr}, the molecular mass above which molecular entanglements play a role in the flow of a molten polymer is calculated from the relationship:

$$K_\theta M_{cr}^{1/2} \approx 13 (cm^3/g) \tag{11.45}$$

The influence and importance of molecular entanglements on flow properties of polymer melts will be treated in detail in Chapter 17.

Once K_θ is known or has been estimated, $[\eta]_\theta$ can be evaluated as follows:

$$[\eta]_\theta = \Phi_o \frac{\langle h^2 \rangle_o^{3/2}}{M} = K_\theta M^{1/2} \tag{11.46}$$

where:

$\Phi_o \quad$ = universal constant

$\approx 2.5 \times 10^{23} mol^{-1}$

The intrinsic viscosity can also be estimated from the Stockmayer-Fixman equation, Equation 11.43 for any given molecular weight, $[\eta] = K_\theta^{1/2} + BM$, providing the interaction parameter B has been determined independently.

Van Krevelen gives additional relationships which allow the estimations of other quantities, including the intrinsic viscosity.

Table 11.1 provides comparisons between calculated and literature values for the selected list of polymers.

Table 11.1

Comparison of Estimated Polymer Unperturbed Viscosity Coefficients
with Literature Values

Polymer	Unperturbed Viscosity Coefficient, K_q (cm^{3}xmol$^{1/2}$/g$^{3/2}$)	
	Calculated	Literature
Poly(ethylene)	0.218	0.20-0.26
Poly(propylene)	0.135	0.12-0.18
Poly(isobutylene)	0.113	0.085-0.115
Poly(vinyl chloride)	0.149	----
Poly(vinyl fluoride)	0.275	0.095-0.335
Poly(chloro trifluoro ethylene)	0.041	----
Poly(vinyl acetate)	0.081	0.077-0.103
Poly(cis butadiene)	0.165	0.13-0.185
Poly(cis 1,4-isoprene)	0.128	0.11-0.15
Poly(styrene)	0.070	0.07-0.09
Poly(amethyl styrene)	0.069	0.014-0.084
Polycarbonate	0.218	0.16-0.28
Poly(ethylene terephthalate)	0.178	0.15-0.20
Poly(butylene terephthalate)	0.183	----
Poly(3,3-bis chloromethyl oxacyclobutane)	0.120	----
Poly(methylene oxide)	0.131	0.132-0.38
Poly(vinyl methyl ether)	0.072	----
Poly(phenylene oxide)	0.055	----
Poly(hexamethylene adipamide)	0.226	0.19-0.23
Poly(acrylamide)	0.241	----
Poly(acrylonitrile)	0.257	0.20-0.25
Poly(methyl methacrylate)	0.069	0.043-0.090
Poly(cyclohexyl methacrylate)	0.045	----

11.12 ABBREVIATIONS AND NOTATION

A	area
a_i	activity of species i

B	an interaction parameter
F	frictional force
$\overline{G}_i$	partial molal free energy
G_m	Gibbs free energy of mixing
$\overline{H}_i$	partial molal enthalpy of species i
$\langle h^2 \rangle$	mean-square end-to-end distance
$\langle h^2 \rangle_o$	unperturbed mean-square end-to-end distance
h_{max}	end-to-end distance for a fully extended chain
J	molar intrinsic viscosity function
K_θ	unperturbed viscosity coefficient
K	empirical parameter
l	length of a polymer chain bond
M_w	molecular mass (weight)
M_{su}	molecular mass/structural unit
M_{cr}	critical molecular weight
n_i	number of moles
p	number of structural units per chain
$\langle R^2 \rangle$	mean-square end-to-end distance
$\langle R^2 \rangle_o$	unperturbed mean-square end-to-end distance
S	partial molal entropy of species i
$\langle S^2 \rangle$	mean square radius of gyration
$\langle S^2 \rangle_o$	unperturbed mean-square radius of gyration
$u^{\cdot}$	velocity at a specific point
v	velocity of flow
V_i	molar volume of species i
x_i	mole fraction
W	work
Z	number of backbone atoms/structural unit
α_R^2	expansion factor related to end-to-end

	distance
α_s^2	expansion factor related to radius of gyration
χ	Flory-Huggins interaction parameter
$\dot{\gamma}$	shear rate
η	viscosity
Ω	angular velocity
η_{rel}	relative viscosity
η_{sp}	specific viscosity
η_{inh}	inherent viscosity
η_{red}	reduced viscosity
$[\eta]$	intrinsic viscosity
ν	empirical parameter
ϕ	angle of restricted group rotation
ϕ	volume fraction
σ_c	polymer chain conformation factor
σ	external stress
θ'	polymer chain backbone bond angles
θ	$180° - \theta'$
Φ	universal constant $\approx 2.5 \times 10^{23} \, \text{mol}^{-1}$
ξ	chain blob size

11.13 SPECIFIC REFERENCES

1. Hildebrand, J.H., *J. Am. Chem. Soc.*, **51**, 66(1929).
2. Hildebrand, J.H., *Discuss. Faraday Soc.*, **15**, 9(1953).
3. Flory, P.J., *Principles of Polymer Chemistry*, Cornell University Press, Ithaca, 1953.
4. Yamakawa, H., *Modern Theory of Polymer Solutions*, Harper and Row, New York, 1971.
5. Flory, P.J., *Statistical Mechanics of Chain Molecules*, Interscience, New York, 1966.

6. Bershtein, T.M., and Pritsyn, O.B., *Conformations of Macromolecules*, Interscience, New York, 1966.
7. Kuhn, W., *Kolloid-Z.*, **68**, 2(1934).
8. Porod, G., *Monatsh. Chem.*, **80**, 251(1949).
9. Kratky, O. and Porod, G., *Rec. Trav. Chim.*, **68**, 1106(1949).
10. Wilson, K.G. and Kogut, *J. Phys. Rep.*, **12C**, 75(1975).
11. DeGennes, P.G., *Scaling Concepts in Polymer Physics*, Cornell University Press, Ithaca, N.Y., 1979.
12. Mark, H., in *Der feste Körper*, R. Sänger (ed), Hirzel, Leipzig 1938.
13. Houwink, R., *J. Prakt. Chem.*, **157**, 15(1940).
14. Kurata, M. and Stockmayer, W.H., *Adv. Polym. Sci.*, **3**, 196(1963).
15. Kurata, M., Tsunashima, Y., Iwama, M., and Kamada, K., in *Polymer Handbook*, J. Brandup and E.H. Immergut, Eds., Wiley, New York, 2nd ed., 1975.
16. Stockmayer, W.H. and Fixman, M., *J. Polymer Sci.* **C1**, 137(1963).

11.14   GENERAL REFERENCES

1. Rodriguez, F., *Principles of Polymer Systems*, McGraw-Hill Book Company, New York, 1970.
2. Van Krevelen, D.W. and Hoftyzer, P.J., *Properties of Polymers*, 2nd Ed. Elsevier, New York, N.Y., 1976.
3. Van Krevelen, D.W., *Properties of Polymers*, Elsevier, New York, N.Y., 1990.
4. Bohdanecký, M. and Kovář, J., *Viscosity of Polymer Solutions*, Elsevier, New York, N. Y., 1982.
5. Morawetz, H., *Macromolecules in Solution*, R. E. Krieger Co., 2nd Ed., 1983.
6. Kurata, M., *Thermodynamics of Polymer Solutions*, Harwood, New York, N.Y., 1982.

CHAPTER 12

Optical Properties of Polymers

12.1   INTRODUCTION

Polymeric materials offer an array of optical properties based on the chemical nature of the polymer and its physical state, namely amorphous or crystalline. A purely amorphous polymer has only one refractive index and hence is clear in appearance. Similarly, a purely crystalline polymer also exhibits only one refractive index and thus is clear. On the other hand, polymers that exhibit crystallinity are generally semi-crystalline with distinctively different refractive index values for the crystalline and amorphous phases and thus are generally hazy, translucent, or opaque in appearance. In the case where the crystalline entities that scatter light are much smaller than the wavelength of light, the polymer presents a clear appearance. These small crystalline entities, however, can be detected by optical techniques which use a smaller wavelength incident beam, such as in X-ray scattering.

Optical properties play an important part in the characterization of polymer blends. A blend of two amorphous polymers with complete miscibility would generally be clear. However, immiscibility or partial miscibility (compatibility) leads to a system with two phases and if the refractive indices of two phases are different the blend would appear hazy. As in the case of semi-crystalline polymers, the blend would be clear if the size of the dispersed phase domains are considerably smaller than the wavelength of light. Similar effects are seen in block and graft copolymers when two immiscible components form two phases.

Another interesting situation occurs when a polymer contains absorbed solvent. If the solvent forms domains by coalesence then these would scatter light to make the systems hazy.

Amorphous glassy polymers offer several possibilities in specialized optical applications because they are lighter than silicate glass, and products with precision can be made in one single operation not requiring grinding or polishing. They have higher impact strength and when they shatter, their fragments are less damaging than silicate glass, making them suitable for applications such as lenses for spectacles. By suitably controlling the glass transition and water absorption, polymers can be used as soft contact lenses - an application which would not be conceivable with any other class of materials. The materials used for these applications are generally acrylic copolymers. Amorphous glassy polymers or copolymers with a high impact strength, such as polycarbonate, may be used for window panes or car windshields. However, the disadvantage of poor abrasion and/or solvent resistance of such polymers prevents their use in such applications, although coatings suitable for protecting the surfaces of such materials are currently being used. Improved, high technology, coatings are under development using plasma deposition which will undoubtedly allow broader applications.

12.2 INTERACTION OF LIGHT WITH POLYMERS

When a light beam impinges on a polymer the following optical phenomena occur: refraction, reflection, absorption and scattering. The first three depend upon the molecular nature of the polymer which would impart average optical properties to the medium. Scattering, on the other hand, is dependent upon the local fluctuation in optical properties within the medium. Systems which are completely isotropic, such as amorphous polymers, would not exhibit scattering. The interaction of light or electromagnetic radiation with a material, is determined by the materials specific conductivity, electrical inductive capacity or dielectric constant, and its magnetic inductance capacity (magnetic permeability). These in turn are related to the refractive index and extinction index of the medium.

In an isotropic and homogeneous system, the transmittance of light through the material is the ratio of intensity of light passing through the sample to that of the incident light. This is dependent upon the amount of light that is reflected, absorbed or scattered. In the absence of absorption and scattering, the material would be transparent. The percent transmittance is thus dependent on the reflection. Materials that have a high level of scattering and thus have virtually no transmittance would be opaque. As an example, pure poly(methyl methacrylate) (PMMA) has a high level of transmittance, but when it is mixed with a small amount of an organic filler it becomes opaque. Materials with intermediate levels of transmission and some scattering would be translucent. Absorption is also an important phenomenon and is related to the frequency of the incident light. At some critical frequencies, the radiation is in resonance with the internal vibration of the material. The material would be a strong absorber at these frequencies of radiation while it would be transparent at other frequencies, i.e., higher or lower than the critical frequencies. In essence any material may appear opaque at some frequencies and transparent at others.

Heterogeneous systems are those which contain two or more species of different refractive index values which results in a point to point variation of refractive index. The dispersion of inorganic filler in an amorphous polymer would be an example. This would give rise to scattering, which would be significant when the size of the heterogeneity in the dispersed phase is comparable to the wavelength of light.

As a special case, if the refractive index of the polymer and the dispersed phase are identical, then practically no scattering would be observed. Another phenomenon that occurs is that of anisotropy. This occurs when a material has two different refractive index values in two different directions. An oriented polymer in which chains are aligned in the stretching direction would exhibit anisotropy. This is quantified as birefringence, which is the difference of refractive index in the two directions. Polymer crystals are anisotropic and thus birefringent. A semi-crystalline polymer, consisting of spherulites, represents another kind of heterogeneous system in which the scattering spheres are anisotropic. Theories of light scattering have been developed to consider scattering due to isotropic as well as anisotropic particles.

When light travels through a medium it changes its phase and some light may be absorbed by the system, i.e., lost by extinction. A complex form of refractive index can illustrate both of these aspects, illustrated by an expression as given by:

$$n^* = n' - in'' \tag{12.1}$$

where, n^* represents the complex refractive index, n' is the in phase component representing the phase lag caused by the material and n'' is the out of phase component representing extinction. Consider a simple harmonic wave traveling in the z direction whose amplitude in the complex form is:

$$U = U_o \exp(-ikz) \tag{12.2}$$

where:

$U_o =$ amplitude of the incident light

$k =$ the wave number $\left(2\pi/\lambda = 2\pi n^*/\lambda_o\right)$

$\lambda_o =$ the wave length of light in vacuum

The wave number k can be written as:

$$k = k\left(2\pi\lambda_o\right)n^* = k_o\left(n' - in''\right) \tag{12.3}$$

Substituting Equation 12.1 and 12.3 in 12.2 gives

$$U = U_o \exp\left[-k_o z(in' + n'')\right] \tag{12.4}$$

This is a harmonic wave with decreasing amplitude as it advances. The intensity, I, for this wave is determined from $|U^2|$ and is given by Lamberts' law as:

$$I = I_o \exp\left(-2k_o n'' z\right) \tag{12.5}$$

where, I_o is the incident beam intensity. An extinction coefficient, E, can be defined as:

$$E = 2K_o n'' = \sigma + K \qquad (12.6)$$

where, σ is the scattering coefficient or turbidity and K is the absorption coefficient or absorptivity.

In anisotropic systems the two components of the complex refractive index are dependent on the state of polarization of light. The anisotropy in the real component is referred to as birefringence while the anisotropy in the imaginary component is referred to as dichroism. Furthermore, the nature of anisotropy may be linear or circular, or a combination of the two. Uniaxially stretched polymers, such as fibers or films oriented in one direction exhibit linear anisotropy. In linear anisotropy the refractive index depends on the polarization direction of the incident light. Unoriented crystalline polymers or biaxially oriented polymers exhibit circular or elliptical anisotropy. The optical activity in polymers is circular birefringence and the extinction counterpart is circular dichroism.

12.3   REFRACTION OF LIGHT BY POLYMERS

Snell's law describes the refraction of light passing through a medium of refractive index n by the expression:

$$n = \sin i / \sin r \qquad (12.7)$$

where, i and r are the angles of incident and refracted light respectively. Refractive index is also known to be the ratio of the velocity of light in the two mediums and in turn the ratio of wave length of light in the two mediums. Thus Snell's law becomes,

$$n = \sin i / \sin r = v_1 / v_2 = \lambda_1 / \lambda_2 \qquad (12.8)$$

If the medium, 1, is vacuum, v becomes v_o and if medium, 2, is transparent in which the velocity of light is v, then

$$\Delta v / v = \left(v_o - v \right) / v = \left(v_o / v \right) - 1 = n - 1 \qquad (12.9)$$

The term $(n-1)$ is referred to as the brake power of the material and $(n-1)/\delta$ becomes the specific brake power. For organic substances specific refraction may be defined as $(n-1)/\delta$ when determined for a standard wave length of incident light.

The dielectric constant, ε, is related to n. Based on the unified theory of electricity, magnetism and light, the dielectric constant, ε, is related to n by the relationship,

$$\varepsilon = n^2 \qquad (12.10)$$

Subsequently, Lorentz and Lorentz related ε and n to molecular polarizability, α, of a chemical group by the equation.

$$(\varepsilon - 1)\cdot M/(\varepsilon + 2)\cdot P = (4n/\ 3)\cdot N_A \cdot \alpha = P \qquad (12.11)$$

where, M is the molar mass, N_A is Avogadro's number and P is the molar polarization. The molar refraction, R_{LL}, can then be expressed as:

$$R_{LL} \equiv P = (n^2 - 1)\cdot M/(n^2 + 1)\cdot \rho = (n^2 - 1)\cdot V/(n^2 + 1) (12.12)$$

where, V is the specific (molar) volume. Group contributions to molar refraction at a wavelength of 589 nm are available in the literature. From these the refractive index of organic materials and polymers can be determined by the additivity of R_{LL}.

12.4 MOLCALC ESTIMATIONS OF THE REFRACTIVE INDEX, n_D, AND COMPARISON TO LITERATURE VALUES FOR A SERIES OF REPRESENTATIVE POLYMERS

The quantity Molcalc uses for estimating the dielectric constant, ε, is the molar polarization of a dielectric based on the estimation method of Van Krevelen and Hoftyzer. Rearranging Equation 12.12 we obtain:

$$n_D = \left(\frac{1 + 2\dfrac{R_{LL}}{V}}{1 - \dfrac{R_{LL}}{V}} \right)^{1/2} \qquad (12.13)$$

Table 12.1 lists the calculated as well as the experimental values available in the literature. It can be seen that for all the polymers the two sets of values match remarkably well. It may be noted that to achieve a high refractive index the polymer must contain chemical groups that have high polarizability and yet do not occupy too high a volume. For polymers that exist in amorphous and crystalline forms, the refractive index increases with increasing crystallinity due to increase in density.

Table 12.1
Comparison of Estimated Refractive Indices with Literature Values

Polymer	Refractive Indices, n_D	
	Calculated	Literature
Poly(ethylene)	1.50	1.49
Poly(propylene)	1.48	1.49
Poly(isobutylene)	1.49	1.508
Poly(vinyl chloride)	1.54	1.539
Poly(vinyl fluoride)	1.42	1.42
Poly(chloro trifluoro ethylene)	1.44	1.39-1.43
Poly(vinyl acetate)	1.47	1.467
Poly(cis butadiene)	1.52	1.516
Poly(cis 1,4-isoprene)	1.52	1.520
Poly(styrene)	1.61	1.591
Poly(αmethyl styrene)	1.59	1.587
Polycarbonate	1.59	1.585
Poly(ethylene terephthalate)	1.57	1.64
Poly(butylene terephthalate)	1.55	----
Poly(3,3-bis chloromethyl oxacyclobutane)	1.53	----

continued

Table 12.1 continued

Polymer	Refractive Indices, n_D	
	Calculated	Literature
Poly(methylene oxide)	1.49	1.510
Poly(vinyl methyl ether)	1.48	1.467
Poly(phenylene oxide)	1.59	----
Poly(hexamethylene adipamide)	1.50	1.53
Poly(acrylamide)	1.53	----
Poly(acrylonitrile)	1.53	1.514
Poly(methyl methacrylate)	1.51	1.49
Poly(cyclohexyl methacrylate)	1.51	1.507

12.5   REFLECTION OF LIGHT BY POLYMERS

Light reflects when it impinges the bonding plane between two non-absorbing media and the level of reflection, r, depends upon the refractive index values of the two media, n_1 and n_2 respectively. If the light impinges perpendicular to the plane boundary, the reflectance is given by Fresnel's equation as follows:

$$r = (n_2 - n_1)^2 / (n_2 + n_1)^2 \qquad (12.14)$$

The level of reflectance would depend on the direction of polarization of light. If the incident light strikes the plane at an angle Fresnel's equation changes and is given by:

$$r = \tan^2(\phi_i - \phi_r) / \tan^2(\phi_i + \phi_r) \qquad (12.15)$$

where, ϕ_i and ϕ_r are the angles of incidence and refraction respectively. From this equation it can be observed that when, $\phi_i + \phi_r = \pi/2$, there is no reflection. The angle of incidence at this condition is the polarizing angle or Brewster's angle, ϕ_B, and is related to n_2 by the Brewster equation, as follows:

$$\tan \phi_B = n_2 \qquad (12.16)$$

The analysis when the media is non-absorbing is more complicated.

12.6   BIREFRINGENCE

A material is birefringent when it has different refractive indices in different directions, due to anisotropy of the material. Birefringence in a material results in the rotation of the plane of polarized light. Birefringence is defined as the difference of refractive index between the maximum and minimum values, given as:

$$\Delta n = n_2 - n_1 \qquad (12.17)$$

In oriented polymers the maximum and minimum refractive indices would generally be in directions parallel and perpendicular to stretching (see Chapter 3, Section 3.2.2). Birefringence may occur in materials naturally or may be induced by perpendicularly orienting the molecules. Crystalline materials would generally be naturally birefringent. Artificial birefringence may be introduced by external forces such as mechanical stretching, electric, and magnetic fields. Birefringence in a polymer gives rise to optical rotation of the plane of polarized light. Several polymer applications require orientation, such as in fibers and films. The mechanical stretching of polymer molecules results in orientation of the polymer chains and causes birefringence to be exhibited. This allows birefringence to be used to determine the degree of orientation. In amorphous polymers the birefringence is entirely due to the orientation of molecules in the amorphous state. On the other hand, in crystalline polymers the contribution to birefringence comes from crystalline as well as amorphous phases and may be written as:

$$\Delta n = x_c f_c \Delta n_c^o + \left(1 - x_c\right) f_a \Delta n_a^o \qquad (12.18)$$

where:

$\quad x_c \qquad\qquad$ = degree of crystallinity

$\Delta n_c^o, \Delta n_a^o$ = birefringence of perfectly oriented crystalline and amorphous phases

f_c, f_a = orientation factors for the crystalline and amorphous phases

The values of Δn_c^o and Δn_a^o are available in the literature for most polymers. The values of f_c and f_a range from 0 for no orientation to 1 for perfect orientation. Generally the strength of an oriented polymer increases in the direction of stretching which may allow a correlation to be made between birefringence and mechanical properties.

A relationship has been developed between birefringence and retractive force for stretched elastomers (cross-linked rubbers). The level of orientation increases birefringence as well as the retractive stress, σ and the relationship between the two for uniaxial stretching is given as:

$$\Delta n = C \cdot \sigma \qquad (12.19)$$

where, C is defined as the stress optical coefficient and its value is dependent on the chemical nature of the polymer. Combining this with the theory of elasticity gives the relationship,

$$C = \frac{2\pi(\alpha_2 - \alpha_1)}{45} \cdot \frac{(n^{-2} + 2)^2}{n_{av} kT} \qquad (12.20)$$

where, α_2 and α_1 are the polarizabilities of a polymer chain parallel and perpendicular to the direction of the chain respectively, and n_{av} is the average refractive index. If the polymer undergoes stress relaxation or creep, the ratio remains the same, and the value of C is thus a constant.

In glassy uncross-linked polymers, orientation is achieved by stretching above the glass transition temperature T_g and then cooling rapidly to below the T_g to preserve the orientation and birefringence. Keeping the polymer at a temperature above T_g after stretching would cause the molecules to relax and decrease in orientation and birefringence. Some polymers are capable of being stretched in their glassy state, by the phenomenon of cold drawing, and thus become birefringent.

Photo-elastic techniques have been developed to study the stresses developed in polymers during conventional processing operations such as extrusion, injection molding and thermoforming. In these the birefringence patterns are observed and are correlated to stresses built in during processing. Such an analysis is useful for improved product design and optimization of processing parameters. The device used for photo-elastic analysis is called a polariscope.

Highly oriented polymers that are transparent would act as polarizing filters. Unpolarized light passing through them would get plane polarized. Polymer melts and solutions undergoing shear or elongational flow would exhibit birefringence, which is referred to as flow birefringence. The stress birefringence relationship for melts would follow Equations 12.19 and 12.20. Flow birefringence measurements can be used to determine values of segmental anisotropy of polymer chains.

12.7   ABSORPTION OF LIGHT BY POLYMERS

When a light beam passes through a medium it experiences a loss of intensity due to absorption by the medium. If the path length is l then the intensity of transmitted light is given by Lambert's relationship, which gives the transmitted intensity as:

$$I = I_o \exp(-4\pi nkl/\lambda) \qquad (12.21)$$

where:

I_o = intensity of incident light

k = characteristic function of the wavelength

Polymers generally do not absorb any specific wavelength selectively in the visible region of the light spectrum and thus are colorless and transparent in their amorphous state. In the crystalline state they appear milky due to the scattering of light caused by the presence of two phases.

In the infrared part of the light spectrum polymers exhibit selective absorbance of specific wavelengths. The determination of infrared absorbance thus becomes a tool for the characterization of polymers. Specific structural groups and atomic vibrations have their own characteristic infrared absorption bands, which provide information about the chemical structures present in polymers. The infrared absorption in semi-crystalline polymers also provides information about the crystallinity since the ordered regions give different absorption bands due to differences in conformations.

12.8   OPTICAL APPEARANCE OF POLYMERS

The optical appearance and properties of polymers such as surface gloss, clarity, and haziness are commercially important. These properties are not dependent on the chemical structure but are determined by physical properties of polymers. The optical appearance can be divided into surface appearance and bulk appearance. Gloss refers to the reflected light intensity as a percentage of that from an ideal surface. Clarity is defined as the degradation of image of a standard printed scale observed through a sheet of plastic. Transparency refers to the transmitted intensity for light that deviates by an angle less than $2.5°$ as a percentage of the incident light intensity.

The main bulk appearance properties of a polymer are its light transmittance and color. As mentioned earlier, polymers do not absorb any particular wavelength and hence are colorless. In their amorphous state they are clear while in the semi-crystalline state the appearance ranges from translucent to hazy or opaque. The transmittance is close to 100% for totally clear materials and 0% for opaque materials. Polymers with a transmittance of greater than 90% are considered to be transparent. At lower values of transmittance the polymers become translucent and are considered hazy and at considerably low transmittance levels they are opaque. The decrease in transmittance is caused by the presence of crystalline structures or external additives. Color is imparted by the addition of pigments or dyes.

The surface appearance depends upon the reflectance of light from the surface. Total reflectance is from mirror like surfaces while almost no reflection is obtained from matte surfaces. In between the two, the surface may be called glossy. Glossy surfaces give a lustrous appearance. The surface of a polymer depends upon surface produced in the processing operation. The mold surface also has a major effect on the surface appearance.

12.9   LIGHT SCATTERING

When a beam of light passing through a medium encounters optical inhomogeneities, the light deviates and this phenomenon is referred to as internal light scattering. Similarly deviation of light from a reflecting surface is referred to as surface light scattering. In both the cases there is no reduction in the total intensity of light.

For internal scattering the intensity of scattered light per unit volume, i_θ integrated over all angles, θ, gives the turbidity as:

$$tur = \int_0^\pi 2\pi i_\theta \sin\theta d\theta \qquad (12.22)$$

The total intensity of the scattered beam of light, I, is related to the intensity of incident beam, I_o, by the expression,

$$I = I_o \exp(-tur \cdot l) \qquad (12.23)$$

where:

l = thickness of the medium

In polymer solutions the scattering of light occurs due to fluctuations in refractive index from point to point in the solution. The relationship between the incident and scattered light is given by the Rayleigh's ratio as follows:

$$R_\theta = r^2 i_\theta / I_o \left(1 + \cos^2\theta\right) \qquad (12.24)$$

where, r is the distance of scattering molecule. Analysis of this leads to a relationship between R_θ and molecular weight of the polymer, M, given as:

$$Kc/R_\theta = 1/M = 2A_2 c \qquad (12.25)$$

where:

c = polymer concentration
A_2 = the second virial coefficient
K = a constant

K is given as:

$$K = 2\pi^2 / \lambda_o N_A \cdot (ndn/dc)^2 \qquad (12.26)$$

Equation 12.24 holds true for small molecules. For large molecules, such as in the case of polymers, a factor for disymmetry has to be introduced. This is because scattered light from different parts of the same particle interfere with each other resulting in a reduction in the observed scattered intensity. The disymmetry factor, $P(\theta)$, is given as:

$$P(\theta) = 1 - 1/3 \cdot h^2 R_g^2 \qquad (12.27)$$

where:

h = $4\pi(n/\lambda_o)\sin(\theta/2)$
R_g = radius of gyration of the polymer molecule

The expression between R_θ and M thus changes to,

$$Kc/R_\theta = 1/M \, P(\theta) + 2A_2 c + ... \qquad (12.28)$$

Further analysis of the light scattering data is done by plotting $(Kc/R\theta)$ versus $(\sin^2(\theta/2) + Kc)$. A grid is obtained which is then extrapolated to $c = 0$ and $\theta = 0$. Molecular weight is the intercept on the y-axis. R_g is obtained from the slope of the $c = 0$ line, and A_2 is obtained from the slope of the line, $\theta = 0$. Light scattering from polymer solutions is one of the few methods to determine absolute molecular weight of a polymer and gives the weight average molecular weight.

12.10 SOLID STATE LIGHT SCATTERING

Solid state light scattering deals with the internal scattering of light as it passes through a polymer, generally in the form of a film. Light will not be scattered by isotropic polymers which have only one phase present such as in unfilled amorphous polymers. Light is scattered by the presence of additives or fillers in amorphous polymers, which makes them appear hazy. Light is also scattered in block copolymers, graft copolymers, or blends with amorphous components, which contain two phases with different refractive index values.

Such systems will also have a hazy appearance as in the case of styrene butadiene graft copolymers (high impact polystyrene), acrylonitrile butadiene styrene terpolymer (ABS), styrene butadiene styrene block copolymers and blends of immiscible or partially miscible components. In semi-crystalline polymers, light is scattered by the difference in the refractive index of the crystalline and amorphous phases. In crystalline polymers that have spherulitic superstructures the scattering takes place due to the anisotropy of the spherulites. In all cases the scattering of light would take place when the inhomogeneities that lead to scattering have a size comparable to that of the wavelength of incident light.

The scattering of light normally takes place at small angles and hence this is often referred as small angle light scattering (SALS). SALS is a useful method that provides information about the structure that has size of the order of wavelength of visible light, i.e., about 1000-100,000 Å. It gives information about the fluctuation in density of the scattering material as well as fluctuations in anisotropy or refractive index. It also gives information about the orientation of the anisotropic entities. The range of sizes that can be measured by SALS overlap with sizes measurable by optical microscopy provide a statistical evaluation of the sizes and shapes of scattering entities. This is not easily done using microscopy. The other advantage SALS offers is that the scattering angle increases with a decrease in the size of scattering entities, thus making it possible to see smaller structures more conveniently. Lastly, SALS can be used to follow rapid changes in structure thus allowing a time dependent evaluation of the structure.

12.10.1 Refractive Index Fluctuations in Glassy Polymers

An amorphous polymer does not contain any regular crystal lattice so it will not scatter light by itself. If, however, it contains another material which has a different refractive index then this would lead to light scattering. Assuming fluctuations in the refractive index, which can be described by a radial correlation function $\gamma(r)$, then this will result in a local deviation, η, in the polarizability from its mean value α_{av}. If the deviations are η_i and η_j at points i and j which are separated by a distance r, then the correlation function is defined as:

$$\gamma(r) = \frac{\langle \eta_i \eta_j \rangle_r}{\langle \eta^2 \rangle_{av}} \qquad (12.29)$$

where, the brackets represent on average for all pairs of points separated by r. Furthermore, for a correlation length, a, the correlation function is given as:

$$\gamma(r) = \exp(-r/a) \qquad (12.30)$$

Debye and Bueche developed the scattering theory to predict the scattering intensity at angle θ to the incident beam given as:

$$I \approx \langle \eta^2 \rangle_{av} \int_0^\infty \gamma(r) r^2 \left(\sin kr / kr \right) dr \qquad (12.31)$$

where:
$$k = 4\pi/\lambda \cdot \sin(\theta/2)$$

This method can be used to determine the correlation length a and subsequently estimate the size of the scattering entities.

12.10.2 Refractive Index and Orientation Fluctuations in Semi-Crystalline Polymers

In semi-crystalline polymers there is a marked difference between the densities of the crystalline and amorphous phases with the density of the crystalline phase being higher. Polymer crystals have covalent bonding in the chain direction (c direction) and Van der Waals interactions in the perpendicular directions. This makes them birefringent materials. The crystals form part of the spherulitic superstructure with the crystals aligned along the radial direction hence the c axis is in the tangential direction of the spherulite. The light scattering from spherulites can be predicted using a model that considers them to be spheres of radius r with polarizabilities being α_r and α_t in the radial and tangential directions respectively. The polymer in the form of a film is viewed between two polarizing filters using monochromatic light. Scattered light with the polarization directions of the polarizers being parallel to each other is referred to as Vv scattering and when they are perpendicular to each other it is referred as Hv scattering. In systems completely filled with spherulites, the respective intensity of scattered light in the two modes depends upon the anisotropy, i.e., $\left(\alpha_r - \alpha_t\right)$ which is given as:

$$I_{Vv} \propto \cos^4 \mu \left(\alpha_t - \alpha_r\right)^2 \tag{12.32}$$

$$I_{Hv} \propto \sin^2 \mu \cos^2 \mu \left(\alpha_t - \alpha_r\right)^2 \tag{12.33}$$

where, μ is the azimuthal angle in the scattered pattern. Hv scattering gives a four leaf clover pattern with no intensity along the polar directions and maximum intensity is the $45°$ directions. Scanning along the $45°$ direction one finds that the angle of maximum intensity is θ_m. From this the average spherulitic radius can be determined from the following relationship:

$$R_s = \left(1.25\lambda/\pi\right) \cdot \sin\left(\theta_m/2\right) \tag{12.34}$$

It can be seen that θ_m has higher values for lower radius and this makes the SALS technique very suitable for studying small superstructures. Similar model approach has been extended to study superstructures of other shapes such as sheafs and rods. Lastly, models have been developed for analyzing scattering behavior from oriented systems containing ellipsoidal spherulites or other oriented superstructures.

12.11 APPLICATIONS

Polymeric materials are used in a number of applications which require control of optical properties. Some of the common applications include barrier films or sheets, horticulture glazing, headlight lenses, aircraft windows, plastic lenses, electronic and information technology. When plastic sheets of film are used as weather barriers they must provide high light transmission making them transparent with high clarity. Glassy polymers are ideal choices as they are clear and more flexible than glass, the traditional material. The processing of sheets by extrusion becomes critical for obtaining optical clarity since sheets must be free of air bubbles and any other foreign particles. Horticulture glazings are used to replace glass greenhouses. For this translucent films of a semi-crystalline polymer is preferred because of the requirements of diffused light. In fact to obtain a real greenhouse effect it is preferred to have a glazing that is transparent in the visible and near infrared regions allowing for photosynthesis, opaque in the far infrared region to minimize loss of radiated energy and resistant to ultraviolet degradation.

In recent times plastics are used in automotive headlight lenses since they are tougher materials and would also result in weight saving. They must be made by precision injection molding to give a minimal of optical distortion. They must also have a high service temperature so that they do not deform during usage. Another important application is that of aircraft windows manufactured from PMMA and used commercially. This results in a considerable weight saving and requires careful manufacture.

Polymers are being extensively used for making optical lenses. These are either used in spectacles or as contact lenses.

Acrylic polymers and polycarbonate are the traditional materials used as optical lenses, generally with structural modifications, that introduce bulky side-chain groups in order to increase the necessary oxygen permeability of contact lenses without adversely affecting optical properties. Hydrogel polymers (water containing polymers), such as, 2-hydroxyethyl methacrylate (HEMA) are used in the manufacture of soft contact lenses. Lastly, the use of plastics is increasing in the area of electronic and information technology. As an example, polycarbonate is extensively used for a storage of information in the form of video and audio disks. This again requires precision molding to produce disks without any optical stresses.

The study and understanding of optical properties of polymers is valuable in a large number of applications.

12.12 ABBREVIATIONS AND NOTATION

A_2	second virial coefficient
c	polymer concentration
C	stress optical coefficient
E	extinction coefficient
f	orientation factor
I	intensity of incident beam
I	intensity of deviated light
I_{Vv}	scattered intensity with parallel polars
I_{Hv}	scattered intensity with perpendicular polars
k	wave number
K	constant factors
l	thickness of the medium
M	molar mass
n	refractive index
N_A	Avogadro's number
P	molar polarization
$P(\theta)$	di-symmetry factor
r	reflectance

R_g	radius of gyration
R_s	spherulitic radius
tur	turbidity
U_o	amplitude of incident light
U	amplitude of refracted light
v	velocity of light
x_c	degree of crystallinity
α_r	radial polarizability of a spherulite
α_t	tangential polarizability of a spherulite
δ	polarizability
ε	dielectric constant
ϕ_i	angle of incidence
ϕ_r	angle of refraction
$\gamma(r)$	correlation function
η	local deviation of polarizability
λ	wavelength of light
λ_o	wavelength of light in vacuum
μ	azimuthal angle
θ	scattering angle
σ	retractive stress

12.13 GENERAL REFERENCES

1. Van Krevelen, D.W., *Properties of Polymers*, third edition, Elsevier, New York, N.Y., 1990.
2. Lenz, R.W. and Stein, R.S., *Structure and Properties of Polymer Films*, Plenum Press, 1973.
3. Kerker, M., *The Scattering of Light and other Electromagnetic Radiation*, Academic Press, 1969.
4. Mark, H.F., Bikales, N.M., Overberger, C.G. and Menges, G., Eds., *Encyclopedia of Polymer Science and Technology*, vol. 10, Wiley Interscience, 1976.
5. Elias, H.G., in *Polymer Handbook*, 3rd Edition, edited by J. Brandrup and E.H. Immergut, Wiley, New York, 1989.

6. Mark, J.E., ed. *Physical Properties of Polymers Handbook*, AIP Press, Woodbury, N. Y., 1996.

7. Seymour, R.B. and Carraher, Jr., C.E., *Polymer Chemistry*, pgs. 144-147, Marcel Dekker, Inc., New York, N.Y., 1988.

CHAPTER 13

Electrical Properties

13.1 INTRODUCTION

The electrical properties of materials are directly related to their structures, assuming the materials are pure and are considered free of contaminants. This is not surprising considering the structural differences between materials discussed in both Chapters 1 and 2.

Materials are generally subdivided from the broad categories of structures discussed in Chapter 1 for consideration of electrical properties. Thus electrical properties of materials are determined by their performance as: conductors (metals), insulators (ceramics, glasses, polymers, e.g.), and semi-conductors (doped materials or others with a narrow band gap) (cf. Chapter 1, section 1.10).

The metallic bond was characterized in Chapter 1 as consisting of mobile electrons that are able to move throughout the metal like an "electron gas" with no net transport in the absence of an electrical field. Semiconductors and insulators were differentiated on the basis of the size of their forbidden energy gaps. In semiconductors, the energy gap is such that usable numbers of electrons are able to jump the gap from the filled valence band to the empty conduction band. These electrons carry a charge toward the positive electrode while the holes in the valence band become available for conduction because they allow electrons present in the material to move into those vacated holes.

Insulators have a large band gap which makes it difficult for electrons to jump this gap. However, it should be realized that the electrical behavior of insulators, particularly polymers, is influenced by temperature, time, moisture, and other contaminants, part geometry, mechanical stress, and the frequency and magnitude of the applied voltage.

13.2   CONTRASTS BETWEEN MATERIALS

The conductivity, σ, and resistivity, ρ, for a material depend upon the number n of charge carriers, the charge q on each, and their mobility, μ, as given by:

$$\sigma = \frac{1}{\rho} nq\mu \qquad (13.1)$$

In those types of materials that conduct electricity the charge is carried in modules of 0.16×10^{-18} coul., this being the charge on an individual electron. In metals the electron moves whereas in ionic materials, the charge is carried by diffusing ions and their charge is simply an integer number of electron charges (− or + for anions and cations, respectively).

Differences in the conductivity between materials covers a large range of values and is quantifiable based on experimental measurements. Differences between opposite ends of the scale as well as examples of materials of each category are shown in Figure 13.1.

In order to provide a better appreciation of the differences between materials, the following summary of contrasts are drawn between different types of materials and the corresponding factors that affect their electrical properties.

13.2.1 Conductivity

Factors affecting the conductivity are as follows:

1. Temperature -
Bulk or lattice diffusion is:

$$D_i = D_o e^{-Q/RT} \qquad (13.2)$$

where:

D_o = pre-exponential diffusion value
Q = experimental activation energy
R = gas constant
T = temperature in Kelvin

This equation, relating diffusion to temperature, shows that as the

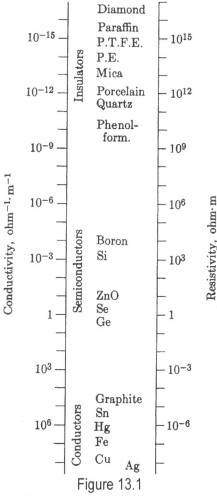

Figure 13.1
Range of Materials Conductivity and Resistivity
Semiconductors Lie Between
10^4 and 10^{-4} ohm^{-1}m^{-1}

temperature is increased, the diffusion should increase. Since the mobility of ions/lattice components increases according to:

$$\mu_i = \frac{qD_i}{kT} \qquad (13.3)$$

Therefore, the conductivity would be expected to increase.

- High Temperatures - Conductivity is a lesser factor at high temperatures, therefore, $\sigma \neq f$ (impurities).
- Lower Temperatures - Around ambient temperatures to those slightly above, $\sigma = f$(impurities, lattice defects, effects of solid solutions, etc.). These factors are significant and for avoidance, defect concentrations should be kept $\leq 10^{-10}$.

2. Defects

In general, the greater the amount of defects present, the greater will be the conduction, especially in the case of vacancies.

3. Ion Size

The presence of ions leads to increased conduction as might be expected. The effect of ion size is such that the smaller the ion, the smaller is its charge and the higher is its coordination number. This leads to increased mobility, as in the case of Na^+ ions whose presence leads to increased conduction.

13.2.2 Other Considerations

13.2.2.1 Glasses

Ionic conduction is controlled by the presence and type of modifier. The smaller the modifier, the greater the conduction.

      a) The presence of RO ions can lead to strange effects.

b) The presence of mixed alkali components have little effect on conduction but a strong influence on viscosity. The addition of alkali modifiers tends to decrease viscosity and increase conduction.

13.2.2.2 Ceramics

Insulators - In general, ceramics provide excellent insulation properties and are the best choice among materials that are classed as insulators. Insulating properties can be enhanced by selecting materials displaying the following characteristics.

a) Materials that possess a high intrinsic viscosity will display the best insulating properties.

b) Minimizing defects by controlling the presence of impurities present will increase the insulation properties.

c) Reducing the glass content will act to increase insulation properties.

d) Controlling the grain boundaries by promoting the presence of large grains and fewer boundaries is preferred over smaller grains and the possibility of increased impurities at the boundaries.

13.2.2.3 Polymers

Polymers are, in general, classed in the family of insulators and some of the above considerations also apply to polymers.

a) These materials should be used at much lower temperatures, vis-à-vis, ceramics and glasses.

b) Materials should be chosen that have the highest possible breakdown voltage.

c) The presence of impurities should be minimized, especially those impurities leading to conducting ions.

13.2.3 Dielectric Properties (Linear Dielectrics)

There are four basic mechanisms that influence the dielectric constants of materials:

1. Electronic polarization from an electric field
2. Ionic polarization which is most important in ceramics and arises from the displacement of ions of opposite

sign from their regular lattice sites due to an applied field

3. The permanent dipole orientation of the material
4. Space charge polarization which results from mobile charges which are present because they are impeded by interfaces since they are not supplied at an electrode or discharged at an electrode or because they are trapped

The following factors affect the dielectric properties of linear dielectrics.

1. Frequency - No appreciable effect until the frequency $> 10^{10}\,Hz$.
2. Temperature - A rise in temperature increases ion mobility and defect concentration which lowers the dielectric properties as we have seen above.
3. The loss tangent must be controlled to provide a good dielectric. This is discussed below. For example, the presence of Na_2O in ceramics increases σ markedly and tan δ at high temperatures.
4. The porosity of a ceramic directly reduces the dielectric constant.

13.2.4 Ferroelectric Materials (Non-Linear Dielectrics)

These are materials that can undergo spontaneous alignment of their electric dipoles by their mutual interaction. They can be grouped into two types.

- non-centro symmetric materials that undergo spontaneous polarization
- non-centro symmetric materials (crystals) that produce a voltage due to a change of dipole moment resulting from the application of a mechanical force

The following factors affect the dielectric properties of non-linear dielectrics.

1. Structure - Perovskite, for example, with a highly charged small cation, Ti^{+4} or Zr^{+4}, can undergo

displacement from the center of the cell and give rise to a permanent dipole.

2. Composition - This factor affects structure and therefore greatly affects performance.

3. Temperature - These materials are greatly affected by temperature since they undergo sharp polymorphic transitions.

13.2.5 Ferromagnetic Materials

Many of the magnetic properties of ceramics are analogous to their dielectric characteristics. Ceramic magnets are often used in high-frequency devices in which the greater resistivity of the ferromagnetic oxide gives them a decisive advantage over metals.

Permanent magnetic moments arise in systems in which unpaired electrons are present. These include metals with conduction electrons, atoms and molecules containing an odd number of electrons, and atoms and ions with partially filled inner electron shells. For the most part the types of materials which constitute this class are the transition elements, rare earth elements, and actinide elements.

The factors which affect the dielectric properties of ferromagnetic materials are:

1. Ceramic ferrites - These materials exhibit low AC losses.

2. Structure - Ferrites and spinels are prime examples of structure which have ferromagnetic properties.

3. Cation - This component exerts a large influence.

4. Composition - This factor controls paired or unpaired electrons/electron spin and therefore properties.

5. Density - This factor has a large influence on the magnetic flux.

6. Grain size - The grain size must be optimized and similar consideration are applicable here as discussed above.

13.3 THEORETICAL ASPECTS

When an electrical voltage or driving force is applied to a

material, current flows if the material is able to conduct. If the voltage is steady (direct), the current is defined by Ohm's law.

$$I = \frac{E}{R}$$

(13.4)

where:

  I = current
  E = voltage; V
  R = resistance in ohms, Ω

We can calculate the drift velocity, $\bar{v}$, of a charge carrier by:

$$\bar{v} = \mu E$$

(13.5)

Equation 13.5 shows that the drift velocity is proportional to the electric field, or volts/m. As mentioned above, in the absence of a field, there is no net drift velocity which can be seen from Equation 13.5.

Voltage, however, is never absolutely steady and usually varies in cyclic fashion or alternating fashion with a repetition rate of $0.1 - 10^{11}$ Hz. The d.c. resistance depends on the dimensions of the material and a number of other factors such as ambient conditions, e.g., heat induced by the current flow, and the characteristics of the material.

The capacitance of a material is given by:

$$C = \frac{Q}{V}$$

(13.6)

where:

  Q = charge
  V = voltage

An insulating material (dielectric) can be represented by a combination of resistance and capacitance as illustrated in Figure 13.2.

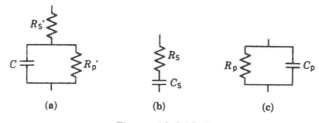

Figure 13.2 [G5]
Schematic Representation of an Insulating Material (Dielectric)

   (a) Complex but realistic representation
   (b) Simplification to an equivalent series circuit
   (c) Simplification to an equivalent parallel circuit

If voltage is suddenly applied to the series circuit, a current flows to charge the capacitance. As the voltage builds up across the capacitance, the current decreases. In the case of a good dielectric, the series resistance is low. In the parallel representation, the higher the resistance, the better the dielectric and less current is by passed or leaked.

   When an alternating voltage is applied to a perfect dielectric, a current flows which is displaced in time in such a way that it is 90° out of phase with the voltage. In actual cases, the voltage will lead the current with time as shown in Figure 13.3. The smaller the phase angle θ, the better the dielectric. The quality of the dielectric is given by the ratio of the resistive component to the capacitive component of the current which is often referred to as the dissipation factor (Equation 13.7).

$$\tan \delta = \frac{I_{R_p}}{I_c} = \text{dissipation factor} \qquad (13.7)$$

   Different materials with the same dimension may have different capacitance values. The relative dielectric constant or permittivity is:

$$\varepsilon' = \frac{C_p}{C_v} \qquad (13.8)$$

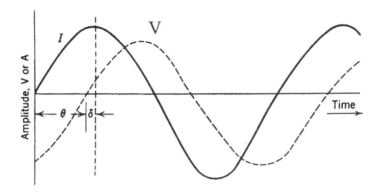

Figure 13.3 [G5]
Relationship of Current to Voltage as a Function of Time
V = Voltage
I = Current
θ = Phase angle
δ = Loss angle

where:

C_p = parallel capacitance of the material

C_v = capacitance of a dimensionally equivalent vacuum

The loss factor ε'' is given by:

$$\varepsilon'' = \varepsilon' \tan \delta = \text{relative loss index} \qquad (13.9)$$

The capacitance C contains both a geometric and a material factor. For a large plate capacitor of area A and thickness d, the geometric capacitance in vacuo is:

$$C_o = \frac{A}{d}\varepsilon_o \qquad (13.10)$$

where ε_o is the permittivity of a vacuum. If an insulator of permittivity, ε' is inserted between the capacitor plates, the capacitance is given by:

$$C = C_o \frac{\varepsilon'}{\varepsilon_o} = C_o \varepsilon = C_o \kappa' \tag{13.11}$$

and ε or κ' is, as before, the relative permittivity or dielectric constant and the material property of interest which determines the capacitance of a circuit element. It can be seen that Equation 13.11 takes into account the effects of design and/or measurement.

It is important to note the difference between an ideal dielectric and a real dielectric which is better represented by the more complex schematic in Figure 13.4(a). The rate of heat generation is proportional to the product of voltage and current and is zero for an ideal dielectric with a finite response time.

Figure 13.4 illustrates that the variation of polarization with time parallels the variation of charge with time. For an alternating field the time required for polarization shows up as a phase retardation of the charging current. This is best represented by expressing the electric field E, and displacement D, or flux density in complex notation.

$$E = E_o e^{i\omega t} \tag{13.12}$$

$$D = D_o e^{i(\omega t - \delta)} \tag{13.13}$$

Since the displacement is equal to $\varepsilon^* E / \varepsilon_o$, the complex dielectric constant is:

$$\varepsilon^* = \varepsilon_s e^{-i\delta} = \varepsilon_s (\cos\delta - i\sin\delta) \tag{13.14}$$

where:

$$\varepsilon_s = \frac{D_o}{E_o} = \text{static dielectric constant}$$

In terms of a complex dielectric constant,

$$\varepsilon^* = \varepsilon' - i\varepsilon'' = \frac{1}{\varepsilon_o}(\varepsilon' - i\varepsilon'') \tag{13.15}$$

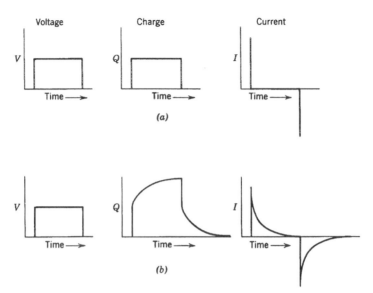

Figure 13.4 [G1]
Schematic Behavior of Charge Buildup and Current Flow in Dielectrics
(a) Ideal dielectric
(b) Real dielectric

and from Equation 13.14,

$$\varepsilon' = \varepsilon_s \cos\delta \tag{13.16}$$

$$\varepsilon'' = \varepsilon_s \sin\delta \tag{13.17}$$

Thus the loss tangent, as before, is given by:

$$tan\ \delta = \frac{\varepsilon''}{\varepsilon'} \tag{13.18}$$

For a simple capacitor with a sinusoidal applied voltage, the charging current is:

$$I_c = i\omega\varepsilon'E \tag{13.19}$$

and the loss current is:

$$I_l = \omega e'' E = sE \tag{13.20}$$

where σ is the dielectric conductivity. These components are shown in Figure 13.5 and are combined as:

$$I = (i\omega\varepsilon' + \omega\varepsilon'')\frac{C_o}{\varepsilon_o}V = i\omega_o C_o\varepsilon^* V \tag{13.21}$$

The corresponding power loss per cycle at applied V, and maximum voltage V_o, is:

$$W_{loss} = 2\pi\varepsilon'\frac{V_o^2}{2}\tan\delta \quad \text{per cycle} \tag{13.22}$$

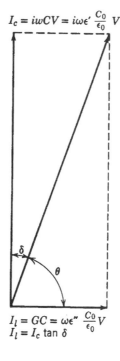

Figure 13.5 [G1]
Charging and Loss Current for a Capacitor

For polymeric insulating materials, not all of the conductance observed derives from the migration of charge carriers in the material. An insulating material may contain bound charges or, under the influence of an electrical stress, develop additional bound charges of various types which are displaced to a limited extent by the action of the electric field. This displacement, discussed above, is an energy consuming process dependent upon both frequency of the applied voltage and the temperature.

13.4   ELECTRICAL BREAKDOWN

Electrical breakdown occurs when the applied voltage can no longer be maintained in a stable fashion without excessive flow of currents, The voltage gradient at failure is called the dielectric strength or electric strength and is usually expressed as kV/mm.

13.4.1 Types of Breakdown

13.4.1.1 Intrinsic

The intrinsic breakdown of a solid is a function of the fundamental properties of the material itself in a pure and defect free state under test conditions which produce breakdown at the highest possible voltage. One theory accounts for intrinsic breakdown by assuming that increased random motion of electrons gain energy faster than they lose it to the material [G1].

13.4.1.2 Thermal Breakdown

Many plastics are noticeably warm or hot after electrical breakdown. At high frequencies (above 1 MHz), failure occurs at relatively low voltages and is usually accompanied by heat emission. Adsorbed moisture may cause polymers to fail at a low voltage with accompanying heat.

Many factors influence the breakdown voltage. Known factors that have an effect on the breakdown voltage are: geometry and size of the part, the thermal conductivity and specific heat of the polymer, rate of voltage increase, magnitude of tan δ, magnitude of ε' and its change with temperature, and the thermal stability of the polymer.

13.4.1.3 Influence of Physical Defects

The presence of cracks, fissures, voids, foreign inclusions, and other types of inhomogeneties influence the breakdown of polymers.

13.4.1.4 Discharge-Dependent Breakdown

Electrical discharges (voltage breakdown in gas) may occur on the surface or in voids or discontinuities of electrical insulation under a wide variety of conditions and *r-f* interference but, more importantly, may cause slow degradation of the surface or within the bulk.

13.4.1.5 Breakdown by Treeing

High electrical stress can also cause breakdown which may initiate a point failure within the bulk at a defect which will produce many fine erratic degradation paths, of "dendritic" character.

13.5   MEASUREMENT TECHNIQUES

There are three commonly used methods that are used for the measurement of resistance (conductance). These are shown schematically in Figure 13.6. Electrode design is very important in the measurement of electrical resistance and is sometimes specified in the test methods for specific materials.

13.6   ALTERNATING CURRENT CHARACTERISTICS

As was mentioned previously, the electrical properties of polymers are largely dependent upon molecular structure [S2]. Examples of this can be seen in the dielectric constant and dissipation factors as a function of temperature and frequency which are very different for polar and nonpolar materials. The values of both properties for nonpolar materials, such as PTFE, are nearly constant over a wide frequency range and at temperatures below the softening point while for polar materials, e.g., phenolic resins and plasticized PVC, the dielectric constant decreases with increasing frequency, and the dissipation factor increases and decreases in a cyclic manner.

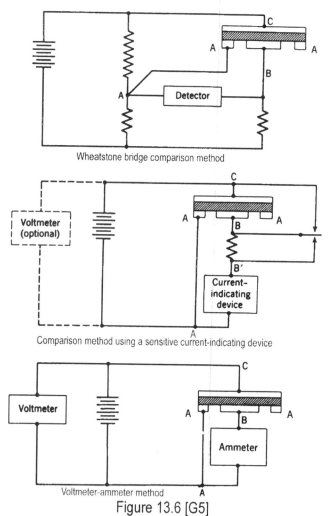

Figure 13.6 [G5]

Typical Circuits for Measuring Electrical Resistance of Insulating Materials

  A. Guard Electrode
  B. Measuring Electrode
  C. High or Voltage Electrode

When dielectric polarization, permittivity, and absorption are plotted as a function of frequency, absorption peaks(maxima in dissipation factor or loss index) occur at the frequencies at which the dielectric constant changes most rapidly as illustrated in Figure 13.7.

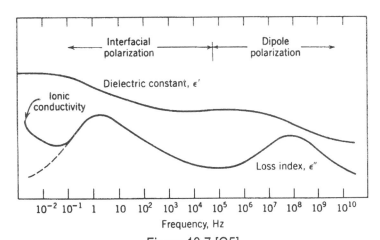

Figure 13.7 [G5]
Representative Example of the Dielectric Constant and Dielectric Loss
Index as a Function of Frequency

At low frequencies the absorption peak (interfacial polarization) is caused by the alternating accumulation of charges at interfaces between different phases of the material and may be strongly influenced by moisture absorption. At high frequencies, the absorption peak is caused by the polarization with dipole orientation. This dipole polarization is related to the frictional losses caused by the rotational displacement of molecular dipoles under the influence of an alternating electric field. At low frequencies, total polarization is indicated but as the frequency is increased the molecular orientation lags behind. At a frequency of about 10^{12} c/s, the dipoles are unable to follow the oscillations of the field.

In molecularly symmetrical materials like PTFE an electric dipole does not exist in the pure (non-polar) material. A replacement of a fluorine atom with a chlorine results in an unsymmetrical structure and the introduction of dipole losses.

Additional types of polarization, evidenced by small atomic and electronic absorption, produces dielectric loss peaks at higher frequencies in the infrared and visible range. The dielectric constant at frequencies higher than those of dielectric absorption is approximately equal to the square of the optical index of refraction as given by Maxwells' relationship:

$$\varepsilon = n^2 \qquad (13.23)$$

For valid comparison, both ε and n should be measured at the same frequency but this is generally not the case since visible light is used for the refractive index measurement $\left(5 - 7 \times 10^{14}\ c/s\right)$. A large variation of ε and n_D^2 may be an indication of semi-conduction or permanent dipoles in the dielectric.

For an ideal nonpolar polymer, PTFE e.g., the dissipation factor, tan δ, and the loss index, ε'', are essentially constant. This is not the case for polar polymers. Figure 13.8 shows that at high frequencies, crystalline barriers or the viscosity of a glassy polymer hinder dipole movement and the dipoles cannot orient in the rapidly reversing field. Consequently, the dipoles do not influence the a.c. characteristics $\left(\varepsilon'_\infty$ and $\varepsilon''_\infty\right)$ at very high frequencies. A decrease in the frequency results in greater dipole movement with an associated loss of energy. The loss peak decreases when the dipole reversals are slow enough to encounter less rotational resistance. Finally, a point is reached where the dielectric constant has its highest value, ε'_s.

The dielectric relaxation time, τ, is approximately the mean time for the molecular dipoles to turn 180° against the viscous resistance in a glassy polymer or the mean time for dipoles in a crystalline structure to jump from one position to another by 180°. The relationships shown in Figure 13.8 can be expressed in terms of a single relaxation time [S3].

$$\varepsilon' = \varepsilon'_\infty + \frac{\left(\varepsilon'_s - \varepsilon'_\infty\right)}{1 + \omega^2 \tau^2} \qquad (13.24)$$

$$\varepsilon'' = \frac{\left(\varepsilon_s - \varepsilon'_\infty\right)\omega\tau}{1 + \omega^2 \tau} \qquad (13.25)$$

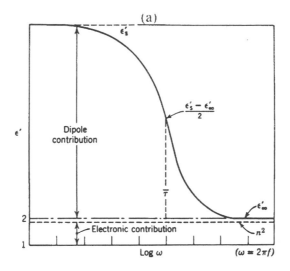

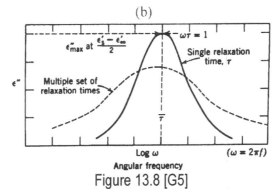

Figure 13.8 [G5]
Dielectric Behavior of an Ideal Polar Material

ε_s' = Static dielectric constant

ε_∞' = High frequency dielectric constant

τ = Dielectric relaxation time

n = Optical index of refraction

(a) Effect of frequency on the static value of the dielectric constant
(b) Debye curves representing a set or distribution of relaxation times

where these quantities are the same as in the nomenclature of Figure 13.8.

The effects of temperature on the static dielectric constants of crystalline and glassy polymers is shown in Figure 13.9. At high temperatures, thermal agitation of the molecules largely prevents

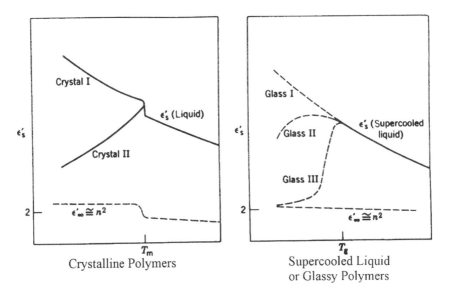

Figure 13.9 [G5]
Effect of Temperature on the Dielectric Constant of Polar Compounds
Static   = Solid Line
High Frequency = Dashed Line

dipole orientation. As the temperature is decreased, structural freezing or demobilization to a crystal or glassy state occurs which partially locks the dipoles.

The dielectric relaxation time decreases as the temperature increases as given by:

$$\tau = A\exp\left(H^*/RT\right) \qquad (13.26)$$

where:

A = pre-exponential constant
H^* = activation energy involved in dipole rotation
R = gas constant
T = absolute temperature (K)

Since the dielectric loss is a maximum when $\omega\tau = 1$ (see Figure 13.8), the loss peaks must shift toward higher frequencies as the temperature increases. This is illustrated in Figure 13.10.

13.7   STATIC CHARGE AND TRIBOELECTRIFICATION

When polymers are brought into intimate contact, e.g. by rubbing, one becomes positively charge while the other becomes charged on separation. In other words they act as electron donors or acceptors. The level to which this charging behavior occurs defines a sequence of arrangement known as the triboelectric series.

The word electric is derived from the Greek word for amber since it was known for a long time that the rubbing of amber with fur produced powers of attraction. As early as 1757, it was observed that materials could be arranged in a triboelectric series.

No correlation of triboelectrification with other properties, such as surface or volume resistivity, has been found although Coehn (1898) [S4] derived the general rule that if two substances

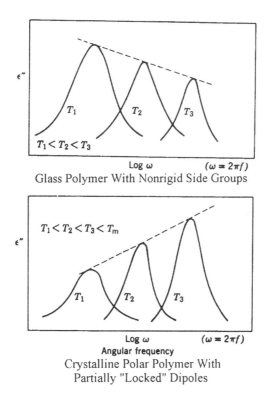

Figure 13.10 [G5]
Loss Index at Different Temperatures as a Function of Frequency

become charged by mutual contact, the substance with the highest dielectric constant will obtain a positive charge. This is illustrated in Figure 13.11.

The presence of moisture, which is always a concern in hydrophilic polymers, will have a large effect as might be expected. Since water has a high dielectric constant, hydrophilic polymers will be antistatic under humid conditions but static under dry conditions.

The charging of polymers is often an important consideration during polymer processing, particularly in the case of polymer fiber and sheet extrusion at high speeds. Although the highest charges observed (500 e.s.u./cm^3) are caused by the transfer of relatively few charges (one electron/10^4 Å), voltages as high as 30,000V or higher have been observed to build up during certain types of processing operations where sheets of polymer products are moved at high speeds. This can result in severe dis-

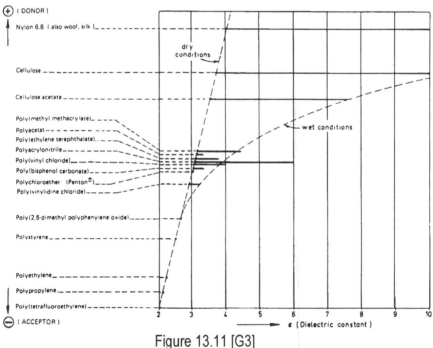

Figure 13.11 [G3]
Triboelectric Series of Polymers

comfort at the least, and serious safety hazards, at the most, to the operators of such processes.

Molecular structure is the prime determinant. Shashoua [S5] found that for a given structural group, the fastest charge decay occurs when a group capable of donating or accepting electrons is present as a side chain substituent rather than in the main chain of the polymer. Also small quantities of ionic impurities can also influence the ability of a polymer to bind ions on its surface and influence its charge selective ability.

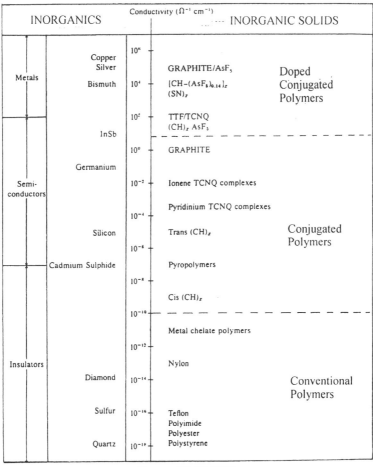

Figure 13.12 [G3]
Electrical Conductivity of Different Classes of Materials Including Conducting Polymers

13.8   ELECTRICALLY CONDUCTIVE POLYMERS

Polymers with conjugated π-electrons backbones display unusual electronic properties such as low energy optical transitions, low ionization potentials, and high electron affinities. The result is a class of polymers which can be oxidized or reduced more easily and more reversibly than conventional polymers. Charge transfer agents or dopants affect this oxidation/reduction and convert insulating polymers to conducting polymers with near metallic conductivity in many cases. The relative ranges of conductivities of doped electrically conductive polymers compared with conventional materials are shown in Figure 13.12.

Polyacetylene has generated the most interest both from theoretical and experimental viewpoints. A significant break-through occurred in 1979 with the discovery [S6] that poly(p-phenylene) could also be doped to high conductivity. This demon-

Polymer	Structure	Typical methods of doping	Typical conductivity (ohm-cm)$^{-1}$
Polyacetylene	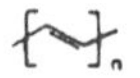	Electrochemical, chemical (AsF_5, I_2Li, K)	500 (2000 for highly oriented films)
Polyphenylene	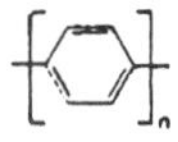	Chemical (AsF_5, Li, K)	500
Poly(phenylene sulfide)	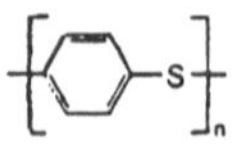	Chemical (AsF_5)	1
Polypyrrole	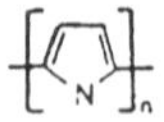	Electrochemical	600
Polythiophene	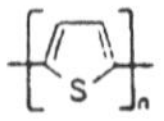	Electrochemical	100
Poly(phenyl-quinoline)	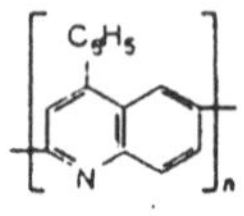	Electrochemical, chemical (sodium naphthalide)	50

Figure 13.13 [G3]
Doped Conjugated Organic Polymers

strated that polyacetylene is not unique and other polymers could be prepared that exhibit high conductivity. This led to the development of a number of new systems, such as shown in Figure 13.13.

13.9 PIEZO- AND PYROELECTRIC POLYMERS

Most piezo- and pyroelectric polymers consist of polymer electrets which are dielectrics in which semipermanent polarization is related to a semipermanent surface charge.

Polymers containing specific structural groups, such as poly (vinylidene fluoride), poly (tetrafluoro ethylene), poly (vinylidene chloride), and poly(vinylidene cyanide) are examples of materials capable of exhibiting piezoelectric properties. When these polymers are placed, while hot, in a d-c electric field, polarized or "poled", and cooled, while the d-c field is still applied, stable trapped charges remain. The resulting materials that display piezoelectric properties are useful in microphone and small speakers.

The extent to which piezoelectricity develops in the polymer depends principally on a combination of factors of morphology, thermal history, and poling The polar β form of PVDF, for example, is most amenable to the generation of strong piezoelectricity compared to the α form which exhibits low piezoelectric response [S7].

13.10 APPLICATIONS

Commercial applications of conducting polymers are based on the promise of a novel combination of light weight, processability, and electronic conductivity. Key applications include rechargeable batteries, circuitry elements, either as conductive traces or Shottky/p-n junction devices, or in electromagnetic shielding. The weight advantages can be visualized readily.

Although the viability has been demonstrated, the number of successful applications depend on the solution of a number of important problems which include processing, consistent quality, reliability, and usable lifetimes.

13.11 MOLCALC ESTIMATIONS OF THE DIELECTRIC
       CONSTANT, ε, AND COMPARISON TO
       LITERATURE VALUES FOR A SERIES OF
       REPRESENTATIVE POLYMERS

The quantity Molcalc uses for estimating the dielectric
constant, ε, is the molar polarization of a dielectric based on the
estimation method of Van Krevelen and Hoftyzer and is given by:

$$\varepsilon = \left(\frac{P_V}{M} \right)^2 \qquad (13.27)$$

where:

P_V = molar polarization

M = M_w of the structural repeat unit

The dielectric constant is related to the refractive index by
Maxwell's relationship, $\varepsilon = n^2$ (Equation 13.23) and the solubility
parameter by:

$$\delta \approx 7.0\varepsilon \qquad (13.28)$$

As was the case in other instances these secondary relationships are
not utilized by Molcalc since this would entail two separate
estimations, the second one of which is dependent upon the
previous. This method can cause serious errors in the second
estimate. For this reason, only estimates of ε, based upon primary
data are used.

The Maxwell relationship has an additional drawback in
that ε is generally measured at relatively low frequencies
$\left(10^2 - 10^9 \, \text{Hz} \right)$, whereas n is measured in the range of visible light,
$5 - 7 \times 10^4 \, \text{Hz}$ usually at the sodium D line). The relationship of ε
to δ is approximate at best.

In the simplest case, i.e. when the dielectric is a pure
compound and the dipole moment, μ , is small ($\mu < 0.6$ Debye
units), Debye's equation can be employed. Debye's equation is
given by:

$$\mu = \left[\frac{\left(\dfrac{\varepsilon - 1}{\varepsilon + 2} - \dfrac{n^2 - 1}{n^2 + 2} \right) \dfrac{M}{\rho}}{20.6} \right]^{1/2} \qquad (13.29)$$

Table 13.1 provides comparisons between calculated and literature values for the selected list of polymers.

Table 13.1

Comparison of Estimated Values of the Dielectric Constant, ε, with Literature Values

Polymer	Dielectric Constant, ε	
	Calculated	Literature
Poly(ethylene)	2.2	2.3
Poly(propylene)	2.2	2.2
Poly(isobutylene)	2.2	2.2
Poly(vinyl chloride)	3.1	3.1
Poly(vinyl fluoride)	8.1	8.5
Poly(chloro trifluoro ethylene)	2.4	2.3-2.8
Poly(vinyl acetate)	3.3	3.3
Poly(cis butadiene)	2.3	2.5
Poly(cis 1,4-isoprene)	2.3	2.4
Poly(styrene)	2.6	2.6
Poly(αmethyl styrene)	2.5	2.6
Polycarbonate	3.1	2.6-3.0
Poly(ethylene terephthalate)	3.5	2.9-3.2
Poly(butylene terephthalate)	3.3	3.2-3.3
Poly(3,3-bis chloromethyl oxacyclobutane)	3.1	3.0
Poly(methylene oxide)	2.9	3.1
Poly(vinyl methyl ether)	2.5	----
Poly(phenylene oxide)	2.9	2.6

continued

Table 13.1 continued

Polymer	Dielectric Constant, ε	
	Calculated	Literature
Poly(hexamethylene adipamide)	4.1	4.1
Poly(acrylamide)	5.7	----
Poly(acrylonitrile)	3.2	3.1
Poly(methyl methacrylate)	3.1	2.6-3.7
Poly(cyclohexyl methacrylate)	2.9	----

13.12 ABBREVIATIONS AND NOTATION

C	capacitance
C_o	geometric capacitance in vacuo
C_p	capacitance, parallel
C_v	capacitance of an equivalent vacuum
D_i	bulk or lattice diffusion
D	displacement
E	electric field strength
H^*	activation energy for dipole rotation
I	current
n	index of refraction
n_D	index of refraction measured using the sodium D line
P_v	molar polarization
Q	charge
Q	experimental activation energy
q	charge carried by a particular species
R	gas constant
R	resistance
R	atom or group
T	temperature, K
$\tan \delta$	dissipation factor
V	voltage

$\bar{v}$	drift velocity of a charge carrier
W	energy
δ	solubility parameter
ε''	loss factor
ε_o	relative permittivity or dielectric constant of a vacuum $\left(\varepsilon'/\varepsilon\right)$
ε'	dielectric constant or permittivity
ε^*	complex dielectric constant
ε_s	static dielectric constant
κ'	permittivity or dielectric constant $\left(\varepsilon'/\varepsilon_o\right)$
σ	conductivity
ρ	resistivity
μ	mobility of ions or lattice components
ω	frequency
μ	dipole moment

13.13 SPECIFIC REFERENCES

1. Frolich, H., *Theory of Dielectrics,* Clarendon Press, Oxford, UK, 1950.
2. Hoffman, J., *IRE Trans. Component* Pts. C **P4(2)**, pp 42-69, June 1957.
3. Debye, P., *Polar Molecules*, Dover Publications, Inc., New York, 1945.
4. Coehn, A., *Ann. Phys.* **64**, 217(1898).
5. Shashoua, J., *J. Polymer Sci.* A-1, **1** 169(1963).
6. Heeger, A.J., MacDiarmid, A.G.., et al., *Phys. Rev. Letter* **39**, 1098(1977); **45**, 1123(1980).
7. Douglass, D.C., McBrierty, V.J., and Wang, T.T., *J. Chem. Phys.* **77**(11), 5826(1982).

13.14 GENERAL REFERENCES

1. Kingery, W.D., Bowen, H.K., Uhlmann, D.R., *Introduction to Ceramics*, 2nd Ed., John Wiley & Sons, New York, N.Y., 1976.
2. Van Krevelen, D.W. and Hoftyzer, P.J., *Properties of Polymers*, 2nd Ed. Elsevier, New York, N.Y., 1976.
3. Van Krevelen, D.W., *Properties of Polymers*, Elsevier, New York, N.Y., 1990.
4. Ku, C.C. and Liepins, R., *Electrical Properties of Polymers*, Hanser, New York, 1987.
5. Mathes, K.N., in *Electrical and Electronic Properties of Polymers: A State of the Art Compendium*, J. I. Kroschwitz, Ed., J. Wiley, New York, 1988.
6. Cusack, N., The Electrical and Magnetic Properties of Solids, Longmans, London, 1967.
7. Brandup, J. and Immergut, E.H., *Polymer Handbook*, Wiley-Interscience, New York, 1977.

CHAPTER 14

Magnetic Properties

14.1 INTRODUCTION

This chapter will serve as a brief introduction to the magnetic properties of materials. For more complete and detailed discussions on this topic, the reader is referred to the general references at the end of this chapter.

It should be noted that unlike the preceding chapters, polymer group contributions and their additivity to polymer diamagnetic susceptibilities are not given. Correlations of estimated diamagnetic susceptibilities of polymers with experimental values are limited and also tend to fall within a narrow range (0.5-0.8) which lessens their usefulness. For those readers, who are interested in the estimation of polymer diamagnetic susceptibilities from additivities, Van Krevelen should be consulted [S1, G1]. In addition, diamagnetic susceptibility has no direct correlation with chemical shift factors, the resolution of which is highly dependent upon the experimental technique, and the instrumental conditions, (frequency, field strength, pulse conditions, etc.). Although we refer the reader to standard tables of chemical shifts of principal structural groups contained in detailed texts and also furnished by manufacturers of nuclear magnetic resonance (NMR) instruments, we have included a listing of the chemical shifts for polymers in section 14.4.3 [G2].

14.2 ORIGIN AND CLASSIFICATION OF MAGNETISM

The magnetic properties of solids originate in the motion of the electrons and in the permanent magnetic moments of atoms and electrons. The fundamental magnetic phenomenon is the Coulomb interaction between magnetic poles. For two magnetic poles whose strengths are m_1 and m_2 (webers) respectively, separated by the distance r, in meters, the force F (newtons) exerted on one pole by the other is:

$$F = \frac{m_1 m_2}{4\pi\mu_o r^2} \tag{14.1}$$

where: μ_o = permeability of a vacuum $\left(4\pi \times 10^{-7}\ \text{H/m}\right)$

H = henry

m = meter

A magnetic field can be generated by magnetic poles or by an electric current. When a current of i (amperes) flows in wind-

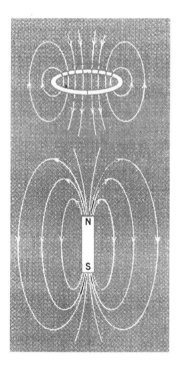

Figure 14.1
The Magnetic Field Distribution Around a Current Loop
and a Bar Magnet

ings having n turns per meter, the intensity of the field, H, is defined by:

$$H = ni \qquad (14.2)$$

The unit of the magnetic field is the ampere per meter which is equivalent to $4\pi 10^{-3} Oe = 0.012 Oe(1\ Oe = 79.6\ A/m)$. The field surrounding a bar magnet is similar to the field distributions surrounding a current loop as shown in Figure 14.1.

Magnetic poles are analogous to electric dipoles which consist of two opposite and separated charges. When a magnetic pole of strength m (Wb) is brought into a magnetic field of intensity $H(A/m)$, the force $F(N)$ acting on the magnetic pole is:

$$F = mH \qquad (14.3)$$

If a bar magnet of length l(m), which has magnetic poles m and -m at its ends, is placed in a uniform magnetic field H, each pole is acted upon by forces as indicated in Figure 14.2, giving rise to a couple of force whose moment is:

$$L = -mlH \sin\theta \qquad (14.4)$$

where θ is the angle between the direction of the magnetic field H

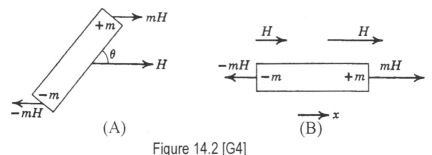

(A) (B)

Figure 14.2 [G4]
Schematic Representation of Magnets in Force Fields

A. Coupled Forces in a Uniform Field
B. Translational Forces in a Gradient Field

and the direction of the magnetization $(-m \rightarrow +m)$ of the magnet. Thus a uniform magnetic field can rotate the magnet but cannot translate it from its original position unless there is a gradient of the field $\dfrac{\partial H_x}{\partial x}$ which provides a translational force given by:

$$F_x = ml \frac{\partial H_x}{\partial x}$$
(14.5)

14.3   CLASSIFICATION OF TYPES OF MAGNETIZATION

There are various kinds of magnetization, each of which is characterized by its own magnetic structure. The magnetic moment per unit volume (m^3) of a magnetic substance is called the intensity of magnetization and is denoted by the vector I. The unit of I is the weber per square meter $(Wb/m^2$ of $I = 1/4\pi \times 10^4$ gauss$)$. The magnetic flux density B is commonly used in engineering applications to describe the magnetization and is given by:

$$B = I + \mu_o H$$
(14.6)

The relation between the intensity of magnetization I and the magnetic field H is:

$$I = \chi H$$
(14.7)

where χ is the magnetic susceptibility. The unit of χ is the henry per meter, which is the same as that of μ_o. The susceptibility is generally measured by the relative susceptibility and is given by:

$$\overline{\chi} = \frac{\chi}{\mu_o}$$
(14.8)

The value of $\overline{\chi}$ is a dimensionless quantity, and its value is 4π times larger than the susceptibility of χ in the cgs system.

The relative susceptibility ranges from 10^{-5} for very weak magnetism to 10^6 for very strong magnetism. In some cases it takes a negative value. The relation between I and H is not always linear which means that $\bar{\chi}$ depends on the intensity of the magnetic field.

The behavior of $\bar{\chi}$ is dependent upon the structure of the material and the various types of magnetism can be classified as follows:

- Diamagnetism
- Paramagnetism
- Antiferromagnetism
- Metamagnetism
- Ferrimagnetism
- Ferromagnetism
- Parasitic ferromagnetism

Carbon based polymers not containing impurities are diamagnetic. Diamagnetism is a universal property of matter. Paramagnetism occurs in only two classes of organic substances: those containing metals of the transition groups of the periodic system, and those containing unpaired electrons in the free radical or the triplet state.

There are two possible atomic origins of magnetism which lead to the magnetization of magnetic substances, viz., orbital motion and spin of electrons. An atom which has a magnetic moment caused by spin or by orbital motion of electrons or by both is generally called a magnetic atom. The magnetic moments of the important magnetic atoms such as iron, cobalt, and nickel are due to the spin motion of the electrons.

There is a magnetic moment, lm_B, associated with an electron's atomic quantum state, where l is the azimuthal quantum number of the state, and μ_B is called the Bohr magnetron which is equal to $\mu_o e\hbar / 2m_e$ $\left(1.165 \times 10^{-29}\,\text{Wb} - \text{m}\right)$. The component of the magnetic moment parallel to the applied field is also quantized and is equal to $m_l m_B$ where m_l is the magnetic quantum number, which may have any integral value between $\pm l$. In any solid, the orbital motions of the atomic electrons give rise to permanent atomic magnetic moments. In the case of filled atomic shells, however, all the moments cancel.

A second source of permanent moments is the electron spin. The term refers to the fact that electrons possess intrinsic angular moments and magnetic moments like those expected of a charged body. The spin angular momentum of the electron is $\pm 1/2$ $(\hbar/2)$, and the spin magnetic moment of a completely filled state (two electrons of opposite spin) or a full atomic shell, is zero.

Diamagnetism is a weak magnetism in most solids in which a magnetization is exhibited opposite to the direction of the applied field. This is schematically represented in Figure 14.3. In diamagnetic solids such as Bi, Cu, Ag, and Au, the magnetic moments oppose the magnetic flux density, B, causing the field inside the solid to be less than the field outside. The susceptibility is negative and the order of magnitude of $\overline{\chi}$ is usually about 10^{-5}. The origin of this magnetism is an orbital rotation of electrons about the nuclei induced by the electric field. From Lenz's law, we know that the induced current produces a magnetic flux which opposes any change in the external field. It can be observed only when all other types of magnetism are totally absent.

Many solids that have small but positive magnetic suscepti-

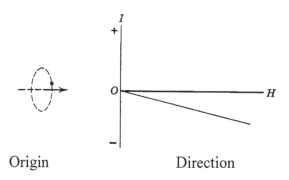

Origin   Direction

Figure 14.3 [G4]
Schematic Representation of Diamagnetism

bilities are called paramagnetic. The source of paramagnetism is the permanent magnetic moments possessed by the atoms and electrons in the solid. The magnetization I is proportional to the magnetic field H [Figure 14.4(a)]. The order of magnitude of $\overline{\chi}$ is 10^{-3} to 10^{-5}. In most cases, paramagnetic substances contain magnetic atoms or ions whose spins are isolated from their magnetic environment and can more or less freely change their directions [Figure 14.4(b).

Under ambient temperatures, the spins take random orientations. The application of a magnetic field produces a slight change in the spins and produces a weak induced magnetization parallel to the applied magnetic field. Here, susceptibility is inversely proportional to the absolute temperature (the Curie law) [Figure (14.4c]. These aspects are illustrated in Figure 14.4.

Conduction electrons which form an energy band in metallic crystals exhibit paramagnetism since in this case, the excitation of minus spins to the plus spin band is opposed by an increase of kinetic energy of electrons irrespective of temperature. In this instance, the susceptibility is independent of temperature (Pauli paramagnetism).

Antiferromagnetism is a weak magnetism which is similar to paramagnetism in the sense of exhibiting a small positive susceptibility. In this case a kink occurs in the $1/\chi - T$ curve at the Néel temperature. Below the Néel temperature l[Figure 14.5(b)] an antiparallel spin arrangement is established in which plus and

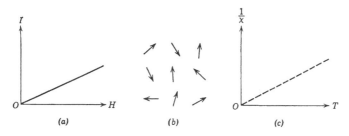

Figure 14.4 [G4]
Schematic Representation of the Effect of the Magnetic Field (a), Spin Directions (b), and Temperature (c), on Susceptibility for Paramagnetism

minus spins completely cancel each other. This is illustrated in Figure 14.5(a).

Ferrimagnetism is the term proposed by Néel [S2] to describe the magnetism of ferrites. In ferrites, magnetic ions occupy two kinds of lattice sites, A and B, and spins on A sites point in the plus direction, while those on B sites point in the minus direction because of a strong negative interaction between the two spin systems on A and B. Since the number of magnetic ions and also the magnitude of spins of individual ions are different on the A and B sites, such an ordered arrangement of spins produces magnetization. At the Curie point, the arrangement of spins becomes completely random and the spontaneous magnetization vanishes. These aspects are illustrated in Figure 14.6. Above the Curie point, the substance exhibits paramagnetism, and the susceptibility decreases with increasing temperatures. The $1/\chi$ vs. T plot is curved as shown in Figure 14.6(b) with the linear extrapolation usually intersecting the temperature axis on its negative side.

In ferromagnetism, the spins are aligned parallel to one another as a result of a strong positive interaction between the neighboring spins. This is illustrated in Figure 14.7. Above the Curie point, $1/\chi$ rises from zero linearly with temperature accord-

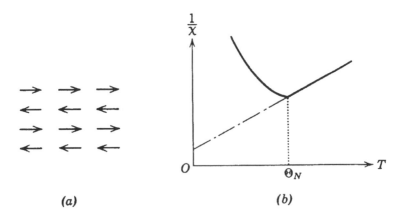

(a)

(b)

Figure 14.5 [G4]
Schematic Representation of Antiferromagnetism

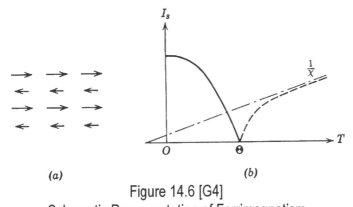

(a)   (b)

Figure 14.6 [G4]
Schematic Representation of Ferrimagnetism

ing to the Curie-Weiss law. Ferromagnetism is exhibited mostly by metals and alloys and a few oxides (Cr_2O and EuO).

In spite of spontaneous magnetization, a ferromagnetic or ferrimagnetic substance is usually not spontaneously magnetized but exists in a demagnetized state because the interior is divided into many magnetic domains each of which is spontaneously magnetized. Since the direction of domain magnetization varies from domain to domain, the resultant magnetization can be changed from zero to the value of spontaneous magnetization.

When an external field *(H)* is applied, the magnetization reaches a saturation level equal to the spontaneous magnetization. These substances display hysteresis and remain magnetized even when the field is removed as is illustrated in Figure 14.8.

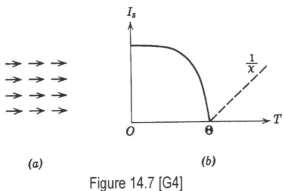

(a)   (b)

Figure 14.7 [G4]
Schematic Representation of Ferromagnetism

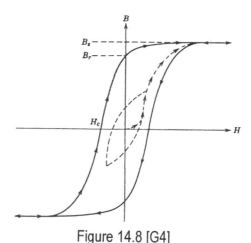

Figure 14.8 [G4]

B Versus *H* Hysteresis Loop for a Ferromagnetic Material

The dashed line indicates the behavior on initially increasing the applied field *H*. If the field is increased to saturation, and then decreased to zero and increased in the reverse direction, the solid curve will result. The dotted curve represents a hysteresis curve which does not reach saturation.

Parasitic ferromagnetism is the name of a weak ferromagnetism which accompanies antiferromagnetism such as is observed for αFe_2O_3. The spontaneous magnetization of this form of magnetism disappears at the Néel point, where the antiferromagnetic arrangement of spins disappears.

14.4 MAGNETIC RESONANCE

The absorption of energy by a material placed in a steady oscillating magnetic field is termed magnetic resonance. The nature of the absorption emanates from transitions between energy eigen states of magnetic dipoles.

Magnetic resonance is attributed to two kinds of transitions:

1. Transitions between energy states of the magnetic moment of the nuclei in the steady magnetic field which is called nuclear magnetic resonance (NMR).

2. Transitions between energy states of the magnetic moment of the electrons in the steady magnetic field

which is called electron spin resonance (ESR).

ESR occurs at a much higher frequency than NMR because the magnetic moment of an electron is about 1800 times that of a proton. ESR is observed in the microwave region (9-38 GHz), while NMR occurs at radio frequencies (10-600 MHz).

14.4.1 NMR

NMR spectroscopy is an exceedingly powerful tool used for structural investigations. The nuclei of some elements possess a net magnetic moment (μ), viz. in the case that the spin (I) of the nucleus is non-zero. This condition is met if the mass number and the atom number of the nucleus are not both even, e.g., Carbon-12 and Oxygen-16. Initially NMR was restricted to nuclei of high natural abundance and high magnetic moment $\left({}^{1}H, {}^{19}F, \text{ and } {}^{31}P \right)$. With modern pulsed Fourier-Transform (FT) spectrometers, spectra can be collected rapidly and less abundant nuclei, in particular, ${}^{13}C$ NMR is routine.

For polymers, the two most important nuclei are the hydrogen isotope ${}^{1}H$ (protonium) and the carbon isotope ${}^{13}C$. Proton NMR gives information on the outer environment of the molecule (mainly hydrogen atoms) while ${}^{13}C$ NMR provides information on the carbon skeleton of the molecule. Contrasts between these two nuclei are listed in Table 14.1.

The nuclear spin I, in a magnetic field H_o, generates $2I + 1$ energy levels, so a spin $I = \frac{1}{2}$ results in two levels. These energy states are characterized by a magnetic quantum number m_I, and are separated in energies ΔE, which is field dependent.

$$\Delta E = h\frac{\gamma H_o}{2\pi} = h\nu_o \qquad (14.9)$$

where:

γ = magneto-gyric ratio ($2\pi\mu / Ih$)

H_o = magnitude of the applied magnetic field

Table 14.1

Comparison of ^{1}H and ^{13}C Properties Relative to the NMR Response [S1]

Isotope	^{1}H	^{13}C
Atomic Number	1	6
Mass Number (atomic mass)	1	13
Natural Abundance (%)	99.98	1.1
Nuclear Spin Quantum Number	½	½
Magnetic Moment	2.79	0.70
Magneto-Gyric Ratio	267.4	67.2
Relative Sensitivity (^{1}H = 1)	1	0.016
Resonant Frequency in a Tesla Field	42.6	10.7

v_o = resonant frequency (Hz)

For the lower state, the magnetic moment is parallel to the applied field, H_o, corresponding to $m_I = + ½$ (α-state), and for the upper state the magnetic moment is antiparallel to B_o, corresponding to $m_I = - ½$ (β-state).

The bulk magnetization (M) is generated from the nuclear spins and behaves in a magnetic field in much the same way as a gyroscope behaves in a gravitational field. The force, generated by $\vec{H}_o$ or $\vec{M}$, will cause $\vec{M}$ to precess about $\vec{H}$ at a frequency $\gamma H_o / 2\pi = v_o$(Hz) (the Larmor precession). After a perturbation, the system will start to return to equilibrium, i.e., the populations of the states will return to their Boltzmann distribution and magnetization returns to an equilibrium value M_o.

The effect of γ and H_o can be realized from Table 14.1. This table shows that at a given field strength, the frequency at which a proton signal occurs is roughly four times that for ^{13}C. The splitting of magnetic energy levels of protons is dramatically affected as the frequency and field strength are increased.

Present NMR techniques use powerful instruments of the

pulsed Fourier Transform type. These instruments employ a powerful constant external magnetic field and pulse the sample repeatedly with strong *RF*. Between the pulses, the magnetic response from the sample is measured. This shortens the measuring time at least a hundredfold. For ^{13}C NMR where the signals are very weak owing to the low natural abundance of ^{13}C isotopes, computer accumulation of the responses, obtained after each *RF* pulse, gives spectra of sufficient quality in a relatively short time.

The magnetic field experienced by the nuclei is slightly modified by the presence of surrounding electrons (chemical shift) and by the presence of interacting neighboring nuclei (spin-spin coupling, causing multiple structure). In liquids, where the molecules can rotate freely and the spectral lines are extremely narrow, these multiplet structures are observed with High Resolution NMR. In solids, where all the nuclei are more or less fixed in position, dipole-dipole interaction causes a broadening of the resonance absorption (Wide-Line NMR). In this case line narrowing can be obtained by suitable NMR techniques and pulse sequence programs.

14.4.2 High Resolution NMR

The phenomena of chemical shift, spin-spin coupling, and time-dependent effects constitute the major applications of NMR. These are observed in gaseous or liquid form materials.

14.4.3 Chemical Shift

The most significant difference that influences the chemical shift of ^{13}C nuclei arise from changes in electronegativity of substituents. Other differences, especially in proton magnetic resonance, arise from long-range effects due to movements of electrons in multiple bonds under the influence of the applied magnetic field. This overall effect of the chemical environment can be described in terms of a shielding constant σ, which is related to the field by:

$$H_{at\ nucleus} = H_o(1 - \sigma) \tag{14.10}$$

The magnetic sweep ΔH is only a tiny fraction (10^{-5} for PMR and 10^{-4} for CMR) of the total strength and therefore the position of the various resonance signals which excludes measurement of their chemical shift in absolute units. Tetramethyl silane $\left((CH_3)_4 Si\right)$ is used as a reference compound and the chemical shift is:

$$\delta = \frac{H - H_R}{H_R} \times 10^6 \qquad (14.11)$$

where:

   H = magnetic field strength of the sample nucleus resonance frequency

   H_R = magnetic field strength of the reference nucleus

For polymers, the chemical shifts have been tabulated by McBrierty and Packer [G2] and are listed in Figure 14.9.

14.4.4 Spin-Spin Interactions

   These signals often show a fine structure with a characteristic splitting pattern due to spin-spin interactions with neighboring nuclei which are mostly one to four bonds away from each other in the molecule. The splitting corresponds to the slightly different energy levels of nuclei whose spins are parallel and anti-parallel to those of their neighbors and can lead to complex spectra. These spin-spin interactions which are called coupling constants can be experimentally determined.

   In proton spectra of polymers, exact knowledge of the coupling constants and chemical shifts makes it possible to simulate the spectra and therefore interpret complex spectra, in terms of structure.

14.4.5 Time Dependent Effects

   NMR signals are sometimes influenced by time-dependent phenomena such as conformational or prototropic changes, which take place at a rate faster than the line width and comparable to the inverse of the differences between the frequencies of the transitions of the different sites.

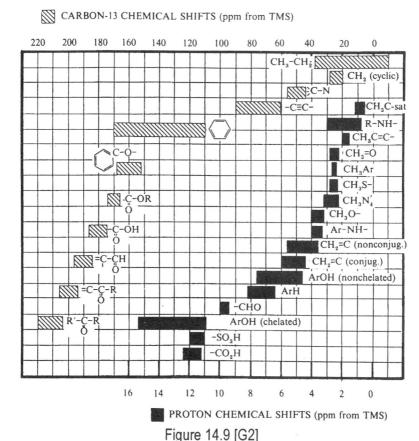

Figure 14.9 [G2]

^1H and ^{13}C Chemical Shifts (ppm from TMS)

14.5   METHODS OF BAND AND LINE NARROWING FOR NMR OF SOLID POLYMERS

The development of pulse techniques to obtain narrow resonance line of solid specimens is responsible for widespread use of contemporary NMR spectroscopy.

There are four major factors responsible for band and line broadening in the spectra of solids.

1. Strong, direct, dipolar coupling
    In solids, the averaging effects which occur in liquids due to proton abundance, and which contribute to local magnetic fields, are absent.

2. Weaker indirect coupling of the spins via intervening covalent bonds

   This is a scalar coupling, similar to the coupling in HR-NMR and is known as J-coupling.

3. Chemical shift anisotropy (CSA)

   The chemical shift, the most important tool for structural determinations, is caused by the shielding (or de-shielding) of nuclei by electrons. A lack of spherical distribution of electrons around the nucleus results in an anisotropic shift in solids accompanied by line broadening due to an angular distribution of the bonds with respect to the magnetic field.

4. Influence of time dependent effects (relaxation).

These four factors can be eliminated or minimized by: 1) the use of high power decoupling techniques 2) the scalar J-coupling (smaller than dipolar coupling) is also removed by high power decoupling 3) the influence of chemical shift anisotropy can be removed by the "Magic Angle Spinning" technique which consists of mechanically rotating the sample about an axis of 54.7° with respect to the magnetic field direction.

The rotation frequency of the sample must be at least the order of CSA line width, usually about 3 kHz. At the magic angle the local field at a ^{13}C nucleus caused by a proton becomes zero and the isotropic chemical shift is obtained, identical with that in solution.

Other techniques, such as Cross Polarization, can be used to enhance the sensitivity of ^{13}C solid state NMR.

14.6   ABBREVIATIONS AND NOTATION

A	ampere
B	magnetic flux density
E	energy
F	force
H	henry

H	magnetic field strength
H_R	magnetic field strength of a reference
H_o	magnitude of applied magnetic field
h	Planck's constant $\left(6.6256 \times 10^{-34}\ \text{J - sec}\right)$
$\hbar$	Planck's constant $\left(\hbar = h/2\pi\right)$
i	ampere
I	intensity of magnetization
I	nuclear spin
m	magnetic moment or strength
M	bulk magnetization
m_e	mass of a single electron
r	distance
RF	radio frequency
χ	magnetic susceptibility
δ	chemical shift
γ	magnetogyric ratio
μ_o	permeability of a vacuum
σ	shielding constant

14.7   SPECIFIC REFERENCES

1. Van Krevelen, D.W. and Hoftyzer, P.J., *Properties of Polymers*, 2nd Ed. Elsevier, New York, N.Y., 1976.
2. Néel, L., *Ann. Physique* **3**, 137 (1948).

14.8   GENERAL REFERENCES

1. Van Krevelen, D.W. and Hoftyzer, P.J., *Properties of Polymers*, 2nd Ed. Elsevier, New York, N.Y., 1976.
2. McBrierty, V.J., and Packer, K.J., *Nuclear Magnetic Resonance in Solid Polymers*, Cambridge University Press, Great Britain, 1993.

3. Van Krevelen, D.W., *Properties of Polymers*, Elsevier, New York, N.Y., 1990.

4. Chikazumi, Sōshin, *Physics of Magnetism*, John Wiley & Sons, New York, 1964.

5. Williams, D.E.G., *The Magnetic Properties of Matter*, Longmans, Green and Co. Ltd., London,1966.

6. Thompson, J.E., *The Magnetic Properties of Materials*, Newnes Books, John Wright & Sons, Stonebridge Press, Bristol, 1968.

7. McBrierty, V.J., "Nuclear Magnetic Resonance Spectroscopy of Polymers in the Solid State", Chap. 19 in *Comprehensive Polymer Science*, ed. G. Allen, Pergamon Press, 1989.

CHAPTER 15

Viscoelasticity and Mechanical Properties

15.1 INTRODUCTION

Elastic solids are those that undergo reversible deformation when subjected to a stress. Metals are elastic solids and generally follow Hooke's law. Ceramics, although brittle materials, can be considered as another class of elastic materials, albeit, over a very narrow range. Viscous liquids, on the other hand undergo irreversible deformation when subjected to a stress. Viscoelastic materials are those that show a combination of the two and polymers fall in this category. Their viscoelastic behavior depends upon the rate of deformation as well as the temperature of testing. As an example, if a polymer above its glass transition temperature T_g is stretched rapidly and then the stress is released, it recovers fully exhibiting elastic behavior. While if the same polymer is stretched slowly, it exhibits permanent deformation like that of a viscous liquid. At rates and duration (time) of deformation between the two extremes, such a polymer above its T_g exhibits viscoelastic properties, i.e., a combination of elastic and viscous behavior. In a typical creep experiment, a weight is suspended by a strip of polymer for a finite period of time after which the weight is removed. The polymer strip will exhibit some permanent deformation and some recovery at the removal of the weight. Similarly in a typical stress relaxation experiment, a fixed amount of strain is applied to a strip of polymer and over a period of time the stress decreases. Thus, in the measurement of modulus it becomes important to specify the time and rate of strain. For example, a 10 sec. modulus would have a different value than a 1000 sec. modulus. The origin of viscoelasticity comes from the molecular nature of polymers which comprise long entangled chains. If the polymer is cross-linked, such as cross-linked rubber, then of course the behavior is like that of an elastic solid.

The understanding of viscoelastic response of polymers forms part of rheology as well as mechanical properties. The study of viscoelasticity also aims to develop models, both mechanical and molecular, to describe the time dependent responses such as creep, stress relaxation, and dynamic mechanical properties.

The mechanical properties of polymeric materials are very important in most application areas, especially when used in load bearing or structural applications. Polymeric materials in general are useful because they provide desirable mechanical properties at an economical cost. Therefore, it is important to understand this mechanical behavior and how it is related to various structural features. Polymers, as a class of materials, have the widest variety and range of mechanical properties. With room temperature as a reference, polymers vary from liquids to soft rubbers to very hard and rigid solids depending on the chemical structure. Further structural variations take place due to crystallinity and orientation. Thus mechanical properties of the polymers are closely related to their structure.

Viscoelasticity and mechanical properties of polymers are intimately related and thus form the basis of this chapter. The first part deals with the viscoelasticity and the second part deals with mechanical properties.

15.2   DEFINITIONS

The mechanical properties of polymers can be described by three moduli values and the Poisson ratio. Modulus is defined as the ratio of applied stress and corresponding deformation. The reciprocal of modulus is compliance. The value of the modulus of a material depends upon the mode of deformation. The three modes of deformation, the specific mode of which determines the modulus of a material, are uniaxial extension, simple shear, and isotropic compression. These are illustrated in Figure 15.1 and described in the following sections.

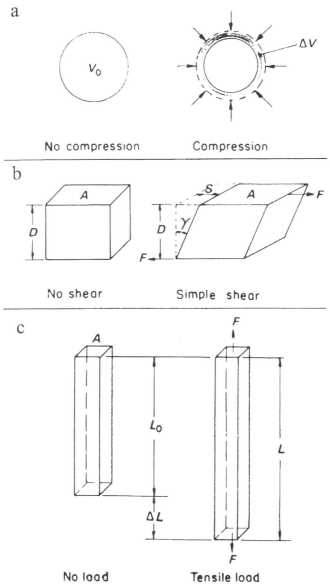

Figure 15.1
Illustration of Three Types of Deformation :
(a) Compression (b) Simple Shear and (c) Uniaxial Extension

- Uniaxial extension gives tensile modulus or Young's modulus, E, expressed as:

$$E = \frac{\text{Force per unit cross-sectional area}}{\text{Strain per unit length}}$$

$$= \frac{\text{Tensile stress}, \sigma}{\text{Tensile strain}, \varepsilon} = \frac{F / A}{\ln(L / L_o)} \qquad (15.1)$$

where, F is the applied force, A is the cross-sectional area, L_o is the original length, and L is the extended length. The tensile compliance, D is given by:

$$D = 1 / E = \text{Strain} / \text{Stress} \qquad (15.2)$$

It may be noted that the true stress would be the load divided by the instantaneous cross-sectional area, which will decrease upon extension. Similarly, nominal stress is the load divided by original cross-sectional area. The measured strain $\Delta L / L_o$ is the nominal strain and the true strain would be an integral of the nominal strain as given by:

$$\varepsilon_{tr} = \int_{L_o}^{L} \frac{dL}{L} = \ln(L / L_o) \qquad (15.3)$$

However, for small deformations the change in area is small, so nominal stress and strain may be used.

- Simple shear gives shear modulus or rigidly, G, expressed as:

$$G = \frac{\text{Shear force per unit area}}{\text{Shear per unit distance between shearing surfaces}}$$

$$= \frac{\text{Shear stress}, \tau}{\text{Shear strain}, \gamma} = \frac{F / A}{\tan \theta} \qquad (15.4)$$

where:

F = the applied force

A = the cross-sectional area

θ = the angular deformation as shown in Figure 1.

The shear compliance, J is given by:

$$J = 1/G \qquad (15.5)$$

- Isotropic (hydrostatic) compression gives bulk modulus, K, expressed as:

$$K = \frac{\text{Hydrostatic pressure}}{\text{Volume change per unit volume}}$$

$$= \frac{P}{\Delta V / V_o} = \frac{PV_o}{\Delta V} \qquad (15.5)$$

where P is the pressure applied, V is the initial volume, and ΔV is the change in volume. Bulk compliance or compressibility $\kappa = 1/K$.

The units of modulus of all three cases in the S.I. system are N/M^2 or Pa. In addition to the above three modes, deformations can also be described in the bending mode which gives flexural modulus and in the torsion mode which gives rigidity.

Poisson's ratio, ν, is dimensionless and ranges from 0 to 0.5, the extremes of which are the values for incompressible solids and viscous liquids, respectively. The interrelations between the three types of moduli using the Poisson's ratio are given as follows:

$$E = 2G(1+\nu) = 3K(1-2\nu) \qquad (15.6)$$

$$E = 3G / \left(1 + 1/3 \cdot G/K\right) \qquad (15.7)$$

$$\nu = 1/2 \cdot E/G - 1 \qquad (15.8)$$

Thus it can be seen that from the three moduli and Poisson's ratio, two are sufficient to describe the other two.

The three modes described could be used to describe viscoelastic response except that a parameter has to be introduced for the viscous part. For example in simple shear the viscous part may be defined by viscosity η. The viscoelastic polymer would now have a relaxation time, λ, associated with it. For shear deformation λ is given as:

$$\lambda = \eta/G \qquad (15.9)$$

The significance of the relaxation time will be discussed later.

15.3 REGIONS OF VISCOELASTICITY

The modulus of polymers exhibits characteristic behavior as a function of temperature. A typical log modulus versus temperature curve for an amorphous polymer is shown in Figure 15.2. This illustrates five different regions of viscoelasticity in different temperature zones which are as follows:

1. At $T \leq T_g$, the glassy region where $G \approx 10^9 \, \text{N}/\text{m}^2$
2. At $T = T_g$, the transition region where G decreases from 10^9 to $10^5 \, \text{N}/\text{m}^2$
3. At $T \geq T_g$, the rubbery plateau region, where G is almost constant at $10^5 \, \text{N}/\text{m}^2$
4. At even higher temperature, rubbery flow begins where G decreases from 10^5 to $10^3 \, \text{N}/\text{m}^2$
5. Finally at T, considerably higher than T_g, viscous flow begins where $G \leq 10^3 \, \text{N}/\text{m}^2$

Chain entanglements are responsible for the elastic properties in the rubbery plateau, even though truly speaking, the polymer is in a liquid state. Higher molecular weight of the polymer leads to larger entanglements and thus a longer rubbery plateau region. Below a critical molecular weight M_{cr} there are

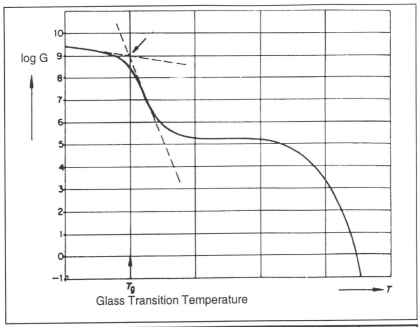

Figure 15.2

Mechanical Behavior	Glassy	Leathery	Rubbery-Elastic	Rubbery Flow	Liquid Flow
Molecular behavior	Only vibrations of atomic groups	Short-range diffusional motion(chain segments)	Rapid short-range diffusional motions. Retarded long-range motions	Slippage of long-range entanglements	Long-range configuratio nal change (whole molecules)

Figure 15.2

Plot of Log Modulus vs. Temperature Illustrating the Regions of Viscoelasticity for an Amorphous Polymer

virtually no entanglements and hence no rubbery plateau. The modulus of an amorphous polymer in the plateau region, G_r, may be estimated by the relation,

$$G_r = \rho R T_g / M_{cr} \qquad (15.10)$$

where, ρ is the density and R is the universal gas constant.

    Cross-links in a rubbery polymer stabilize the rubbery plateau and extend it till the polymer degrades. The permanent cross-linking points replace the temporary chain entanglements. Highly crystalline polymers show practically no transition effect at T_g and flow at the crystallite melting point. In semi-crystalline polymers the crystallites do not undergo a transition while the amorphous regions do and hence show a less pronounced rubbery plateau with a higher modulus value. As compared to an amorphous polymer flow takes place at the melting point for a crystalline polymer. Thus for semi-crystalline polymers both T_g and T_m are of importance.

15.4   TIME TEMPERATURE SUPERPOSITION

The log modulus versus temperature curve for a polymer is generally prepared by choosing a time scale (say 10 sec) for measuring the modulus. Similar log modulus versus log time curves can be made by choosing a reference temperature. The time scale for each measurement is changed in the range from 10 to 10^4 seconds. Thus it means that the same modulus value can be obtained at different combinations of temperature and time. Above T_g, amorphous polymers obey the time temperature superposition principle. Modulus versus time data at one temperature can be superimposed with data at another temperature by shifting the curves. William, Landel and Ferry developed a procedure to convert stress-relaxation data at widely different temperatures to a single curve covering several decades of time at a chosen reference temperature. The time temperature principle can be illustrated by first determining modulus at several temperatures over a range of time scale, as illustrated in Figure 15.3. A reference temperature, say T_3, is then chosen to prepare a master curve. The curves for other temperatures are shifted horizontally one at a time to the left or the right, until portions of the shifted curve superimpose on that of the reference curve. The amount each curve has to be shifted along the x-axis of log time is called the shift factor, a_T, which can be plotted as a function of temperature. The composite of the shifted curves produce a master curve at the reference temperature as shown in Figure 15.3. Similar procedure can be carried out by choosing different reference temperatures. It can be seen that this

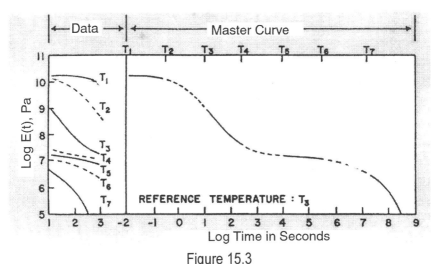

Figure 15.3

Illustration of Time-Temperature Superposition by Preparing a Stress Relaxation Master Curve Using Experimentally Measured Modulus vs. Time Curves at Various Temperatures

procedure gives modulus values over time scales, which are not possible to determine due to practical limitations.

In mathematical form, the time temperature shifting may be expressed as:

$$E(T_1, t) = E(T_2, t/a_T) \qquad (15.11)$$

William, Landel and Ferry showed that if T_g is taken as the reference temperature then the shift factor is given by the WLF equation which is given by:

$$\log a_T = \log \frac{\eta}{\eta_g} = \frac{-17.44(T - T_g)}{51.6 + (T - T_g)} \qquad (15.12)$$

The viscosities, η and η_g, are the values of these components at T and T_g, respectively.

The main significance and usefulness of the time temperature shifting is that by this procedure it is possible to predict the stress relaxation modulus at either very long time scales or at very short time scales at the reference temperature, by using experimentally accessible time scales. Similarly the time temperature superposition may be applied to creep behavior.

15.5   CREEP AND STRESS RELAXATION

Creep and stress relaxation represent time dependent changes in a polymeric material, or manifestation of viscoelastic behavior. In creep, a polymer is subjected to constant stress and it shows time dependent deformation that is irreversible. Higher stress values give larger deformations. Since the stress is constant and the strain is changing, generally creep compliance is determined and plotted as a function of time. Creep behavior depends upon the temperature of the experiment. At temperatures below T_g very little creep occurs and the rate of creep increases as the temperature is raised above T_g.

In a stress relaxation experiment, the polymer is subjected to a fixed constant strain. The stress is found to decrease as a function of time. Since strain is constant, the stress relaxation modulus is determined as a function of time. The extremes of this would give the unrelaxed and relaxed moduli values at very short and at very long times, respectively.

In both, creep and stress relaxation, the relationship between stress and strain is linear at low stress or low strain values. This forms the region of linear viscoelasticity and is generally used to determine properties of a polymer. At large stress or strain values the relationships are not linear and this forms the region of non-linear viscoelasticity. In the linear viscoelastic region creep compliance can be used to determine stress relaxation and vice-versa while this is not possible in the non-linear region.

15.6   DYNAMIC MECHANICAL PROPERTIES

Dynamic mechanical properties of viscoelastic polymers are measured when the applied stress or strain is oscillatory in nature with a specific frequency. For elastic materials the stress and strain would be totally in phase while viscoelastic materials would show that the stress and strain are out of phase.

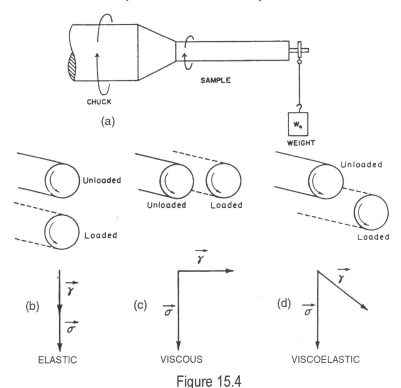

Figure 15.4
Illustration of Elastic, Viscous and Viscoelastic Effects Under Dynamic
Loading Using a Simple Experiment

A simple experiment to illustrate the dynamic response
consists of a rod of a material attached to a chuck and rotated at a
frequency ω. Then a weight W_o is hung from the free end using a
suitable bearing arrangement as shown in Figure 15.4(a). If the
material is purely elastic the rod will deform straight down as
shown in Figure 15.4(b), with strain in phase with the applied
stress. For a purely viscous material the rod will deform in the
horizontal direction as shown in Figure 15.4(c), with strain 90° out
of phase with the applied stress. Lastly for a viscoelastic material
the rod will deform at an angle between 0° and 90° as shown in
Figure 15.4 (d). Let us assume that the strain is of a sinusoidal type
then the stress and strain is given as:

$$\varepsilon = \varepsilon_o \sin \omega t \tag{15.13}$$

$$\sigma = \sigma_o \sin(\omega t + \delta) \qquad (15.14)$$

The strain lags behind the stress by a phase angle δ, the loss angle. This is represented in Figure 15.5. Expanding the expression for σ gives:

$$\sigma = (\sigma_o \cos\omega t)\sin\delta + (\sigma_o \sin\omega t)\cos\delta \qquad (15.15)$$

From this equation it is seen that the stress has two components, one in phase and the other 90° out of phase. Thus the relationship between stress and strain can be written as:

$$\sigma = \varepsilon_o (E' \sin\omega t + E'' \cos\omega t) \qquad (15.16)$$

and:

$$E' = (\sigma_o / \varepsilon_o)\cos\delta \qquad (15.17)$$

$$E'' = (\sigma_o / \varepsilon_o)\sin\delta \qquad (15.18)$$

The quantity E' is the in-phase component and E'' is the component $90°$ out-of-phase component. Such an expression is similar to the relationship between current and voltage and thus

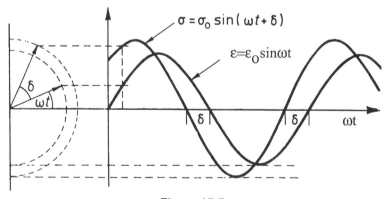

Figure 15.5

Vector Representation of a Sinusoidal Stress Leading a Sinusoidal Strain by the Phase Angle δ

another approach is to use complex representation of strain and stress as shown below:

If the strain is given by:

$$\varepsilon^* = \exp(i\omega t) \qquad (15.19)$$

Then the stress is:

$$\sigma^* = \exp i(\omega t + \delta) \qquad (15.20)$$

The complex modulus would be

$$E^* = \frac{\sigma^*}{\varepsilon^*} = \frac{\sigma_o}{\varepsilon_o}\exp(i\delta) \qquad (15.21)$$

$$= \left(\sigma_o/\varepsilon_o\right)(\cos\delta + i\sin\delta) \qquad (15.22)$$

$$= E' + iE'' \qquad (15.23)$$

where E' is the real part and termed as the storage modulus while E'' is the imaginary part and termed as the loss modulus.

Similarly if the stress is given by:

$$\sigma^* = \sigma_o \exp(i\omega t) \qquad (15.24)$$

Then the strain becomes:

$$\varepsilon^* = \varepsilon_o \exp i(\omega t - \delta) \qquad (15.25)$$

and the complex compliance is given by:

$$D = \left(\varepsilon_o/\sigma_o\right)(\cos\delta - i\sin\delta) \qquad (15.26)$$

$$= D' - iD'' \qquad (15.27)$$

From these expressions the phase angle can be determined as:

$$\tan\delta = \frac{E''}{E'} = \frac{D''}{D'} \qquad (15.28)$$

For purely elastic solids tan δ equals zero since there is no out of phase component. Metals and ceramics are examples of such materials. Polymers have δ values of several degrees. In the glass transition range it may be as high as $30°$.

Complex modulus is related to complex viscosity. In the shear mode it may be expresses as:

$$G^* = i\omega\eta^* \qquad (15.29)$$

and

$$G'' = \omega\eta', \ G' = \omega\eta'' \qquad (15.30)$$

The real part of the complex modulus is the term that determines the amount of recoverable energy stored as elastic energy hence called storage modulus. The imaginary parts consists of damping terms which determine the dissipation of energy as heat when the material is deformed and hence called loss modulus.

Dynamic mechanical properties are often determined to study the polymer structure and its relationship with thermo-mechanical behavior. Thus complex moduli are determined as a function of temperature at a constant frequency. For every transition region there is a sharp decrease in the storage modulus and this is accompanied by a corresponding peak in loss tangent.

15.7 MECHANICAL MODELS

The viscoelastic behavior of viscoelastic polymeric materials is often determined by the use of mechanical models. A simple spring represents an ideal elastic material obeying Hooke's law and in this the elastic deformation takes places immediately upon application of stress. The equation describing the stress strain behavior for the spring is:

$$\sigma = \varepsilon E \qquad (15.31)$$

A dashpot filled with a Newtonian fluid represents an ideal

viscous fluid in which deformation increases linearly with time upon application of stress and this deformation is completely irreversible. The equation describing the stress strain behavior of a dashpot in tensile mode is:

$$\sigma = \eta\left(d\varepsilon/dt\right) \qquad (15.32)$$

The series arrangement of the spring and dashpot form the Maxwell element and is useful in describing stress relaxation, creep and dynamic properties. The parallel arrangement of spring and dashpot form the Voigt element and is useful in describing creep and dynamic properties. These two elements form the basic

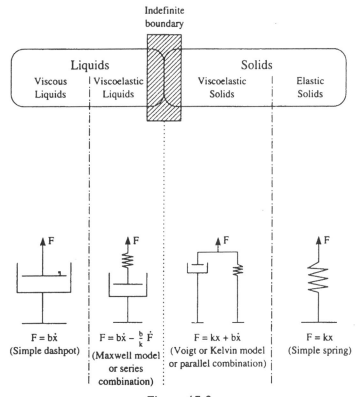

Figure 15.6
Representation of Viscous, Viscoelastic and Elastic Behavior of Polymers Using Simple Mechanical Models

elements for describing the behavior of viscoelastic polymers. These various elements are illustrated in Figure 15.6.

More complicated mechanical models are made by arrangements of dash pots and spring elements. They are used to describe the viscoelastic behavior of polymeric materials more completely.

15.7.1 Maxwell Model

The Maxwell element, a series combination of a Hookean spring and a Newtonian dashpot, is useful in qualitatively illustrating the behavior of a real polymer but is inadequate for quantitative correlation with properties. Stress relaxation for Maxwell element is illustrated in Figure 15.7.

Application of stress σ gives an elongation ε_1 in the spring and a rate of elongation $d\varepsilon_2/dt$ in the dashpot. The stress obviously is the same on both the elements while the total strain is a sum of individual strains. This can be written as:

$$\sigma = \sigma_1 = \sigma_2 \qquad (15.33)$$

$$\varepsilon = \varepsilon_1 + \varepsilon_2 \qquad (15.34)$$

$$\frac{d\varepsilon}{dt} = \frac{d\varepsilon_1}{dt} + \frac{d\varepsilon_2}{dt} \qquad (15.35)$$

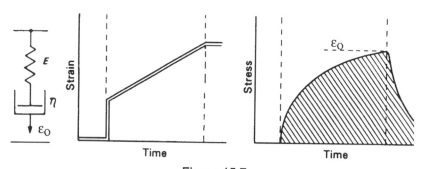

Figure 15.7
Illustration of Stress Relaxation in a Maxwell Element
for a Step Strain of ε_o

and
$$\frac{d\varepsilon}{dt} = \frac{1}{E}\frac{d\sigma}{dt} + \frac{\sigma}{\eta} \tag{15.36}$$

For stress relaxation a fixed elongation ε_o is applied at time $t = 0$ and held constant. Stress and stress relaxation modulus are then determined as a function of time in the following manner:

$$\frac{d\varepsilon}{dt} = 0 = \frac{d\varepsilon_1}{dt} + \frac{d\varepsilon_2}{dt} \tag{15.37}$$

or
$$\frac{1}{E}\frac{d\sigma}{dt} + \frac{\sigma}{\eta} = 0 \tag{15.38}$$

Integrating Equation 15.38, we obtain:

$$\ln\sigma\Big|_{\sigma_o}^{\sigma_t} = -\frac{Et}{\eta}\Big|_{t=0}^{t} \tag{15.39}$$

at
$$t = 0, \ \varepsilon = \varepsilon_o, \ \sigma = \sigma_o = E\varepsilon_o \tag{15.40}$$

and finally
$$E(t) = \frac{\sigma(t)}{\varepsilon_o} = E\exp\left(-Et/\eta\right) \tag{15.41}$$

Since η/E has been defined as the relaxation time, λ,

$$E(t) = E\exp\left(-t/\lambda\right) \tag{15.42}$$

The plot of a step strain and the corresponding stress is given in Figure 15.7. It may be noted at long times stress goes to zero indicating complete relaxation.

In a creep experiment a fixed stress, σ_0, is applied to the element at time $t = 0$ and held constant. Strain and creep compliance are then determined as a function of time in the following manner:

$$\frac{d\varepsilon}{dt} = \frac{d\varepsilon_1}{dt} + \frac{d\varepsilon_2}{dt} \tag{15.43}$$

With the stress being constant (σ_o), ε the spring strain will also be constant and thus, $d\varepsilon_1/dt = 0$. This reduces the above equation to:

$$\frac{d\varepsilon}{dt} = \frac{\sigma_o}{\eta} \tag{15.44}$$

Integrating and noting that at $t = 0$; $\varepsilon = \varepsilon_o = \sigma_o/E$, therefore:

$$\varepsilon(t) = \varepsilon_o + \frac{\sigma_o t}{\eta} = \frac{\sigma_o}{E} + \frac{\sigma_o t}{\eta} \tag{15.45}$$

or
$$J(t) = \frac{\varepsilon(t)}{\sigma_o} = \frac{1}{E} + \frac{t}{\eta} \tag{15.46}$$

    The plot of a step stress and the corresponding strain is given in Figure 15.8. It may be noted that upon removal of stress the spring part recovers while the deformation due to the dash pot is permanent.

    For dynamic response, let us assume that the Maxwell element is subjected to a sinusoidal force resulting in a sinusoidal strain at the same frequency, but out of phase. Alternately the input may be the strain with stress being the output. Let the input be sinusoidal strain with a frequency ω (rad/sec.)

$$\varepsilon = \varepsilon_m \sin\omega t \tag{15.47}$$

which is the real part of:

$$\varepsilon^* = \varepsilon_m \exp(i\omega t) \tag{15.48}$$

    The resulting stress will be directly proportional to E. By analysis it is seen that complex modulus is given as:

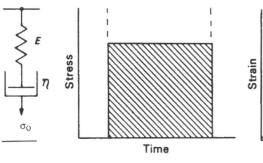

Figure 15.8
Illustration of Creep in a Maxwell Element for a Step Stress of σ_o Applied
for Time, *t*

$$E^* = \frac{E\omega^2\lambda^2}{1+\omega^2\lambda^2} + i\frac{\omega\lambda E}{1+\omega^2\lambda^2} \qquad (15.49)$$

$$= E' + iE'' \qquad (15.50)$$

where E' is the real component or storage modulus and is:

$$E' = \frac{E\omega^2\lambda^2}{1+\omega^2\lambda^2} \qquad (15.51)$$

and E'' is the imaginary component or loss modulus and is:

$$E'' = \frac{E\omega\lambda}{1+\omega^2\lambda^2} \qquad (15.52)$$

and $\tan\delta$ is, as before, given by:

$$\tan\delta = \frac{1}{\omega\lambda} = \frac{E''}{E'} \qquad (15.53)$$

15.7.2 Voigt Element
    A Voigt element is a parallel combination of a Hookean

spring and a Newtonian dashpot and is useful in qualitative illustration of creep but is inadequate for representing relaxation. It serves another useful purpose in that it may be combined with other elements for describing viscoelastic behavior. In the Voigt element, the strain is the same in both the elements while the stress is a sum of stresses in the individual elements.

$$\sigma = \sigma_1 + \sigma_2 \tag{15.54}$$

$$\varepsilon = \varepsilon_1 + \varepsilon_2 \tag{15.55}$$

The dashpot cannot be stretched instantaneously and hence this model is incapable for studying stress relaxation. For creep a constant stress is applied and strain as well as creep compliance are measured as a function of time.

For creep behavior with the application of a constant stress σ_o,

$$\sigma = \sigma_o = \sigma_1 + \sigma_2 = E\varepsilon + \frac{\eta d\varepsilon}{dt} \tag{15.56}$$

Integration gives:

$$D(t) = \frac{\varepsilon(t)}{\sigma_o} = D\left(1 - \exp\left(-t/\lambda\right)\right) \tag{15.57}$$

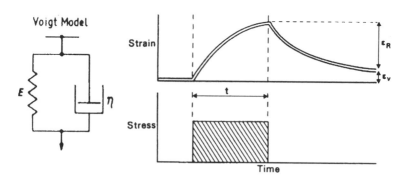

Figure 15.9
Illustration of Creep for a Voigt Element for a Step Stress σ_o Applied for Time, t.

The plot of a step stress and the corresponding stress is given in Figure 15.9. It is seen that the Voigt material creeps up to the limit of deformation of the spring and the entire deformation is recoverable when the stress is removed.

15.7.3 Other Models

Jeffrey Model

   In this a spring element is put in series with the Voigt element as shown in Figure 15.10(a). The extra elements of dashpot results in permanent deformation in the creep experiment as shown in Figure 15.11.

Zener Model

   In this a spring element is put in parallel with the Voigt elements as shown in Figure 15.10(b). The extra spring element helps to explain the immediate deformation upon application of stress as shown in the Zener Model in Figure 15.11.

Berger Model

   This is a series combination of a Maxwell element and Voigt element as shown in Figure 15.10(c). This helps in explaining both the immediate deformation upon application of stress as well as permanent set as shown in Figure 15.11.

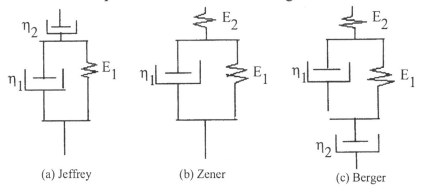

(a) Jeffrey   (b) Zener   (c) Berger

Figure 15.10
Illustration of Other Combination Models
(a) Jeffrey (b) Zener (c) Berger

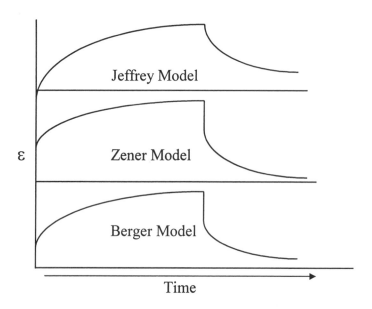

Figure 15.11
Creep Behavior in Jeffrey, Zener and Berger Models as a Result of Step
Strain σ_0 Applied for Time, t

15.7.4 Maxwell-Wiechert Model

In this a number of Maxwell elements are placed in parallel, each with a different combination of E and η values. With different relaxation times as shown in Figure 15.12(a). In all the individual elements the strain in the same and the total stress is the summation of individual stresses in each element. This is expressed as λ.

$$\frac{d\varepsilon(t)}{dt} = 0 = \frac{1}{E_i}\frac{d\sigma_i}{dt} + \frac{\sigma_i}{\eta_i} \qquad (15.58)$$

$$\sigma = \sigma_1 + \sigma_2 + ... + \sigma_i ... + \sigma_N \qquad (15.59)$$

For stress relaxation $d\varepsilon(t)/dt = 0$.

The stress relaxation modulus for a Maxwell-Wiechert

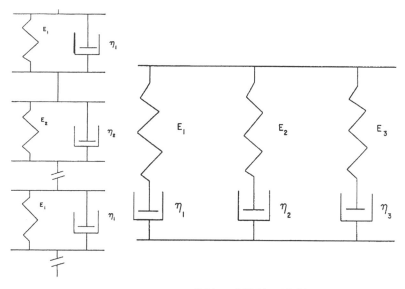

(a) Voigt-Kelvin Model   (b) Maxwell-Weichert Model

Figure 15.12
(a) Voigt-Kelvin Model and (b) Maxwell-Wiechert Model

model with N Maxwell elements is thus given as:

$$E(t) = i\sum_{i=1}^{N} E_i \exp\left(-t/\lambda_i\right) \qquad (15.60)$$

where: $\lambda_i = \eta_i/\lambda_i$

This helps in depicting the master curve which has a glass transition as well as viscous flow behavior following the rubbery plateau.

15.7.5 Voigt-Kelvin Model

In this a number of Voigt elements are placed in series, each with a different combination of E and η values as shown in Figure 15.12(a). The creep compliance for a Voigt-Kelvin model with N Voigt elements is given as:

$$D(t) = \sum_{i=1}^{N} D_i \left(1 - \exp\left(-t/\lambda_i\right)\right) \qquad (15.61)$$

15.8   MOLECULAR MODELS

Several molecular theories have been developed to predict the viscoelastic behavior of polymers. In these the polymer molecule is represented in a manner in which its response to an applied stress can be determined and theory thus developed is useful in predicting the distribution of relaxation times and partial moduli associated with each relaxation time. One of the earlier theories and several modifications of it are based on the bead and spring representation of a polymer molecule. More recently a polymer chain is viewed by the reptation model. Both are discussed briefly in this section. The molecular theories are not based on mechanical models but the results based on them may be presented in forms of the parameters of mechanical models.

15.8.1 Bead and Spring Model

In this model a single polymer chain is sub-divided into z sub-units or sub-molecules, each being long enough to have a Gaussian, end-to-end distribution. The mass of the sub-molecule is concentrated at the beads which are held together by Hookean springs on each side of the beads, as shown in Figure 15.13. The restoring force, f, on each spring when stressed by an amount Δx is given as:

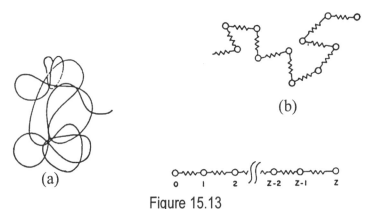

Figure 15.13
Representation of a Real Polymer Chain in Dilute Solution
(a) Polymer Coil Model
(b) Bead and Spring Model

$$f = 3kT\Delta x / a^2 \qquad (15.62)$$

where k is the Boltzmann constant and a^2 is the mean square end-to-end distance of the sub-molecule.

Rouse had first developed a molecular theory based on the spring bead representation of a polymer molecule in dilute solutions. This was then modified by several including Petiolas, Bueche, and Zimm. The analysis of the spring bead model consists of determining the restoring force on each bead when the entire molecule is subjected to a unidirectional deformation. The force on the ith beads in the x-direction is the amount by which the bead i is displaced from its equilibrium position. The deformation of each bead and the corresponding restoring force are affected by the deformation of the adjacent beads. Since the analysis is based on the spring-bead chain in a solution, there is an additional force acting on the sub-molecules due to viscous nature of the solvent they are immersed in. The drag force of each bead can thus be determined by the relationship:

$$f_{ix} = \rho_f \cdot dx_i / dt \qquad (15.63)$$

where ρ_f is the friction factor. The restoring force and the viscous force can now be equated. Detailed analysis using matrices, relaxation times in the stress relaxation of element p is given by the expression:

$$\lambda_p = \frac{1}{2B \cdot Y_p} \qquad (15.64)$$

where:

$$B = 3kT / a^2 \rho_f$$

$$Y_p = 4\sin^2\left(\frac{p\pi}{2(z+1)}\right)$$

This may be compared with the Maxwell-Wiechert model consisting of z elements, each being associated with a relaxation time, λ_p. Thus the stress relaxation for a spring bead model is given as:

$$E(t) = \sum_{p=1}^{z} E_p \exp\left(-t/\lambda_p\right) \tag{15.65}$$

It can shown that at $t = 0$, $E(0)$ is given as:

$$E(0) = \sum_{p=1}^{z} E_p = 3ckTz \tag{15.66}$$

where c is the polymer concentration in terms of molecules per ml.

The spring-bead model can be further analyzed on the basis of viscosity measurements in a manner similar to the Maxwell-Wiechert model. The difference in viscosity of the solution and the solvent is the sum of viscosity of each element and expressed as:

$$\eta - \eta_s = \sum_{p=1}^{z} G_p \cdot \lambda_p \tag{15.67}$$

where:

η = the shear viscosity of the solution
η_s = the shear viscosity of the solvent
G_p = the shear modulus

The relaxation time now becomes:

$$\lambda_p = \frac{6\left(\eta - \eta_s\right)}{ckT\pi^2 p^2} \tag{15.68}$$

This result may now be extended to bulk polymers to give the relaxation times as:

$$\lambda_p = \frac{6\eta}{NkT\pi^2 p^2} \tag{15.69}$$

where N is the number of polymer chains per ml of the bulk material. The shear stress relaxation modulus would be given as:

$$G(t) = Nkt \sum_{p=1}^{z} \exp\left(-t/\lambda_p\right)$$ (15.70)

15.8.2 Reptation Model

Another approach to molecular theories is to consider the motion of polymer molecules in terms of reptation or creeping flow. In this approach, the movement of a polymer molecule is in terms of a motion through the matrix formed by neighboring molecules. Figure 15.15(a) shows reptation view of a single chain that is placed with obstacles around it, which may be formed due to chain entanglements. According to the reptation model, upon relaxation the chain must move through the obstacle course in a snake-like or worm-like fashion. For ease of analysis the movement of the chain through the obstacles can be visualized as the motion of the chain through a constraining tube of a given diameter as illustrated in Figure 15.15 b. The maximum relaxation time for the

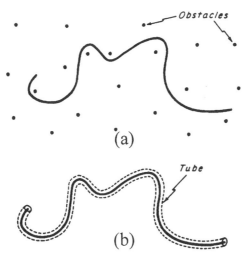

Figure 15.15
Reptation Model View of a Polymer Chain
(a) with obstacles from entanglements
(b) in a constraining tube

chain is then related to the time required for the chain to diffuse through the entire original constraining tube. This is because the original tube would be influenced by the external applied forces and these would be totally relaxed when the molecule is no longer in this tube. The analysis now involves the determination of diffusion times of the polymer molecules in the tube.

DeGennes used scaling concepts to analyze the reptation model. The application of a steady force to the chain results in velocity, v, in the tube and this gives the mobility of the tube, (μ tube, as:

$$\mu_{tube} = v/f \tag{15.71}$$

If the number of repeat units in the molecule is n, then to obtain the same velocity v with molecules of different lengths the force would also be proportional to n:

$$f = \frac{v}{\mu_{tube}} = f \propto n \tag{15.72}$$

or
$$\mu_{tube} = \mu_1/n \tag{15.73}$$

where λ_1 is independent of chain length. Now the diffusing constant through the Nernst-Einstein equation

$$D_{tube} = kTM_{tube} = \frac{kT\mu_1}{n} = \frac{D_1}{n} \tag{15.74}$$

where k is the Boltzmann constant and D_1 is also independent of chain length. The equation for diffusion is given as:

$$D = x^2/2t \tag{15.75}$$

where x is the average distance of movement of molecules in time t in a medium of diffusion constant D. This relaxation time is thus given as:

$$\lambda_{max} = \frac{L^2}{D_{tube}} = \frac{L^2}{D_1} \propto n^3 \tag{15.76}$$

Here, the diffusion time is the maximum relaxation time at the diffusion distance which is the entire tube length L. The tube length is directly proportionate to the polymer chain length hence the maximum relaxation time depends upon molecular weight or chain length to the third power.

$$\eta = \lambda E \qquad (15.77)$$

$$\therefore \eta \propto n^3 \qquad (15.78)$$

Thus this theory predicts that the shear viscosity is proportional to the third power of molecular weight. Experimentally, one often finds an exponent slightly larger than three, which is not fully understood. The theory is further extended to give a more detailed analysis for stress relaxation.

15.9 MECHANICAL PROPERTIES - GENERAL

Mechanical behavior involves the deformation of a material under the influence of applied forces. The mechanical properties of polymers are affected by their chemical composition, surrounding conditions and test conditions. The various factors that affect the mechanical properties are:

1. Molecular weight and molecular weight distribution
2. Cross-linking and branching
3. Crystallinity and crystalline morphology
4. Copolymerization (random, block or graft)
5. Plasticization
6. Fillers, type and amount
7. Blending and related morphology of the blend
8. Molecular orientation

In any given polymer system one or more of the above factors would be operative. The effect of these factors can be correlated with the mechanical behavior, which help in tailoring properties.

In addition several environmental and external variables

that affect the mechanical behavior of polymers are:

1. Temperature
2. Time, frequency, rate of stressing /straining
3. Pressure
4. Stress and strain amplitude
5. Type of deformation
6. Thermal history
7. Nature of surrounding atmosphere, such as moisture level, ozone level, etc.

Processing methods and the conditions of processing play an important role in governing the polymer morphology and hence the resulting mechanical properties. The micro-structure produced during processing affects the viscoelastic nature and thus the response to applied stress during testing. It is, therefore, important to also correlate processing with structure and properties.

Polymer properties show a much stronger dependence on temperature and time as compared to other materials like metals and ceramics. This time temperature dependence is primarily due to the viscoelastic nature of polymer. Viscoelasticity has been discussed earlier along with the time dependent properties such as creep, stress relaxation and dynamic mechanical properties. This section deals with the ultimate mechanical properties of polymers. A number of test methods have been developed to study the mechanical behavior of polymers and some of the important ones are discussed.

The mechanical properties are characteristic parameters for mechanical behavior and subsequent failure under use conditions. Broadly they can be divided into two categories.

1. Properties associated with high stress levels for short periods.
2. Properties associated with low stress levels for long periods.

This can be further classified on the basis of mechanical end

Table 15.1
Mechanical End Use Properties

Class	Short-Term Behavior	Long-Term Behavior
Deformation properties	*Stiffness*	*Creep*
	Stress-strain behavior Modulus Yield stress	Uniaxial and flexural deformation Creep behavior
Durability properties	*Toughness*	*Endurance*
Bulk	Ductile and brittle fracture Stress and elongation at break (Impact strength)	Creep rupture Crazing (and cracking) Flexural resistance Fatigue failure at cyclic stress
	Hardness	*Friction and Wear*
Surface	Scratch resistance Indentation hardness	Coefficient of friction Abrasion resistance

use properties as described in Table 15.1.

15.10 TENSILE PROPERTIES

The tensile properties of polymers are normally determined by studying the stress-strain behavior at relatively high strains. A typical plot of stress strain curve is shown in Figure 15.16. The initial part of the curve has a linear stress-strain relationship exhibiting elastic deformation of the polymer. The slope in this linear region gives the tensile modulus of the material. The point at which it begins to deviate from linearity is called the proportionality limit. At slightly higher strains, the yield point is reached and after this the polymer deforms in a plastic manner and all the strain is not recoverable. The stress at this point is called the yield stress with a corresponding elongation at yield. Beyond the yield point, the material is permanently deformed. If the stress is removed after the yield point the polymer exhibits some recovery and some permanent deformation. At a higher strain level the polymer breaks giving the ultimate tensile strength and the corresponding strain at break. Another feature of importance is that during the plastic deformation there is an increase in stress which is called strain hardening.

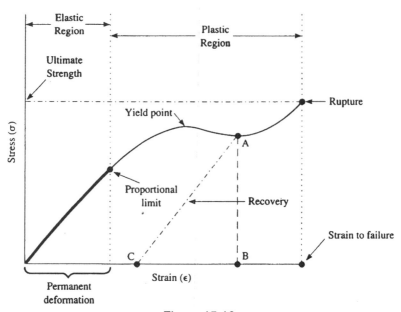

Figure 15.16
Stress-Strain Behavior of Polymeric Materials

The common tensile test involves elongation of a dumbbell shaped sample held in jaws that pull the sample at a constant rate and the load required for this is measured as a function of time. The load-elongation curve obtained on a uniaxial tensile testing machine is plotted and then converted to stress-strain curve.

Ductile materials have a relatively higher elongation at break while brittle materials have a lower elongation at break. Similarly stronger materials have higher yield stress and ultimate strength values. Stiffer materials have a higher modulus as compared to softer materials. Amorphous polymers in their glassy state are generally brittle and in the rubbery state they are ductile. Another feature of importance is that during the plastic deformation there is an increase in stress, which is called work hardening.

The stress-strain relationship at large values of deformation gives an idea about the type of behavior a polymer has. A typical set of classifications on the basis of stress-strain curves are shown in Figure 15.17 and described below.

1. A material with low modulus, low yield stress and moderate elongation at break is soft and weak and is generally represented by uncross-linked rubbers.
2. A material with low modulus but high elongation and high stress at break is soft and tough, generally represented by cross-linked rubbers.
3. A material with high modulus and low elongation at break is hard and brittle generally represented by amorphous glassy polymers.
4. A material with high modulus, high yield stress, high breaking stress, and high elongation at break, is hard and tough and generally represented by crystalline glassy polymers.

Polymer	Characteristics of Stress-Strain Curve			
	Modulus	Yield Stress	Ultimate Strength	Elongation at Break
Soft&Weak	Low	Low	Low	Moderate
Soft&Tough	Low	Low	(Yield Stress)	High
Hard&Strong	High	High	High	High
Hard&Tough	High	High	High	High
Hard&Brittle	High	None	Moderate	Low

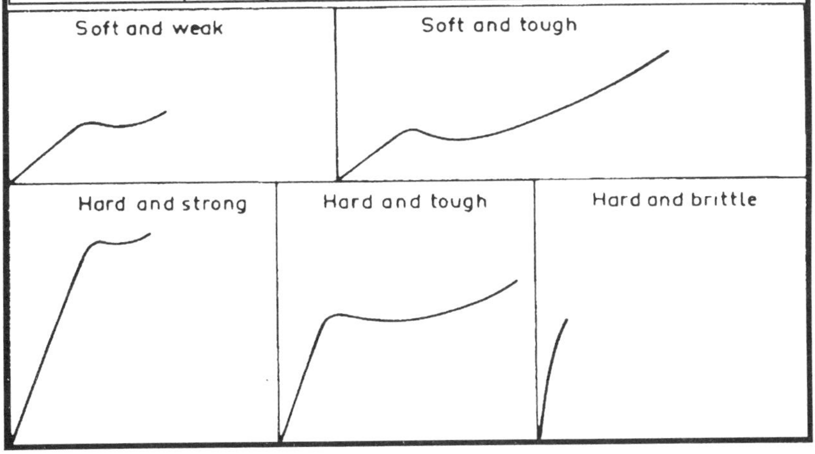

Figure 15.17
Typical Tensile Stress-Strain Curves of Different Types
of Polymeric Materials.

5. A material with high modulus, high yield stress, high breaking stress, and moderate elongation at break, is hard and strong, generally represented by oriented crystalline polymers.

It may be noted that the stress-strain curves are dependent on the rate of elongation and the temperature of testing. The addition of fillers or other polymers may change the behavior significantly. In rupture above the T_g, the polymer chains have a chance to rearrange and flow before the material fails, while in rupture well below the T_g, the chains are immobilized within the period of test and hence failure occurs at small strain values.

15.10.1 Ultimate Strength

The theoretical values for the brittle strength of a material is approximated by the following equation.

$$\sigma_{th} = E/10 \tag{15.79}$$

where E is the Young's modulus.

The observed brittle strength is generally 10 to 100 times lower than the theoretical value. This is mainly due to the presence of cracks and flaws in the material, especially at the surface. These flaws act as stress multipliers. The ultimate strength of ductile materials is nearly the maximum stress the material can support which is its yield stress, if there is no work hardening. In polymers with work hardening the ultimate strength would be higher.

15.11. TOUGHNESS AND IMPACT STRENGTH

15.11.1 Toughness

Toughness of a material is its ability to absorb energy without breaking. The area under the stress strain curve is often taken as a measure of toughness since energy absorption is the sum of all the force resistance effects within the system. This can be obtained by integrating the equation describing the stress-strain curve.

Toughness has units of joules per cubic meter when stress is in pascals and strain is in meters per meter. The toughness is in inch-pounds per cubic inch if the stress is in pounds per square inch and strain as inch per inch. This also represents the energy per unit volume. This may be stored as elastic energy in a cross-linked rubber or it may be dissipated as heat during the permanent deformation of a crystalline material. As an example a glassy polymer may have higher strength and thus greater ability to bear load than a rubber. However, the rubber would have a much higher elongation, thus the energy that the rubber can absorb before breaking would be many times higher than the glassy polymer. As another example, a rubber when cooled to glassy state becomes stronger but much more brittle.

15.11.2 Impact Strength

From a practical view, the real relevance of toughness comes when the force is applied suddenly over a short period of time, such as in an impact. Thus impact toughness or impact strength relates the maximum force that a material can withstand upon sudden impact or in other words the resistance to break under high velocity impact. Impact strength or toughness is highly dependent on the ability of the material to deform in order to accommodate the sudden application of force applied during impact. This deformation is basically strain and hence materials that show high elongation at break in a stress-strain curve have better impact strength. Thus materials with low modulus and high elongation would be tougher than those with high modulus and low elongation, the latter being more brittle.

The molecular factors and crystallinity have a significant effect on impact strength. Higher molecular weight helps in increasing toughness. Generally, crystallinity increases tensile modulus and strength but decreases impact strength. Cross-linking of a brittle polymer generally decreases impact strength but cross-linking of a rubbery polymer increases impact strength.

Impact strength is difficult to define in scientific terms and to measure its value different types of impact methods have been developed. Some of these have become standard test methods while others are specially developed for specific end-use applications. Some of the standard tests measure the energy required to break a sample of specified dimensions under specified conditions. The most commonly used tests are Izod and Charpy. (See Chapter 8, section 8.7, for a description of the test).

The test in which the rigid sample is held vertically and the pendulum swings to impact the free edge, is called the Izod Impact strength test. On the other hand in a Charpy test the pendulum impacts the middle portion of a sample held in a horizontal position. In both cases the impact strength value is given in terms of energy absorbed per thickness with units of joules per mm. In Izod and Charpy tests, the pendulum will continue to swing after breaking the samples and the amount of the swing will depend upon the energy absorbed by the sample from the potential energy of the swinging pendulum. Thus higher the reduction of height of the pendulum, higher the impact strength. The energy used in breaking the sample is proportional to the height of the pendulum before and after failure of the sample. It is given as:

Impact strength = energy to break the sample per unit thickness

$$\left(h_1 - h_2\right)w/d \quad \text{J/mm or ft.lb/in} \tag{15.80}$$

where:

h_1 = height the pendulum swings without the sample
h_2 = height the pendulum swings with the sample in place
w = width of the sample
d = the thickness of the sample

A more practical oriented approach is to determine impact strength by the falling dart test method. In this generally a flat circular sample is supported such that its center is unsupported. A metal dart of a specific weight and shape is dropped from a specific height, and the height is increased in increments until the sample breaks. At the failure point, height times the weight of the dart is related to the impact toughness. Variations of this test can be adopted for specific end use testing.

15.12 FLEXURAL STRENGTH

The determination of flexural strength normally involves a three point bending test with the load applied in the middle on a sample supported at the ends. Young's modulus is calculated by applying the standard equations for a beam undergoing small elastic deflections. For a rectangular beam of length L, width b, and thickness d, the modulus, E, maximum tensile stress, σ, and strain, ε, are given as:

$$E = \left(L^3 P\right)/\left(4bd^3 y\right) \qquad (15.81)$$

$$\sigma = \left(3PL\right)/\left(2bd^2\right) \qquad (15.82)$$

$$\varepsilon = 6dy/L^2 \qquad (15.83)$$

where P is the applied load and y is the deflection.

15.13 COMPRESSIVE STRENGTH

Polymeric materials are subjected to compressive stresses in several application areas. In uniaxial compression the macroscopic phenomenon of collapse are the shear and kink bands. In polymers they are caused by buckling of chains along with changes in chain conformation. The resistance against buckling is expressed by the yield strength under axial compression. An empirical correlation relates σ_c to T_g by the expression:

$$\sigma_c = 1.85 \log T_g - 2.75 \qquad (15.84)$$

or $$\sigma_c \approx T_g^2 \qquad (15.85)$$

This approximation applies to polymers as well as to other materials such as carbon fiber, glass, fused quartz and diamond. There is no simple relation between compressive and tensile strength. However, compressive strength and modulus may be related by the equation.

$$\sigma_c \approx 7.10^{-2} E^{1/2} \tag{15.86}$$

The value of σ_c depends on the combined effect of the boundary stiffness and the lateral cohesive binding forces.

15.14 CREEP RUPTURE

Creep test in tensile or flexural mode provides information on the long-term dimensional stability of the material in a load bearing application. Creep tests generally use small loads so that the sample does not break due to the loading. With time, as the load approaches the breaking strength, rupture will occur. In a creep experiment the time for creep rupture can be given as:

$$\ln t_r = A + \left(E_{act} - B\sigma\right)/RT \tag{15.87}$$

where:

σ = the applied stress
E_{act} = is the activation energy for the fracture process
A and B = constants

The creep test may be combined with a variable temperature, which would measure the deflection temperature at which creep rate would increase significantly. Creep rate may also be conducting in a hostile environment, such as in an organic solvent or soap solution, to give environmental stress cracking. These tests in general are useful in determining the service life of a polymer.

15.15 FAILURE

Rate of testing and temperature of testing affect the stress strain curve. For rigid polymers the modulus and ultimate strength increase while elongation at break decreases with increasing rate of strain. For very brittle polymers the effects are small while for rigid ductile polymers and elastomers the effects are large when the speed of testing varies over several decades. Moduli and yield strength obtained on rigid polymers at different temperatures can be superimposed by using shift factors.

For elastomers, stress-strain curves can be obtained over a large range of temperature and speed of testing. Results can be superimposed by time temperature shifts. If the elastomer undergoes no crystallization during orientation the stress-strain failure envelope can be made as illustrated in Figure 15.18.

Lowering the temperature or increasing the speed of testing has similar effect. Lines *OA*, *OB* and *OC* represent stress-strain curves at different temperatures or different strain rates. The points of rupture fracture are the points on the failure envelope. The lower line *OA* is the equilibrium elastic line. Such curves are also quite useful in determining creep or stress relaxation. If the material is stretched to point *D* and held at constant stress, it would creep to point *F*. Similarly if it is stretched to point *D* and held at constant strain the stress would relax to point *E*. On the other hand if the material is stretched to *G* and held at constant stress or strain, the diagram predicts eventual failure in either condition.

The mechanism of failure depends upon the temperature of testing. At temperatures well below the glass transition temperature the polymers exhibit almost linear stress-strain responses and generally fail by brittle feature. Shear band yielding and crazing are competing mechanisms in such failure. By increasing the temperature the mode changes from brittle failure to ductile failure

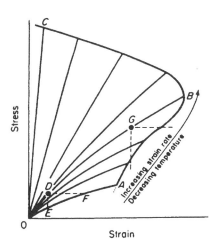

Figure 15.18
Failure Envelope Illustrating the Dependence of Stress-Strain Curves on Shear and Temperature

giving a yield point before failure. This may be accompanied by necking. A further rise in temperature leads to necking, which may be accompanied by cold drawing, leading to large elongations. Lastly at still higher temperature polymer deforms by viscous flow. Each mechanism has a range of temperature and strain rate.

15.16 FATIGUE

Fatigue is associated with decay in mechanical properties after repeated application of stress or strain. This subsequently leads to failure at a stress level below the ultimate stress at failure in a single deformation. Generally the fatigue life of a material is defined as the number of cycles of oscillation, N, that are required to bring about failure of rupture at a given stress or strain. The lower the stress or strain the larger would be the number of cycles before failure. Fatigue property is useful in applications which experience cycling loading such as in a car tire, shoe sole, refrigerator door and gaskets etc. For some materials, the stress reaches a limit below which failure does not occur in a measurable number of cycles.

There are several types of fatigue testers used to determine fatigue strength. These include constant amplitude of cyclic stress or strain in a tensile or flexural beam test, rotating beam test, constant rate of increase in amplitude of stress or strain and superimposition of oscillation on a static stress or strain. The results are in terms of number of cycles to failure versus stress level used. The limiting stress below which the material does not fail in a finite number of cycles is called the fatigue or endurance limit. Generally for polymers the endurance limit is about one third of the static tensile strength in a tensile test. Therefore, this becomes an important property where the application is subjected to cyclic loading or vibration. Fatigue tests are very important for engineering plastics, and components in load bearing applications where plastics are replacing metals. No other test gives information about fatigue or endurance.

Temperature plays an important factor on the fatigue life, which decreases with increasing temperature. This is because fatigue life is related to the strength of polymers which generally decreases with temperature making the process of crack growth easier. For a rigid polymer, the fatigue life is related to temperature as given by:

$$\log N = A + \left(B/T\right) \qquad (15.88)$$

where N is the fatigue life in number of cycles, A and B are constants and T is the absolute temperature.

At low frequencies fatigue is not significantly affected by the frequency of test. However, at high frequencies the fatigue life decreases due to damping effects. High damping may result in heat build-up thus increasing the temperature of the polymer and in turn reducing the fatigue life. The effect of molecular structure on fatigue life is not well understood.

15.17 HARDNESS

Hardness can be divided into two categories.

1. Scratch resistance
2. Indentation hardness

The Moh's scale of hardness is the oldest method to determine scratch resistance. The Moh's scale can be used to determine the scratch resistance of polymers relative to other materials it comes into contact with. It has limited applicability but is useful in differentiating between polymers. Generally polymers would have Moh's values in the range of 2-3, whereas inorganic materials have values in the range of 7-10. The scratch resistance of polymers is related to abrasion and for rigid polymers it can be correlated with modulus.

15.17.1 Indentation Hardness

In determining indentation hardness, a very hard indenter is pressed under a load into the surface of the polymer. The indenter may be a hard sphere as in the Brinell test or a diamond pyramid as in the Vickers test. In the indentation tests, when the indenter presses the material surface the material deforms elastically. Upon increasing the load, the stresses exceed the elastic limit and plastic flow starts. When further increasing the load the material below the indenter becomes completely plastic. Upon release of load some elastic recovery takes place.

15.17.2 Types of Indentation Hardness Tests

1. Hardness tests determine the resistance of a material to indentation. Examples of such tests are: Brinell hardness, Vickers hardness, and Barcal hardness tests, which measure the residual indentation when the stress is removed. Tests like Shore durometer and ISO durometer determine the indention when the load is applied, which relates to compliance.
2. Some hardness tests measure rebound efficiency or resilience such as the various Rockwell hardness tests. In these hardness tests the load is applied and deformation noted like in a creep test. Hence the rate of loading, time after loading when deformation is observed have to be carefully controlled. This is more important for polymers that exhibit viscoelastic behavior. Rockwell tests have properly specified time scales. Using the theory of stress field around the indenter, it is found that about two thirds of the mean pressure is in the form of hydrostatic component and hence does not contribute to plastic flow.

Thus an approximation expression gives:

$$P_y = W/A = 3\sigma_y \qquad (15.89)$$

where:

W = the load applied

A = the surface area under the indenter

σ_y = the uniaxial yield stress of the material

P_y = the pressure at which plastic flow starts.

Thus the yield stress of a material can be determined by a hardness measurement.

In the Brinell hardness test with a ball indenter, the analysis based on elasticity theory gives:

$$W/E = 3(d_1 - d_2)(d_1 D)^{1/2} \qquad (15.90)$$

where:

W	= the load applied
E	= the Young's modulus
d_1	= the indentation depth
$(d_1 - d_2)$	= the distance of recovery
D	= the diameter of the ball indenter

Thus from Equation 15.90, the modulus can be determined. The hardness is defined as:

$$H_p = W/\pi D \cdot d \qquad (15.91)$$

Hardness may be related to the modulus, E, as:

$$H_p \approx E^n \qquad (15.92)$$

An empirical relationship between H_p and E is:

$$H_p = 10E^{3/4} \qquad (15.93)$$

where both H_p and E are in N/m^2.

For rubbers and elastomers, which are relatively soft, Shore hardness tests are used. In this a steel pencil in the form of a truncated cone (Shore A and C) or a rounded cone (Shore D) is used as the indenter. The pencil is pressed into the material and the indentation depth measured on a scale of 0 to 100.

15.18 FRICTION

Friction properties of a polymer are important when it is

used in applications that require abrasion, wear and scratching. In applications like tires or shoe soles it is desirable to have a high friction for tires against the road and shoe soles against the floor surfaces. On the other hand low friction is required in plastic bearings and plastic bottom skis against snow. Friction is a measure of the force resisting the motion of one surface against another. The coefficient of friction is defined by:

$$\mu = F/W \tag{15.94}$$

where F is the tangential force required to produce motion at the interface between two surfaces when they are pressed together by a normal load W. Friction can be of three kinds: static, dynamic and rolling and they have different values. The external factors that affect coefficient of friction include load, contact area, the surface structure (smooth or rough), velocity of sliding, temperature, lubricants and the type of measuring apparatus. The factors that contribute to the total frictional force are:

1. The relative hardness of the two materials in contact
2. Molecular adhesion which depends on the nature of two surfaces
3. The large scale mode of deformations
4. The mode of deformation of surface aspirates
5. Bulk deformation process

If polymers were perfectly elastic with no damping then a rigid ball or a wheel would have little friction when rolling over a smooth polymer surface. When such a ball rolls, it depresses the polymer in front of it, but the polymer behind the ball immediately snaps back and pushes the ball from the rear. If the polymer has mechanical damping, part of the energy in deforming the polymer is dissipated as heat, leaving less elastic energy available for pushing the backside of the rolling object. Thus rolling friction can be correlated with damping, and the variables that affect damping should in turn affect the rolling friction. Rolling friction has been correlated with dynamic mechanical properties for a hard ball rolling on a polymer surface and is given by the expression:

$$\mu_r = C_1\left(E''/E'\right)\left(W/E'R^2\right)^{1/3} \qquad (15.95)$$

where:

W	= the load on the ball
R	= radius
E'	= the storage modulus
E''	= loss modulus
C_1	= a constant which varies with Poisson's ratio $\cong 0.48$

This shows that when $E''/E'\,(= \tan\delta)$ is high, friction will be large, such as in the transition regions. A similar expression has been developed for rolling wheels on a polymer surface relating frictional force, F, to dynamic mechanical properties by the expression:

$$F = C_2(G')^{1/3}\tan\delta = K(G')^{1/3}\left(G''/G'\right) \qquad (15.96)$$

where G' is the dynamic shear modulus, $\tan\delta$ is the dissipation factor and C_2 is a constant. Due to the close relationship between the coefficient of friction and dynamic mechanical properties, the temperature and velocity variation should follow the time-temperature superposition principle. This has been found to be the case where frictional data obtained at different temperatures and velocity has been superimposed using the WLF equation.

15.19 MOLCALC ESTIMATIONS OF ELASTIC PROPERTIES AND COMPARISON TO LITERATURE VALUES

The quantities Molcalc uses for estimating the elastic properties of materials are based on the additive group contributions of the molar functions, U_H and U_R. The elastic properties are related to the molar functions by:

$$G = \rho\left(\frac{U_H}{V}\right)^6 \qquad (15.97)$$

and

$$K = \rho\left(\frac{U_R}{V}\right)^6 \qquad (15.98)$$

Therefore:

$$\frac{G}{K} = \left(\frac{U_H}{U_R}\right)^6 \qquad (15.99)$$

These equations allow all elastic parameters to be estimated. These relationships allow estimation of the tensile modulus and the Poisson ratio from:

$$E = \frac{3G}{1 + G/3K} \qquad (15.100)$$

and:

$$\nu = \frac{1/2 - G/3K}{1 + G/3K} \qquad (15.101)$$

These equations are also given in Chapter 16 since they are used for estimating sonic properties. Table 15.2 provides comparisons between calculated and literature values of elastic properties for the selected list of polymers.

Table 15.2
Comparison of Estimated Elastic Property Values with Literature Values

Polymer	Elastic Properties							
	K		G		E		ν	
	Calc.	Lit.	Calc.	Lit.	Calc.	Lit.	Calc.	Lit.
Poly(ethylene)	4.57	4.54	0.93	0.91	2.62	2.55	0.40	0.41
Poly(propylene)	4.33	4.37	1.49	1.54	4.02	4.13	0.35	0.35
Poly(isobutylene)	3.70	---	1.29	---	3.46	---	0.34	---
Poly(vinyl chloride)	5.54	5.5	1.71	1.81	4.65	4.90	0.36	0.35
Poly(vinyl fluoride)	3.55	---	0.75	---	2.11	---	0.40	---
Poly(chloro trifluoro ethylene)	3.70	1.4	0.81	0.74	2.27	---	0.40	---
Poly(vinyl acetate)	3.64	---	1.13	---	3.06	---	0.36	---
Poly(cis butadiene)	2.87	---	0.62	---	1.74	---	0.40	---
Poly(cis 1,4-isoprene)	3.11	2.3	0.51	0.43	1.46	1.3	0.42	0.49
Poly(styrene)	4.16	4.21	1.39	1.39	3.75	3.76	0.35	0.35
Poly(αmethyl styrene)	3.83	---	1.29	---	3.48	---	0.35	---
Polycarbonate	4.60	4.72	1.18	1.10	3.27	3.09	0.38	0.39
Poly(ethylene terephthalate)	4.49	>4	1.10	1.1	3.06	~3.0	0.39	0.40
Poly(butylene terephthalate)	4.57	---	1.08	---	3.01	2.3-2.5	0.39	
Poly(3,3-bis chloromethyl oxacyclobutane)	6.27	---	1.34	---	3.76	---	0.40	---
Poly(methylene oxide)	6.53	6.59	1.49	1.43	4.15	4.01	0.39	0.40
Poly(vinyl methyl ether)	5.16	---	1.73	---	4.67	---	0.35	---
Poly(phenylene oxide)	4.79	4.22	1.26	1.09	3.47	3.0	0.38	0.38
Poly(hexamethylene adipamide)	6.51	6.53	1.46	1.43	4.07	4.00	0.39	0.40
Poly(acrylamide)	3.35	---	0.98	---	2.68	---	0.37	---
Poly(acrylonitrile)	6.54	---	1.96	---	5.34	---	0.36	---
Poly(methyl methacrylate)	6.47	6.49	2.33	2.33	6.24	6.24	0.34	0.34
Poly(cyclohexyl methacrylate)	5.46	---	1.47	---	4.05	---	0.38	---

15.20 ABBREVIATIONS AND NOTATION

A	area
C_1, C_2	constants
D^*	complex compliance
D	tensile compliance
D'	storage compliance
D''	loss compliance
E	Young's modulus
E'	storage modulus
E''	loss modulus
E^*	complex modulus
F	force
G	shear modulus or rigidity
H	hardness
J	shear compliance
K	bulk modulus
L	extended length
L_o	original length
P	pressure
τ	shear strain
T_g	glass transition temperature
T_m	melting temperature
t	time
V_o	initial volume
ΔL	change in length
ΔV	change in volume
δ	phase angle
ε	tensile strain
k	bulk compliance or compressibility
λ	relaxation time
λ	time (relaxation time) or time constant
ν	Poisson's ratio
σ	tensile stress
θ	angular deformation
η	viscosity

G_{cr}	critical shear modulus
δ	density
M_{cr}	critical molecular weight
at	shift factor
ϖ	frequency
ε_o	peak strain
σ_o	peak stress
η^*	complex viscosity

15.21 GENERAL REFERENCES

1. Van Krevelen, D.W., *Properties of Polymers*, third edition, Elsevier, 1990.
2. Aklonis, J.J. and MacKnight, W.J., *Introduction to Polymer Viscoelasticity*, second edition, John Wiley and Sons, 1983.
3. Ferry, J.D., *Viscoelastic Properties of Polymers*, Third Edition, Wiley, 1980.
4. Nielson, L.E. and Landel, R.F., *Mechanical Properties of Polymers and Composites*, second edition, Marcel Decker, Inc., 1994.
5. Ward, I.M., *Mechanical Properties of Polymers*, Wiley, 1971.
6. McCrum, N.G., Buckley C.P. and Bucknall, C.B., *Principles of Polymer Engineering*, second edition, Oxford Science Publications, 1997.
7. Brent Strong, A., *Plastics: Materials and Processing,* Prentice Hall, 1996
8. Rodriguez, *Principles of Polymer Systems*, Fourth edition, Taylor and Francis, 1996.
9. Porter, D., *Group Interaction Modeling of Polymer Properties*, Marcel Decker, Inc., 1995.

CHAPTER 16

Acoustic Properties

16.1   INTRODUCTION

Most monographs that address the subject of acoustics are concerned mainly with the control and tailoring of sound waves for various applications to produce a desired effect. This may involve design and control of the sound source, the method and frequencies to be transmitted, and control of the reflections, vibrations, and dampening characteristics at the receiver. These aspects are vital for acoustic engineering applications.

This chapter is mainly concerned with acoustic properties of materials not from their application use, e. g., as a sound barrier or material that screens unwanted frequencies, but from the standpoint of providing acoustic information that can be related to many molecular structural factors. Among these molecular structural factors are: transition temperatures, material morphology, and cross link density. However, this does not preclude in any way the use of fundamental acoustic property data being utilized as a source of engineering data for acoustic applications.

The discussion of acoustic properties in this chapter offers insights into the origin of acoustic waves and the influence of compositional differences on sonic velocities. However, estimation of acoustic properties, and the resulting elastic properties derived from acoustic data, are relegated only to polymers since these are the only materials where molar additivities are available.

16.2   SOUND WAVE FUNDAMENTALS

Sound is an alteration in pressure, particle displacement or

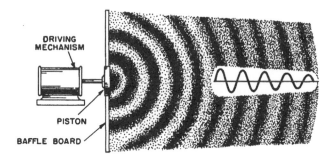

Figure 16.1 [G2]
Example of Production of Sound Waves Using a Vibrating Piston

particle velocity propagated in an elastic material and, of course, is also the sensation produced through the air. Sound is produced when an elastic medium is set into vibration by any means whatsoever. As an example, when a piston is set in rapid oscillating motion by some external means, sound is produced as illustrated in Figure 16.1.

The air in front of the piston is compressed when it is driven forward, and the surrounding air expands to fill up the space left by the retreating piston when it is retracted. This produces a series of compressions and rarefactions of the air as the piston is driven back and forth. The elasticity of the air allows these areas of compression and rarefaction to move outward in all directions. If a line in the direction of sound propagation is measured and plotted with the ordinates representing the pressure, a periodic but dampened wave would result as shown in Figure 16.1 with the compression and expansion (rarefaction) pressures producing the positive and negative points.

The pressure waves that propagate through a medium due to the motion of a vibration are called acoustic or sonic waves. They can be thought of as being similar to electromagnetic waves in the field of light and optics.

The general case of sound propagation involves three dimensions. For understanding the concept of the equation for continuity, consider the influx and efflux through each pair of faces of a cube of dimensions Δx, Δy, and Δz, the difference between the latter and the former for the whole cube is:

$$-\left[\frac{\partial(\rho' u)}{\partial x} + \frac{\partial(\rho' v)}{\partial y} + \frac{\partial \rho' w}{\partial z}\right]\Delta x \Delta y \Delta z \qquad (16.1)$$

where:

x, y, z = coordinates of a particle in the medium
u, v, w = component velocities of a particle in the medium
ρ' = density of the medium

Since the amount of matter which enters the boundaries of a small volume equals the increase of matter inside, the rate of growth of mass is:

$$\frac{\partial \rho'}{\partial t} \Delta x \Delta y \Delta z \qquad (16.2)$$

The rate of growth of mass in the cube must be equal to Equation 16.1, which may be written as:

$$\frac{\partial \rho'}{\partial t} + \frac{\partial(\rho' u)}{\partial x} + \frac{\partial(\rho' v)}{\partial y} + \frac{\partial(\rho' w)}{\partial z} = 0 \qquad (16.3)$$

where:

t = time

The acceleration of momentum parallel to x is $\rho' \Delta x \Delta y \Delta z \frac{\partial u}{\partial t}$. The mean pressures on the faces perpendicular to x are:

$$\left(p_o' - \frac{\partial p_o'}{\partial x}\frac{\Delta x}{2}\right)\Delta y \Delta z \qquad (16.4)$$

and

$$\left(p_o' + \frac{\partial p_o'}{\partial x}\frac{\Delta x}{2}\right)\Delta y \Delta z \tag{16.5}$$

where:

p_o' = pressure in the medium

The difference is a force $\dfrac{\partial p_o}{\partial x}\Delta x \Delta y \Delta z$ in the direction of increasing x.

Equating this to the acceleration of momentum the result is the equations of motion given by:

$$\rho'\frac{\partial u}{\partial t} = -\frac{\partial p_o'}{\partial x} \tag{16.6}$$

$$\rho'\frac{\partial v}{\partial t} = -\frac{\partial p_o'}{\partial y} \tag{16.7}$$

$$\rho'\frac{\partial w}{\partial t} = -\frac{\partial p_o'}{\partial z} \tag{16.8}$$

The equation of motion may be written as:

$$\frac{\partial V_{uvw}}{\partial t} + \frac{1}{r}\,Grad\,p_o' = 0 \tag{16.9}$$

For an adiabatic process:

$$\frac{p_o'}{p_o} = \left(\frac{\rho'}{\rho}\right)^{\gamma} \tag{16.10}$$

where:

p_o = static pressure, i.e., absence of sound waves

p_o' = total pressure (static + excess)

ρ = static or original density

ρ' = instantaneous density (static + change)

γ = ratio of specific heat at constant pressure to that at constant volume

The condensation that occurs is:

$$s' = \frac{\rho' - \rho}{\rho} \qquad (16.11)$$

Combining Equations 16.10 and 16.11

$$\frac{p_o'}{p_o} = \left(\frac{\rho'}{\rho}\right)^{\gamma} = (1 + s')^{\gamma} = 1 + \gamma s' \qquad (16.12)$$

$$\text{or} \quad p_o' = p_o + p_o \gamma s' \qquad (16.13)$$

The instantaneous sound pressure p, which is $p_o' - p_o$ becomes:

$$p = p_o \gamma s' \qquad (16.14)$$

Equations 16.2, 16.9, and 16.10 characterize the disturbances at any amplitude. In general, acoustic waves are of infinitesimal amplitudes, the alternating pressure is small compared with the atmospheric pressure and the wavelength is so long that u, v, w, and s change very little with x, y, and z. Substituting Equation 16.11 in 16.3, and neglecting higher order terms, we have:

$$\frac{\partial s'}{\partial t} + \frac{\partial u}{\partial x} + \frac{\partial v}{\partial y} + \frac{\partial w}{\partial z} = 0 \qquad (16.15)$$

A scalar velocity potential ϕ, is defined as being:

$$u = \frac{\partial \phi}{\partial x}, v = \frac{\partial \phi}{\partial y}, w = \frac{\partial \phi}{\partial z} \qquad (16.16)$$

or

$$V_{u,v,w} = Grad\phi \qquad (16.17)$$

Substituting Equations 16.16 in 16.6, 16.7, and 16.8, and multiplying by dx, dy, and dz, we have:

$$\frac{\partial}{\partial t} d\phi = -\frac{1}{\rho'} dp'_o \qquad (16.18)$$

and integrating

$$\frac{\partial \phi}{\partial t} = -\int \frac{dp'_o}{\rho'} \qquad (16.19)$$

Since the density changes very little, the mean density, ρ, may be used. The excess pressure, which is equal to the integral $\int dp'_o$ in Equation 16.19, then becomes:

$$\rho \frac{\partial \phi}{\partial t} = -p \qquad (16.20)$$

From Equations 16.14, 16.15, 16.16, and 16.20, we have:

$$\frac{\partial^2 \phi}{\partial t^2} - \frac{\gamma p_o}{\rho}\left(\frac{\partial^2 \phi}{\partial x^2} + \frac{\partial^2 \phi}{\partial y^2} + \frac{\partial^2 \phi}{\partial z^2}\right) = 0 \qquad (16.21)$$

This may be written as:

$$\frac{\partial^2 \phi}{\partial t^2} = c^2 \nabla^2 \phi \qquad (16.22)$$

which is the standard D'Alembertian wave equation for ϕ.

The resulting velocity of propagation is:

$$c^2 = \frac{\gamma p_o}{\rho}$$ (16.23)

where:

   c = sound velocity, m/s

This equation demonstrates the importance of density on sound propagation.

   There are two basic types of wave motion, longitudinal and transverse. The longitudinal waves are propagated by the back and forth motion of the molecular particles of the medium in the direc-

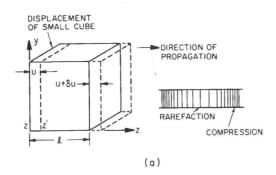

(a)

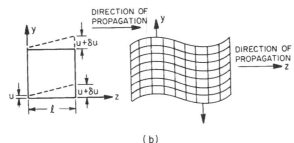

(b)

Figure 16.2 (after Kino [G3])
Propagation of Acoustic Waves
(a) Longitudinal Wave Propagation
(b) Shear Wave Propagation

tion of propagation. This produces expansion or contraction of a particle in the medium. The shear wave is produced when the direction of motion of a particle in the medium is transverse to the direction of propagation. These aspects are illustrated in Figure 16.2. There is no change in the volume or density of the material in a shear wave mode [Figure 16.2(b)].

In general, acoustic waves that propagate through a solid medium may combine shear and longitudinal motion. However, in a crystalline medium with anisotropic elastic properties, the direction of propagation can be chosen to be along one of the principal axes of the crystal: in this case the basic modes can be purely longitudinal or purely shear waves.

16.3   PLANE SOUND WAVES

For a wave being transmitted along the x-axis, ϕ is a function of x and t only and the wave Equation 16.22 reduces to:

$$\frac{\partial^2 \phi}{\partial t^2} = c^2 \frac{\partial^2 \phi}{\partial x^2}$$
(16.24)

A solution of this equation for a simple harmonic wave traveling in the positive x direction is:

$$\phi = A \cos k(ct - x)$$
(16.25)

where:

   A = amplitude of ϕ
   $k = 2\pi / \lambda$
   λ = wavelength, cm
   $c = f\lambda$ = sound velocity, cm/s
   f = frequency, cycles/s

The particle velocity is then given by:

$$u = \frac{\partial \phi}{\partial x} = kA \sin k(ct - x)$$
(16.26)

and it follows from Equations 16.14, 16.20, and 16.23 that:

$$\frac{\partial \phi}{\partial t} = -c^2 s' \tag{16.27}$$

and the condensation in a plane wave from Equations 16.27 and 16.25 is:

$$s' = \frac{Ak}{c} \sin k(ct - x) \tag{16.28}$$

Finally:

$$p = c^2 \rho s' \tag{16.29}$$

which is a different form of Equation 16.23.

From Equations 16.28 and 16.29, the pressure in a plane wave is:

$$p = kc\rho A \sin k(ct - x) \tag{16.30}$$

16.4 RELATIONSHIP OF ELASTIC PROPERTIES TO SOUND WAVE VELOCITIES

16.4.1 Stress

As we have seen from Chapter 15, stress is the force per unit area applied to a solid. A force applied to a solid, of infinitesimal length l, can consist of a longitudinal stress and/or a shear stress. The direction of forces and their resultants are shown in Figure 16.3.

16.4.2 Displacement and Strain

For a one dimensional case, a plane z in a material is displaced in the z direction by a longitudinal stress to a plane $z' = z + u$, as illustrated in Figure 16.2. The parameter u is called the displacement of the material and in general is a function of z.

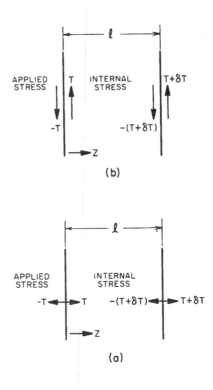

Figure 16.3 (after Kino [G3])
Representation of Stress in a Slab of Length *l*
(a) Stress in the longitudinal direction
(b) Stress in the shear direction

At other points in the material, the displacement u, changes to $u + \delta u$. The change in u in a length l is δu and from a Taylor expansion is:

$$du = \frac{\partial u}{\partial z} l = Sl \qquad (16.31)$$

The fractional extension of the material is:

$$S = \frac{\partial u}{\partial z} \qquad (16.32)$$

where S is the strain.

16.4.3 Hooke's Law and Elasticity

Hooke's law states that for small stresses applied to a one dimensional system, stress is proportional to strain.

$$T = kS \qquad (16.33)$$

where:

k = elastic content of the material

16.4.4 Equation of Motion

The equation of motion is:

$$\frac{\partial T}{\partial z} = \rho \dot{v} \qquad (16.34)$$

where:

v = particle velocity of the material
ρ = mass density, as before

16.4.5 Conservation of Mass

The particle velocity of the material is:

$$v = \dot{u} \qquad (16.35)$$

so that in a small length l, in the one dimensional case the change in velocity is:

$$dv = \frac{\partial v}{\partial z} l \qquad (16.36)$$

and the change of strain with respect to time is:

$$\frac{\partial S}{\partial t} = \frac{\partial v}{\partial z} \qquad (16.37)$$

Equation 16.37 represents another way of writing the equation of conservation of mass of the material for longitudinal waves. The same equation holds for shear waves but since there is no change of density with shear motion, the equation of conservation of mass is not implied.

16.4.6 The Wave Equation and Definition of the Propagation Constant

The previous equations lead to a propagation constant, at a radian frequency ω, given by:

$$\beta = \omega \left(\frac{\rho}{k}\right)^{1/2} \tag{16.38}$$

and:

$$V_a = \left(\frac{k}{\rho}\right)^{1/2} \tag{16.39}$$

where:

V_a is the acoustic wave velocity

For propagation in a forward direction:

$$v = -V_a S = -\frac{V_a}{k} T \tag{16.40}$$

The magnitude of the strain, S, is also the ratio of the particle velocity of the medium to the sound wave velocity of the medium.

16.4.7 Sound Wave Propagation Comparisons

Table 16.1 lists the longitudinal and shear wave propagations for various types of common materials so that they can be compared. A liquid such as water, for example, is relatively easy to compress and has a small mass density. Its elastic constant is $2.25 \times 10^9 \, N/m^2$, so the acoustic wave velocity in water is 1500 m/s. Most metals have longitudinal wave velocities on the order of 5000 m/s, and shear wave velocities approximately half this value.

There are exceptions such as beryllium, 1290 m/s, and lead, 1960 m/s. In the former case the material is light and rigid while in the latter it is heavy and flexible. Since liquids and gases cannot support shear stresses, shear waves cannot propagate through them.

Table 16.1
Longitudinal and Shear Wave Sound Propagation for Isotropic or Polycrystalline Materials*

Material	V_l (km/s)	V_s (km/s)
Aluminum	6.42	3.04
Araldite 5061956	2.62	-----
Bakelite	1.59	1.40
Beryllium	1.289	-----
Bismuth	2.20	1.10
Boron carbide	11.0	-----
Brass, yellow 70%Cu, 30%Zn	4.70	2.10
Butyl rubber	1.80	-----
Cadmium	2.80	1.50
Carbon, Pyrolitic variable properties	3.3	-----
Carbon, vitreous	4.26	2.68
Chromium	6.65	4.03
Copper, Rolled	5.01	2.27
Epoxy, DER332, MPDA 15 phr resin, 60 ° C cure	2.68	1.15
Fused quartz	5.96	3.76
Glass, coming sheet	5.66	-----
Glass, crown	5.1	2.8
Glass, Schott FK3	4.91	2.85
Glass, Pyrex	5.64	3.28
Gold, hard drawn	3.24	1.20
Granite	6.5	-----
Inconel	5.7	3.0
Indium	2.56	-----
Iron	5.9	3.2
Lead	2.2	0.7
Lithium niobate	11.2	-----

continued

Table 16.1 continued

Material	V_l (km/s)	V_s (km/s)
Lithium niobate, crystal-type trigonal 3m, propagation along Z axis	11.2	-----
Lucite or Plexiglas	2.7	1.1
Magnesium, drawn annealed	5.77	3.05
Molybdenum	6.3	3.4
Monel	5.4	2.7
Mylar	2.54	-----
Nickel	5.6	3.0
Niobium	4.92	2.10
Nylon	2.6	1.1
Paraffin wax	1.5	-----
Platinum	3.26	1.73
Polyethylene, low density	1.95	0.54
Polypropylene	2.74	-----
Polystyrene	2.40	1.15
Porcelain	5.9	-----
PVC, gray rod stock	2.38	-----
Quartz, Propagation along Z axis	6.32	-----
RTV-11 rubber	1.05	-----
RTV-577 rubber	1.08	1.35
Rubidium	1.26	-----
Rutile, crystal-type tetragonal, propagation along Z axis	7.90	4.26
Sapphire, crystal-type trigonal 3m, propagation along Z axis	11.1	6.04
Silicon nitride, ceramic	11.0	6.25
Silicone rubber (Sylgard182)	1.027	-----
Silver	3.6	1.6
Steel, mild	5.9	3.2
Stycast 1267	2.57	-----
Tantalum	4.10	2.90
Teflon	1.39	-----
Thorium	2.40	1.56
Tin	3.3	1.7
Titanium	6.1	3.1
Tungsten	5.2	2.9
Uranium	3.4	2.0
Vanadium	6.0	2.78

continued

Table 16.1 continued

Material	V_l (km/s)	V_s (km/s)
Vinyl, rigid	2.23	-----
Zinc	4.2	2.4
Zirconium	4.65	2.25

*Unless otherwise noted

16.5 THEORETICAL RELATIONSHIPS OF THE ELASTIC PARAMETERS TO SOUND VELOCITY

The above discussions provide the basis for understanding why fundamental elastic parameters may be derived from the velocity of acoustic waves.

In liquids only one type of sound wave occurs, viz., the bulk or compressional wave while in isotropic solids both longitudinal wave and shear waves exist, as we have seen. When the lateral dimensions are much less than the wave length, an extensional wave is propagated.

The acoustic properties are related to the bulk modulus (K), the shear modulus of rigidity (G), and the tensile or Young's modulus (E), the terms of which have been defined in Chapter 15, as follows:

$$V_l = \left[\left(K + 4/3G \right)/\rho \right]^{1/2} \tag{16.41}$$

$$V_s = \left(G/\rho \right)^{1/2} \tag{16.42}$$

$$V_{ext} = \left(E/\rho \right)^{1/2} \tag{16.43}$$

where V_l, V_s, and V_{ext} are the sound velocities for longitudinal, shear, and extensional waves, respectively.

In liquids $G = 0$ and $E = 0$ and the sound velocity under these conditions is:

$$V_B = \left(K/\rho \right)^{1/2} \tag{16.44}$$

where V_B is the bulk velocity of sound.

16.6 ADDITIVE MOLAR FUNCTIONS FOR SOUND PROPAGATION

As was the case in previous chapters for a number of other properties, a correlation of the sound velocity with the chemical structure was found. In 1940 Rama Rao [S1, S2] showed that for organic liquids, the ratio $V^{1/3}/\rho$ is nearly independent of the temperature, and multiplied by the molar mass provides a molar function with additive properties.

The Rao function or molar sound velocity function is given by:

$$U_R = MV^{1/3}/\rho \qquad (16.45)$$

This expression is also called the Molar Elastic Wave Function and in general form reduces to:

$$U = V_V V^{1/3} \qquad (16.46)$$

or

$$V = (U/V_V)^3 \qquad (16.47)$$

where:

V_V = volume of the repeat structural unit
V = sound velocity
There are two additive functions:

$$U_R = V_V V_B \text{ (Rao-function)} \qquad (16.48)$$

$$U_H = V_V V_s \text{ (Hartmann-function)} \qquad (16.49)$$

Van Krevelen and Hoftyzer [S3] have utilized a corrected form for the Rao function in their tabulation of values which is given by:

$$U_R = V_V V_l^{1/3} \left[\frac{1+v}{3(1-v)} \right]^{1/6} \qquad (16.50)$$

where: U_R, V_V, V_l are as before, and:

v = Poisson's ratio

16.7   SONIC ABSORPTION

Sonic absorption has not been studied as intensively as sonic speed. Sound absorption is of great interest for applications involving vibration damping and nose reduction. When a material can dissipate energy so that it is not purely elastic, and k is complex, there is a loss term that contributes. The viscous forces between neighboring particles with different velocities are a major cause of acoustic wave attenuation in solids and liquids. There are additional viscous stresses, T_η on the particles in a material through which a plane wave is propagating.

This leads to an attenuation or absorption constant, α, which is given by:

$$\alpha = \frac{\eta \omega^2}{2V_a^3 \rho} \tag{16.51}$$

where:

η = viscosity of the medium

ω = frequency

V_a and ρ are as before

Equation 16.51 shows that the attenuation of the wave due to viscous losses varies as the square of the frequency and inversely to the cube of the velocity. As shear waves typically have velocities of the order of half those of longitudinal waves in the same material, we might expect the shear wave attenuation per unit length to be considerably larger than the longitudinal wave attenuation per unit length, although this is not always the case.

In water at room temperature, the attenuation is $2.2 \times 10^{-3} dB / cm$ at 1MHz. At 1GHz, the attenuation is approximately $2.2 \times 10^3 dB / cm$. This shows the role that frequency plays and that low frequency waves can propagate over long distances in water. For polymers, amorphous stiff (high modulus materials) such as poly(styrene) or poly(methyl methacrylate) would be expected to show less sound absorption than rubbers with a low cross link density and indeed, this is the case.

There are many other sources of loss in real materials. One is thermal conduction. When a material is compressed adiabatically, its temperature increases and decreases when the material expands. Since thermal conduction causes the process to be non-adiabatic and contributes to a loss of energy, it tends to give higher attenuation in metals than in insulators. The attenuation due to thermal conduction also varies as the square of the frequency. Sonic attenuation will also be influenced by grain size and dislocations in crystalline solids.

Acoustic attenuation in common materials varies over a very wide range. Thus a hard single crystalline material such as sapphire can be used for acoustic delay lines at frequencies up to 10GHz. The attenuation at this frequency may be as much as 40dB/cm (10^4 power ratio). Viscous materials, such as rubbers, exhibit relatively high losses at frequencies greater than a few kilohertz, as mentioned above, and therefore are good sound absorbers.

16.8 MOLCALC ESTIMATIONS OF SOUND SPEEDS AND COMPARISON TO LITERATURE VALUES

The quantities Molcalc uses for estimating the sonic properties of materials are based on the additive group contributions of molar functions. The elastic properties are related to the molar functions, U_H and U_R as follows:

$$G = \rho\left(\frac{U_H}{V}\right)^6 \tag{16.52}$$

$$K = \rho\left(\frac{U_R}{V}\right)^6 \tag{16.53}$$

and therefore:

$$\frac{G}{K} = \left(\frac{U_H}{U_R}\right)^6 \tag{16.54}$$

The interrelationships of the various elastic parameters were given in Chapter 15. As we stated previously, these equations allow all elastic parameters to be estimated and allow estimation of the tensile modulus and the Poisson ratio as given in Equations 16.55 and 16.56.

$$E = \frac{3G}{1 + G/3K} \tag{16.55}$$

and:

$$\nu = \frac{1/2 - G/3K}{1 + G/3K} \tag{16.56}$$

Once the elastic properties have been estimated, the sonic properties can be determined from the following relationships.

$$V_l = \left[(K + 4G/3)/\rho\right]^{1/2} \tag{16.57}$$

and:

$$V_s = \left(G/\rho\right)^{1/2} \tag{16.58}$$

It should be noted that estimations for the linear and shear sound velocities, V_l and V_s, are valid only when absorption is low. Table 16.2 provides comparisons between calculated and literature values for the linear and shear sound velocities in the selected list of polymers.

The values in Table 16.2 should be used with a great deal of caution since estimated values and literature values represent only averaged values. From the previous chapters, the reader will appreciate that these values do not take into account sample crystallinity, orientation, nor does it account for other variables introduced during processing and testing of the polymer to obtain the final elastic properties comparisons.

The estimation of mechanical or elastic properties from fundamental data is always fraught with danger and should be used with caution. The prerogative to estimate such properties becomes a valuable tool when an agreed upon and established database is available since under those circumstances valid comparisons between materials can be made.

For estimating the sound absorption properties of polymers, the approximate relationship used by Van Krevelen [S3] is employed in Molcalc for estimating the longitudinal absorption coefficient for non-crosslinked polymers.

$$a_l(\text{dB}/\text{cm}) = 40(n - 0.30) \tag{16.59}$$

The corresponding shear absorption coefficient is estimated from the approximate relationship given by:

$$\frac{\alpha_{sh}}{\alpha_l} \approx 5 \tag{16.60}$$

No comparisons of estimated values vs. literature values are given since these comparisons require knowledge of the frequency and, in addition, there are only very limited data available. The estimated absorption coefficients of selected polymers are listed in the tables in Chapter 5.

Table 16.2
Comparison of Estimated Sonic Property Values with Literature Values

Polymer	Sound Speeds			
	V_l, m/s		V_s, m/s	
	Calc.	Lit.	Calc.	Lit.
Poly(ethylene)	2611	2430	1046	950
Poly(propylene)	2715	2650	1320	1300
Poly(isobutylene)	2540	---	1239	---
Poly(vinyl chloride)	2350	2376	1105	1140
Poly(vinyl fluoride)	1895	---	771	---
Poly(chloro trifluoro ethylene)	1542	---	635	---
Poly(vinyl acetate)	2099	---	982	---
Poly(*cis* butadiene)	2035	---	834	---
Poly(*cis* 1,4-isoprene)	2053	1580	755	---
Poly(styrene)	2292	2400	1102	1150
Poly(αmethyl styrene)	2259	---	1090	---
Polycarbonate	2271	2280	994	970
Poly(ethylene terephthalate)	2139	---	921	---
Poly(butylene terephthalate)	2220	---	943	---
Poly(3,3-bis chloromethyl oxacyclobutane)	2453	---	1002	---
Poly(methylene oxide)	2660	2440	1113	1000
Poly(vinyl methyl ether)	2722	---	1310	---
Poly(phenylene oxide)	2338	2293	1053	1000
Poly(hexamethylene adipamide)	2778	2710	1153	1120
Poly(acrylamide)	1869	---	857	---
Poly(acrylonitrile)	2853	---	1319	---
Poly(methyl methacrylate)	2881	2690	1422	1340
Poly(cyclohexyl methacrylate)	2624	---	1168	---

16.9 ABBREVIATIONS AND NOTATION

A	amplitude of ϕ
c	sound velocity

E	Young's modulus
f	frequency
G	shear modulus of rigidity
k	constant $(2\pi / \lambda)$
k	elastic constant
K	bulk modulus
l	length
M	molar mass per repeating structural unit
p	instantaneous sound pressure
S	strain
t	time
T	stress
U	molar elastic wave function
U_H	Hartmann function
U_R	Rao function
V	sound velocity
V_a	acoustic wave velocity
V_l	longitudinal sound velocity
V_s	shear sound velocity
V_V	volume of the polymer repeat structural unit
V_{ext}	extensional wave sound velocity
V_B	bulk sound velocity
u, v, w	component velocities in x, y, and z directions
x, y, z	directions in the Cartesian coordinate system, respectively
α_l	longitudinal absorption coefficient
α_{sh}	shear absorption coefficient
ρ'	density (instantaneous) of the medium
ρ	static or original density
p'_o	sound pressure in the medium
p_o	static sound pressure
s'	condensation
ϕ	scaler velocity potential

v	particle velocity in the material
$\dot{v}$	rate of change of velocity
ν	Poisson's ratio
η	viscosity
ω	frequency

16.10 SPECIFIC REFERENCES

1. Rao, M.R., *Ind. J. Phys.* **14**, 109, (1940)
2. Rao, M.R., *J. Chem. Phys.* **9**, 682 (1941).
3. Van Krevelen, D.W. and Hoftyzer, P.J., *Properties of Polymers*, Elsevier, New York, N.Y., 1990.

16.11 GENERAL REFERENCES

1. Van Krevelen, D.W. and Hoftyzer, P.J., *Properties of Polymers*, 2nd Ed. Elsevier, New York, N.Y., 1976.
2. Olsen, H.F., *Acoustical Engineering*, D. Van Nostrand Co., New York, N. Y. 1960.
3. Kino, G.S., *Acoustic Waves: Devices, Imaging, and Analog Signal Processing*, Prentice Hall, Inc., Englewood Cliffs, New Jersey, 1987.
4. Blake, M.P. Ed., and Mitchell, W.S., *Vibration and Acoustic Measurement Handbook*, Spartan Books, New York, 1972.
5. Brandup, J. and Immergut, E.H., *Polymer Handbook*, 3rd Ed., J. Wiley & Sons., N. Y. 1989.
6. Mark, J.E., *Physical Properties of Polymers Handbook*, American Institute of Physics, Woodbury, N. Y., 1996.
7. Wylie, C.R., *Advance Engineering Mathematics*, 4th Ed., McGraw-Hill Book Co., N. Y., 1976.
8. Van Krevelen, D.W., *Properties of Polymers*, Elsevier, New York, N.Y., 1990.

CHAPTER 17

Rheological Properties of Polymer Melts

17.1 INTRODUCTION

This chapter deals with the rheological properties of polymer melts. The flow of polymer melts behave differently from other traditional materials and require special theoretical and analytical treatment. Some features of polymer melts that differ, compared to the melts of other materials, are density, low thermal conductivity, and high shear viscosity (typically $10^2 - 10^4$ Ns/m^2). The first two result in long times required for heating and cooling along with non steady state conduction. The densities and viscosities provide low Reynolds numbers which are responsible for producing laminar flow in most situations. Furthermore, most polymer melts show a deviation from Newtonian behavior over the shear rates encountered in common processing operations.

Let us compare the flow involved in processing of a metal and a thermoplastic by melting the material followed by a forming process such as casting for metals and injection molding for plastics. The metal would require a much higher temperature to melt and results in a melt of low viscosity, thus it can easily flow into the mold cavity. The thermoplastic, on the other hand, would melt at a considerably lower temperature but would produce a fluid of very high viscosity requiring high pressures to force it into the mold. The energy required to heat up the metal is higher. Lastly, there are differences in the cooling behavior requiring different suitable designs for flow channels for each of these two materials.

Rheological properties of polymer melts are extremely important for analyzing the flow behavior during processing. Rheology by definition deals with the science of deformation and flow of matter. The rheological behavior of polymers includes several widely different phenomenon which in turn are related to different molecular mechanisms. The discussion in this chapter will be related primarily to the rheological properties of polymers when they are in their molten form, i.e. they are above their glass transition temperature if they are amorphous or above their melting temperature if they are crystalline/semi-crystalline. The various flow phenomena of such melts involves irreversible deformation in which polymer chains slip past each other in an irreversible manner. The resistance to flow when a stress is applied is related to the viscosity of the polymer which depends upon the molecular weight, temperature, and the stress.

Polymeric fluids are unique in their behavior because of their long chain molecular structure. The chains would normally be entangled for normal molecular weights of commercial polymers and the flow phenomenon would involve movement of the chains in an array of entanglements. This gives rise to deviation from ideality and thus polymeric fluids have to be treated as non-Newtonian fluids. Furthermore, there are viscoelastic effects and viscous heating associated with polymeric fluids and these can be quantified.

17.2   VISCOSITY AND SHEAR FLOW

Viscosity can be defined as a measure of the energy dissipated by a fluid in motion as it resists an applied shearing force. There are generally two kinds of flows, viz., pressure and drag flow:

1. Pressure flow occurs due to a pressure gradient along the length of pipe or a channel. In this kind of flow, there are no moving boundaries.
2. Drag flow occurs when a moving boundary drags the fluid along with it resulting in a flow, even though, no external pressure is applied. Some flow situations may be a combination of these two kinds of flows.

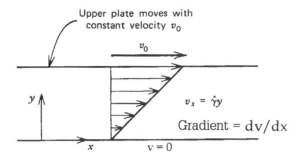

Figure 17.1
Illustration of Simple Steady Shear Flow Between
Semi-Infinite Parallel Plates

Shear flow is one in which the fluid can be divided into layers with a viscosity gradient across the layers, as illustrated in Figure 17.1 for drag flow between two parallel plates. In unidirectional laminar flow between parallel plates, the shear stress τ is given as the force applied for unit area (f/A) which is required to maintain a constant velocity gradient $(\dot{\gamma} = v/y)$. If the fluid follows Newtonian law, then the viscosity is $\eta = \tau/\dot{\gamma}$ and the fluid is called a Newtonian fluid which is independent of the magnitude of τ or $\dot{\gamma}$. Basically the viscosity is the constant of proportionality between τ and $\dot{\gamma}$. However, for polymer melts, the shear stress and shear rate are not proportional over all ranges, hence lead to non-Newtonian viscosity which is not a constant.

For a Newtonian fluid, a plot of τ vs. $\dot{\gamma}$ results in a straight line passing through the origin and having a slope of η. All fluids which do not follow this flow curve are non-Newtonian fluids. These can be further classified as time-independent fluids, time-dependent fluids, and viscoelastic fluids. The time-independent fluids are those in which the shear rate is a function of the applied shear stress but the relationship between the two is non-linear. If the shear rate decreases from proportionality with the shear rate, then the apparent viscosity $(\tau/\dot{\gamma})$ decreases with shear rate. Such fluids are called pseudoplastic. On the other hand, if the apparent

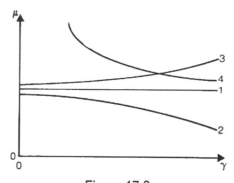

Figure 17.2
Viscosity vs. Shear Rate for Different Kinds of Fluids
1. Newtonian Fluid
2. Pseudoplastic Fluid
3. Dilatant Fluid
4. Bingham Fluid

viscosity increases with shear rate then the fluid is called dilatent. Time dependent fluids are those in for which shear rate depends on both the magnitude and duration of shear. Such fluids can also exhibit yield stress and such behavior is seen in Bingham plastics. In all purely viscous fluids the applied stress is fully dissipated and the deformation is irreversible. However, polymeric fluids also ex-

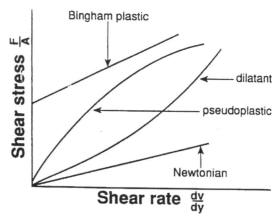

Figure 17.3
Shear Stress vs. Shear Rate Comparison Between Newtonian Viscous
Flow and Examples of Non-Newtonian Viscous Flow

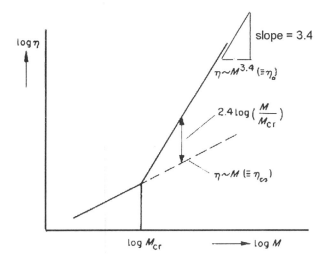

Figure 17.4
Typical Plot of Log η vs. Log M_w

hibit elastic effects which results in recovery of applied stress and thus are called viscoelastic fluids. Different kinds purely viscous fluids are exhibited in Figure 17.2 and 17.3. At low shear rates most polymers would exhibit a Newtonian behavior and the viscosity in this region is referred to as the zero shear viscosity.

17.3   EFFECT OF MOLECULAR WEIGHT

The increase in viscosity of polymer melts is divided into two sections. Generally log viscosity is plotted as a function of log molecular weight and this results in two straight lines of different slopes which intersect at a point as illustrated in Figure 17.4. The line at lower molecular weights has a slope of 1 or slightly greater than one. In this region the effect of entanglements is not very significant and hence the rise of viscosity is almost proportional to the molecular weight. The line at the higher molecular weight has a slope of around 3.4 and in this region entanglements play a major role in determining viscosity. This results in a much greater increase in viscosity with increasing molecular weight. The chains have to flow past the obstacles created by entanglements and hence the resistance to flow is not only dependent on the chain length but also on entanglement density. The point of switch over from the first region to the second region occurs at a critical molecular weight, M_{cr}, the value of which varies from polymer to polymer.

Thus we can write:

$$\eta = K_1 M \qquad \text{for } M < M_{cr} \qquad (17.1)$$

and:

$$\eta = K_2 M^{3.4} \qquad \text{for } M > M_{cr} \qquad (17.2)$$

Table 17.1 lists the critical molecular weight values determined using Molcalc along with literature values. The agreement between the two is very good.

Table 17.1
Comparison of Estimated Values of the Critical Molecular Weight, M_{cr} with Literature Values

Polymer	Critical Molecular Mass, M_{cr} (g/mol)	
	Calculated	Literature
Poly(ethylene)	~3560	3500
Poly(propylene)	~9300	7000
Poly(isobutylene)	~13260	16000
Poly(vinyl chloride)	~7590	6200
Poly(vinyl fluoride)	----	----
Poly(chloro trifluoro ethylene)	----	----
Poly(vinyl acetate)	~26000	25000
Poly(cis butadiene)	~6180	6000
Poly(cis 1,4-isoprene)	~10300	----
Poly(styrene)	~34000	35000
Poly(αmethyl styrene)	~36000	40000
Polycarbonate	~3600	3000
Poly(ethylene terephthalate)	~5400	6000
Poly(butylene terephthalate)	~5000	----
Poly(3,3-bis chloromethyl oxacyclobutane)	~12000	----
Poly(methylene oxide)	~9900	----

continued

Table 17.1 (continued)

Polymer	Critical Molecular Mass, M_{cr} (g/mol)	
Poly(vinyl methyl ether)	~33000	----
Poly(phenylene oxide)	----	----
Poly(hexamethylene adipamide)	~3300	5000
Poly(acrylamide)	----	----
Poly(acrylonitrile)	~2500	1300
Poly(methyl methacrylate)	~38000	30000
Poly(cyclohexyl methacrylate)	~84000	----

17.4 EFFECT OF TEMPERATURE

Melt viscosity of polymers decreases with increasing temperature. The specific temperature dependence of the zero shear viscosity is generally expressed by Williams - Landel-Ferry (WLF) equation which states that

$$\eta(T) = a_t \eta(T_S) \qquad (17.3)$$

where:

$$\log a_t = \frac{C_1(T - T_S)}{C_2 + (T - T_S)} \qquad (17.4)$$

T_S is a characteristic reference temperature while C_1 and C_2 are constants. If T_S is taken to be the glass transition temperature (T_g), then it has been found that C_1 and C_2 are almost universal for all polymers, being 17.4 and 51.6 respectively. This expression is generally valid within $100°C$ of T_g. Doolittle proposed the concept of free volume to derive T_g and viscosity by the expression:

$$\log \eta = A + B \cdot V_o / V_f \qquad (17.5)$$

and

$$V = V_o - V_f \qquad (17.6)$$

where:

V = total volume
V_o = is the occupied volume
V_f = is the free volume

The fractional volume is given as:

$$f_o = f - f_g = \alpha \Delta T \tag{17.7}$$

where α is the difference between the expansion coefficients for melt and glass, f_g is the fractional volume at T_g, and $\Delta T = T - T_g$. Thus in the WLF equation, C_1 and C_2 are given by:

$$C_1 = B / f_g \tag{17.8}$$

$$C_2 = f_g / \alpha \tag{17.9}$$

For most polymers, $\alpha = 4.8 \times 10^{-4}\,K$, hence $f_g = 0.025$

The relationship between viscosity and temperature is often expressed by an Arrhenius type of expression given by:

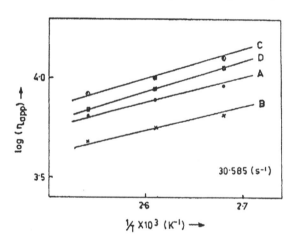

Figure 17.5
Variation of Viscosity with Temperature for a Series of EPDM Polymers
with Carbon Black Additions

$$\eta = Ae^{-E/RT} \qquad (17.10)$$

where E is the activation energy for flow and A is a constant referred to as the collision frequency factor. A plot of η vs. $1/T$ yields the value of E, which is obtained from the slopes of the example curves in Figure 17.5. Table 17.2 lists the calculated and experimental values of activation energy estimated by Molcalc for a number of common polymers (see section 17.10). Good agreement between the estimated and literature values are evident.

Table 17.2
Comparison of Estimated Activation Energies for Viscous Flow with
Literature Values

Polymer	Activation Energy for Viscous Flow, $E_h(\infty)^*$, kJ/mol	
	Calculated	Literature
Poly(ethylene)	26.8	27
Poly(propylene)	43.5	43
Poly(isobutylene)	48.1	48
Poly(vinyl chloride)	85.1	85
Poly(vinyl fluoride)	----	----
Poly(chloro trifluoro ethylene)	----	----
Poly(vinyl acetate)	63.2	63
Poly(cis butadiene)	25.9	26
Poly(cis 1,4-isoprene)	26	26
Poly(styrene)	57.5	58
Poly(αmethyl styrene)	58.2	----
Polycarbonate	84.8	85
Poly(ethylene terephthalate)	47.1	47
Poly(butylene terephthalate)	44.1	----
Poly(3,3-bis chloromethyl oxacyclobutane)	67.7	----
Poly(methylene oxide)	26.9	----
Poly(vinyl methyl ether)	38.4	----
Poly(phenylene oxide)	----	----

continued

Table 17.2 (continued)

Polymer	Activation Energy for Viscous Flow, $E_h(\infty)^*$, kJ/mol	
Poly(hexamethylene adipamide)	36.4	36
Poly(acrylamide)	34.8	----
Poly(acrylonitrile)	----	----
Poly(methyl methacrylate)	63.3	64
Poly(cyclohexyl methacrylate)	----	----

$^*E_h(\infty)$ = value of E_h for $T \gg T_g$

17.5   SIMPLE SHEAR FLOW

The "simple shear flow" is the simplest type of flow that a fluid can experience. It has one non-zero component of velocity changing only in one direction. This results in only one shear rate and one shear component in flow geometry. The respective coordinate system may be designated as 1, 2, and 3 as the flow direction, the direction of velocity variation, and the neutral direction respectively. Typical simple shear geometries are shown in Figure 17.6. These consist of: poisuelle flow, in which pressure flow takes place through a capillary; couette flow, in which drag flow takes place between two concentric cylinders; parallel plate torsion, and cone and plate torsion. The last two are different cases of drag flow. The non-zero components in each case and their dependence on the coordinates are also illustrated in Figure 17.6. The measurement of viscous properties is carried out in viscometers which are designed on simple shear flow behavior.

Thus we can define the velocity and deformation tensor for a simple shear flow as:

$$v = (v_1, 0, 0) \tag{17.11}$$

$$\Delta = \dot{\gamma}(x_2) \begin{pmatrix} 0 & 1 & 0 \\ 1 & 0 & 0 \\ 0 & 0 & 0 \end{pmatrix} \tag{17.12}$$

Flow Geometry	Coordinate Notation
	1 2 3

1. Poiseuille Flow 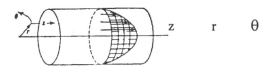   z   r   θ

2. Couette Flow    θ   r   z

3. Parallel Plate Torsion    θ   z   r

4. Cone and Plate Torsion    φ   θ   r

Figure 17.6
Simple Shear Flow Notation

where $\dot{\gamma}(x_2)$ is the scalar function of the x_2 coordinate and $\dot{\gamma}$ is the shear component of the rate of deformation tensor and is called the shear rate. The non-zero components of Δ in simple shear flow are Δ_{12} and Δ_{21} such that $\Delta_{12} = \Delta_{21}$. The corresponding shear stress is τ_{12}. The apparent viscosity, η_a, of a fluid can be defined as the ratio of shear stress to shear rate,

$$\eta_a = \tau_{12} / \Delta_{12} = \tau_{12} / \dot{\gamma}_{12} \qquad (17.13)$$

This serves as a definition for Newtonian as well as non-Newtonian fluids. At low shear rates, η approaches a constant value, η_o, the zero shear viscosity.

17.6   MEASUREMENT OF VISCOSITY

The importance of determining viscosity lies in analyzing the flow behavior in polymer processing operations such as extrusion, calendering and molding, since these operations require the flow of polymeric fluids. The shear rates associated with some manufacturing methods is given below:

- Compression molding, $\dot{\gamma} = 1 - 10 \ \sec^{-1}$

- Calendering, $\dot{\gamma} = 10 - 100 \ \sec^{-1}$

- Extrusion, $\dot{\gamma} = 10 - 1000 \ \sec^{-1}$

- Injection molding, $\dot{\gamma} = 1000-10000 \ \sec^{-1}$

It can be seen that the shear rate requirements vary over a wide range. It is therefore necessary to determine the melt viscosity of polymers over a shear rate range rather than at a fixed shear rate. In general pressure flow produces higher shear rates while drag flow produces lower shear rates. In some instances, viscosity has to be measured by two different instruments to cover the shear rate range of interest.

The instruments developed for the measurement of viscosity are generally based on simple shear flow. Analysis of shear flow is easier since there is only one non-zero component of velocity. Four of such geometries have been illustrated earlier in Figure 17.4 and a brief description of each method follows.

In all the cases some general assumptions are made which are:
a) the fluid is Newtonian
b) the fluid is incompressible
c) there is no viscous heating
d) there are no end effects

e) the flow is steady state
f) the flow is laminar

17.6.1 Capillary Viscometer (Poiseulle Flow)

In this technique, the fluid is forced through a capillary of radius R of length L under a pressure ΔP. The shear stress is given as:

$$\tau_{rz} = \Delta P/2L \qquad (17.14)$$

From this and using dynamic equations, the velocity profile v and the flow rate Q are determined as:

$$v = \Delta P\left(R^2 - r^2\right)/4\eta L \qquad (17.15)$$

$$Q = \pi\Delta P R^4/8\eta L \qquad (17.16)$$

and

$$\eta = \pi\Delta P R^4/8QL \qquad (17.17)$$

Thus knowing the relationship of ΔP and Q, the viscosity η can be determined.

17.6.2 Cup and Bob Viscometer

This consists of a bob of radius R_1 and length L, rotating at an angular speed of Ω inside a cup of radius R_2. The torque on the bob at the speed of rotation Ω is $\Im$. The shear stress is given as:

$$\tau = \Im/2\pi r^2 L \qquad (17.18)$$

From this the $\Im$ is determined as

$$\Im = 4\pi\eta L\Omega\Big/\left[1/\left(R_2\right)^2 - 1/\left(R_1\right)^2\right] \qquad (17.19)$$

Knowing the values of $\Im$ and Ω, η can be calculated.

17.6.3 Cone and Plate Viscometer

In this technique, a cone of radius R rotates over a plate at an angular velocity Ω. The cone angle is ψ which is generally about 1 to 2°. Since the cone angle is so small, the shear rate is estimated to have almost a constant value of

$$\dot{\gamma} = \Omega / \psi \qquad (17.20)$$

The torque on the cone is related to Ω by the relation:

$$3\Im/2\pi R^3 = \eta\Omega/\psi \qquad (17.21)$$

From this equation the viscosity can be determined.

17.7   NORMAL STRESS

Some fluids when subjected to a simple shear flow not only develop shear stress but also normal stresses. These stresses are in the direction normal to the flow directions and are due to the viscoelastic effects in the polymer fluid. Generally, such normal stresses increase with the rate of shear and molecular weight. In the example of flow between parallel plates, the top plate would be pushed upwards due to the presence of normal stresses. The magnitude of these normal stresses can be determined by measuring the force required to maintain the initial gap between these plates. In the flow through a circular capillary, the normal stress relaxation of this elastic energy at the exit of the pipe results in an increase in diameter over that of the capillary and is called extrudate swell. In yet another situation, the viscoelastic fluids being stirred by a rod exhibit climbing due to normal stresses. It is important to study these effects since they have practical implications during processing.

Normal stress coefficients may be defined as the ratio of normal stress to the square of shear rate, given by the expression:

$$\psi_{ij} = \left(\tau_{ii} - \tau_{jj}\right)/\dot{\gamma}^2 \qquad (17.22)$$

17.8   SIMPLE ELONGATIONAL FLOW

Several flow situations may produce elongational flow rather than shear flow, such as in the case of stretching of fibers and films. Simple elongation refers to elongation applied in one direction which results in accompanying deformation in the other two directions. The rate of deformation tensor for simple elongation is given as:

$$\Delta = \begin{pmatrix} \Delta_{11} & 0 & 0 \\ 0 & \Delta_{22} & 0 \\ 0 & 0 & \Delta_{33} \end{pmatrix} \tag{17.23}$$

$$= \varepsilon \begin{pmatrix} 2 & 0 & 0 \\ 0 & -1 & 0 \\ 0 & 0 & -1 \end{pmatrix} \tag{17.24}$$

where ε is the applied strain rate

The stress accompanying the extrusion of the material is of particular importance in defining operating conditions for such processes. This leads to the deformation of elongational viscosity η_e, also referred to as tensile viscosity and is given by the relationship

$$\eta_e = \tau_{ii} / \varepsilon \tag{17.25}$$

where τ_{ii} is the total stress (not dynamic stress) on the system. For a Newtonian fluid η_o and η_e are related by the equation, $\eta_e = 3\eta_o$. This is attributed to Trouton and called Trouton viscosity.

17.9   NON-NEWTONIAN FLUIDS

It has been assumed until now that the viscosity is a constant for any polymer with given molecular weight, temperature and pressure. Two more variables are the rate at which deformation takes place and the length of time at a given rate of shear. The variation of viscosity with rate of deformation can be shown by combination of any two variables. As described previously, if the viscosity increases with shear rate or shear stress, the fluid is called dilatent; and if the viscosity decreases with an increase in shear rate or shear stress it is called pseudoplastic. Most examples of dilatent behavior are exhibited by dispersed systems such as latexes, slurries, and suspensions. The typical behavior of polymer solutions and melts is pseudoplastic. No single mathematical model can explain the behavior of these non-Newtonian fluids. One of the simplest approaches is to approximate sections of the log-log plots by straight lines which is referred to as power-law model and is given as:

$$\tau = K\dot{\gamma}^{n} \quad \text{or} \quad \eta = K\dot{\gamma}^{n-1} \qquad (17.26)$$

For dilatent fluids n has a value greater than 1 while pseudoplastic fluids have n values less than 1. Examples of mathematical models used to describe non-Newtonian behav-ior are described in the following sections.

17.9.1 Truncated Power Law

$$\eta = \eta_{o} \quad \text{for} \quad \dot{\gamma} \subseteq \dot{\gamma}_{o} \qquad (17.27)$$

$$\eta = \eta_{o}\left(\dot{\gamma}/\dot{\gamma}_{o}\right)^{n-1} \qquad (17.28)$$

where η_{o} is the viscosity at shear rates below $\dot{\gamma}_{o}$

17.9.2 Ellis Model

$$\eta/\eta_{o} = 1 + \left(\tau/\tau_{1/2}\right)^{-1} \qquad (17.29)$$

$\tau_{1/2}$ is the shear stress at $\eta = \eta_{o}/2$ and η_{o} is the viscosity at low shear rates.

17.9.3 Carreau Model
The Carreau model is given by the following equation:

$$\eta - \eta_o / \eta_o - \eta_\infty = \left[1 + (\lambda\dot{\gamma})^2\right]^{(n-1)/2} \tag{17.30}$$

where η_o is the zero sheer viscosity, η_∞ is the viscosity at very high shear rates and is a time constant.

17.10 MOLCALC ESTIMATIONS OF THE CRITICAL MOLECULAR WEIGHT M_{cr}, AND THE ACTIVATION ENERGY $E_{\eta}(\infty)$

The basis and background used to calculate M_{cr} in Molcalc have been presented in Section 11.11 of Chapter 11 and will not be repeated here. For the sake of completeness, M_{cr} is given by:

$$M_{cr} = \left(13/K_\theta\right)^2 \tag{17.31}$$

The relationship used to calculate the activation energy for viscous flow when $T \gg T_g$ is based on the additivity relationship found by Van Krevelen and Hoftyzer in 1976 which is given by:

$$E_{\eta}(\infty) = \left(H\eta/M\right)^3 \tag{17.32}$$

where:

H_η = structural group contribution to the molar viscosity-temperature function

M = structural group molecular weight

The structural group contributions and additivity summations for K_θ and H_η are an integral part of Molcalc. The results of the estimation of activation energies for viscous flow were listed previously in Table 17.2.

Comparisons, between Molcalc calculated values and literature values of M_{cr} for the selected list of polymers, were presented previously in Table 17.1.

17.11 ABBREVIATIONS AND NOTATION

a_t	shift factor
B	constant
C_1, C_2	constants in WLF equation
$E_\eta(\infty)$	activation energy for viscous flow for $T \gg T_g$
f_o	fractional free volume
H_η	molar viscosity-temperature function
K_1, K_2	constants
K_θ	unperturbed viscosity coefficient
M	molecular weight
M_{cr}	critical molecular weight
P	pressure
Q	flow rate
R	radius
T_s	characteristic reference temperature
V	volume, total
v	velocity
V_o	occupied volume
V_f	free volume
α	expansion coefficient difference between melt and glass
η	viscosity
η_a	apparent viscosity
η_o	zero shear rate viscosity
η_e	elongational viscosity
$\dot{\gamma}$	shear rate
$\mathfrak{I}$	torque

Ω	speed of rotation
ψ	cone angle
τ	shear stress

17.12 GENERAL REFERENCES

1. Aklonis, J.J. and Macknight, W.J., *Introduction to Polymer Viscoelasticity*, 2nd Ed., John Wiley & Sons, 1983.
2. Bird, R. B., Armstrong, R.C., and Hassager, O., *Dynamics of Polymeric Liquids*, Vol. 1, *Fluid Mechanics*, John Wiley & Sons, New York, 1977.
3. Cheremisinoff, N.P., *An Introduction to Polymer Rheology and Processing*, CRC Press, 1993.
4. Cheremisinoff, N.P., Ed., *Encyclopedia of Fluid Mechanics*, Vol. 9, *Polymer Flow Engineering*, Gulf Publishing Co.,1990.
5. Dealy, J. M., *Rheometers for Molten Plastics*, VNR company, 1982.
6. Middleman, S., *The Flow of High Polymers,* John Wiley & Sons, New York, N.Y,1968.
7. White, J. L. *Principles of Polymer Engineering Rheology*, John Wiley & Sons, 1990.
8. Van Krevelen, D.W. and Hoftyzer, P.J., *Properties of Polymers*, 2nd Ed. Elsevier, New York, N.Y., 1976.

CHAPTER 18

Transfer Properties of Materials

18.1 INTRODUCTION

In this chapter we consider both heat conduction transfer and mass diffusion transport together since both processes originate from molecular activity. The ultimate aim of this chapter is to provide models which can be used to estimate the thermal conductivity of polymers and the permeation of simple gases through these materials.

The rate of heat transport through polymers is important for applications involving thermal conduction or insulation properties of the polymer, or processing requirements, where cooling or heating characteristics must be known.

Permeation of small molecules in polymeric materials depends jointly upon their solubility and diffusivity. The process of diffusion involves mass transport and this must therefore be a prime consideration.

The following sections provide the background for many fundamental concepts in heat and mass transport that are vital in understanding these processes and their relationships to thermal conductivity, diffusion, and permeation.

18.2 FUNDAMENTAL CONCEPTS OF TRANSFER PROCESSES

18.2.1 The Thermal Conduction Model

Both conduction heat transfer and mass diffusion are transport processes that originate from molecular activity or move-

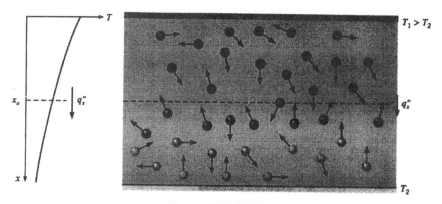

Figure 18.1 [S1]
Illustration of the Relationship of Conduction Heat Transfer with
Diffusion of Energy Due to Molecular Activity

ment. These processes can best be visualized using models where the materials under consideration are gases.

For thermal conduction, the model we use is a gas contained between two surfaces that are maintained at different temperatures. The energy of the gas molecules at any point is proportional to the temperature at that point and is related to the translational, rotational, and vibrational motions of the molecules. This is illustrated in Figure 18.1.

Higher temperatures, of course, are associated with higher molecular energies and when neighboring molecules collide the higher energy molecules transfer or impart a portion of their energies to their less energetic neighbors. In the presence of a temperature gradient, energy transfer by conduction then occurs in the direction of a decreasing temperature gradient. The arbitrary plane x_o, in Figure 18.1, is constantly being crossed by molecules from above and below due to random motion of the molecules. Since those molecules above this plane are exposed to a higher temperature than those below it, a net transfer of energy occurs in the positive x direction.

18.2.2 The Diffusion Transport Process

Like the process of conductive heat transfer, a similar situation exists during mass transfer. In this case, however, a difference in the concentration of some species exists, and this concentration gradient acts as the driving potential for transport of that species.

The mass transport model consists of a chamber in which two different gas species at the same temperature and pressure are initially separated by a partition. When the partition is removed, as illustrated in Figure 18.2, both species will be transported by diffusion. The concentration of species A $\left(C_A\right)$ decreases with increasing x, while the concentration of species B (C_B) increases with x.

Since mass diffusion is in the direction of decreasing concentration, there is a net transport of species A to the right and species B to the left of the imaginary plane x_o. At equilibrium, uniform concentrations of A and B are established, and net transport of species A and B ceases. It can be seen that this situation is analogous to that of thermal transport.

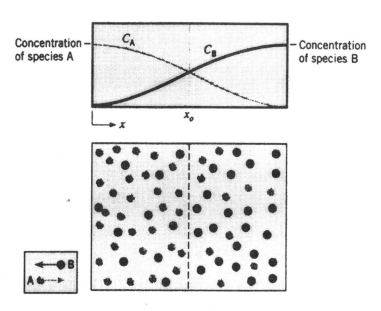

Figure 18.2 [S1]
Illustration of Mass Transfer by Diffusion In a Binary Gas Mixture

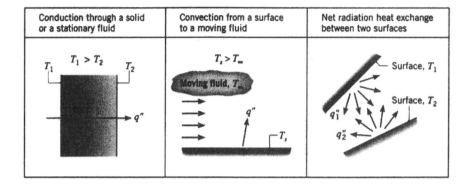

Conduction through a solid or a stationary fluid	Convection from a surface to a moving fluid	Net radiation heat exchange between two surfaces

Figure 18.3 [S1]
Illustration of Conduction, Convection, and Radiation Heat
Transfer Modes

18.3 HEAT TRANSFER MECHANISMS

There are three types of heat transfer mechanisms of which conduction is the process of concern here. The other two processes are convection and radiation. These three mechanisms are illustrated in Figure 18.3.

Convection refers to heat transfer that will occur between a surface and a moving fluid when they are at different temperatures. In the case of radiation, energy is emitted from surfaces in the form of electromagnetic waves and in the absence of an intervening medium, there is net heat transfer between two surfaces at different temperatures. We will not concern ourselves here with the latter two instances.

The mechanism of heat transfer can be quantified by using appropriate models and the equations which define these models. Consider a homogeneous slab, initially at T_0, the upper plane of which is maintained at temperature T_0 and the lower plane of which is suddenly changed to a temperature T_1. Assuming that the absolute temperature does not cause any significant change in the properties of the solid, a steady temperature profile develops in the solid as shown in Figure 18.4.

Experimental evidence shows that there is a direct proportionality, between the temperature gradient and the amount of en-

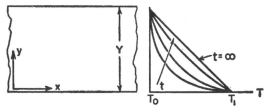

Figure 18.4
Time-Temperature Profile in a Homogeneous Slab with Upper and Lower
Surfaces Maintained at Temperatures, T_0 and T_1

ergy transferred across a unit area of the solid in the y-direction. Again, this direction of transfer is in accord with the Second Law of thermodynamics, and occurs from the region of high temperature to the region of lower temperature.

For the one-dimensional plane slab of area A, the rate equation or the heat flux q'' (W / m^2) is given by:

$$\left(\frac{q''}{A}\right)_y = k\frac{T_1 - T_0}{y} = k\frac{dT}{dy} \tag{18.1}$$

or at any position during the transient is:

$$\left(\frac{q''}{A}\right)_y = -k\frac{\partial T}{\partial y} \tag{18.2}$$

This empirical relation is known as Fourier's equation of heat conduction. The proportionality constant k is a transport property know as the thermal conductivity (J/m·K) or equivalent units and is characteristic of the material. For a more general case of heat flow in three dimensions, the conduction rate equation (Fourier's Law) becomes:

$$q'' = -k\nabla T = -k\left(i\frac{\partial T}{\partial x} + j\frac{\partial T}{\partial y} + k\frac{\partial T}{\partial z}\right) \tag{18.3}$$

where:

$$\nabla = \text{three dimensional del operator}$$
$$T(x,y,z) = \text{scaler temperature field}$$

18.4 THERMAL CONDUCTIVITY

In general terms, for heat flux across a surface to the temperature gradient in a direction perpendicular to the surface, the thermal conductivity is given by:

$$k = -\frac{q_x''}{(\partial T / \partial x)} \qquad (18.4)$$

This relationship shows that the heat flux by conduction increases with increasing thermal conductivity.

Materials vary enormously in their range of thermal conductivities which can be linked directly to the structure of materials as discussed in Chapters 1 and 2. The range of variation can be seen in Figure 18.5 which shows almost five orders of magnitude between the thermal conductivity of gases and those of metals.

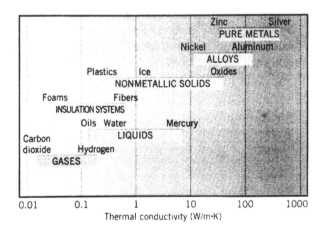

Figure 18.5 [S1]
Thermal Conductivities for Various Materials at Normal
Temperatures and Pressures

The conduction of heat in dielectric solids may be considered either the propagation of anharmonic elastic waves through a continuum or the interaction between quanta of thermal energy called phonons. The thermal conductivity, k, can be represented as:

$$k = 1/3 \int c(\omega) v l(\omega) d\omega \qquad (18.5)$$

where:

$c(\omega)$ = contribution to the specific heat per frequency interval for lattice waves of that frequency

$l(\omega)$ = attenuation length for lattice waves

v = average molecular velocity

The major process giving rise to a finite thermal conductivity and energy dissipation from thermal elastic waves is phonon-phonon interaction corresponding to phonon scattering.

The differences in thermal conductivity of the above wide range of materials can be partly understood by considering a number of factors which influence thermal conductivity.

Factors Influencing Thermal Conductivity in Dielectric Solids

- Use or measurement temperature - The temperature level affects the phonon conductivity in dielectric solids by affecting the phonon mean free path.
- Composition - The lower the mean atomic weight and/or the simpler the structure the higher the thermal conductivity.
- Metal Oxides - The lower the weight of the cation, the higher k becomes.
- Grain Boundary Effects - These act as scattering sites. Single crystals have, in general, higher thermal conductivities.
- Impurities - Solid solutions shorten the mean free path, which is important particularly at lower temperatures.

- Phase Composition - The presence of two or more phases also influences the thermal conductivity.
- Porosity - The porosity greatly reduces the thermal conductivity.
- Thermal Cracking/Spalling - This results from temperature gradients which may be caused by phase transformations or anisotropic expansions.
- Glasses - For glasses, the mean free path, l, is small because there is no lattice network, therefore, at lower temperatures k is smaller than at higher temperatures.

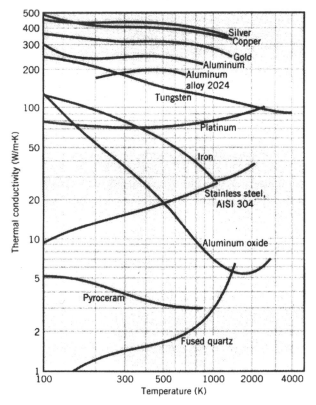

Figure 18.6 [S1]
Temperature Dependence of the Thermal Conductivity of
a Range of Materials

Many of the results of these factors are illustrated in Figure 18.6 which shows the temperature dependence of a number of materials including metallic conductors for comparison.

18.5 ESTIMATION OF THE THERMAL CONDUCTIVITY OF POLYMERS

There is no theory which allows the accurate estimation of the thermal conductivity of polymeric melts or solids. Most of the theoretical or semi-theoretical expressions are based on Prof. Debye's treatment of heat conductivity [S2]. The thermal conductivity was given by:

$$k = \Lambda c_v \rho V L \tag{18.6}$$

where:

Λ = a constant of about 1
c_v = specific heat capacity at constant volume
ρ = density
L = average mean free path length
V = velocity of elastic waves (sound waves)

Kardos [S3] and Sakiades and Coates [S4] proposed an analogous equation:

$$k \approx c_p \rho V_l L \tag{18.7}$$

in which L represents the distance between the molecules in adjacent isothermal layers.

The variation of the thermal conductivity, specific heat at constant pressure, the density, and the sound velocity for polymers are shown in Figure 18.7. If the assumption that ΛL in Equation 18.6 or L in Equation 18.7 are nearly constant and independent of temperature, a direct proportionality exists between the thermal diffusivity $k/c_p \rho$ and the sound velocity, V.

In the next section, previous relationships developed in Chapter 16 are utilized for estimating thermal conductivities.

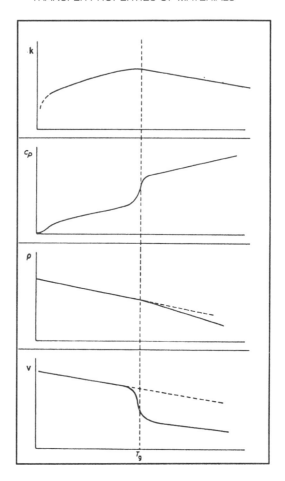

Figure 18.7
Effect of Thermal Conductivity, Specific Heat, Density, and Acoustic
Velocity On Amorphous Polymers

18.6   ADDITIVE MOLAR FUNCTIONS FOR THERMAL
        CONDUCTIVITY AND MOLCALC ESTIMATIONS OF
        THERMAL CONDUCTIVITIES AND COMPARISONS
        TO LITERATURE VALUES

Utilizing the Rao function, Chapter 16, and rearranging
Equation 16.50, we have:

$$V_l = \left(\frac{U_R}{V_v}\right)^3 \left[\frac{3(1-\nu)}{1+\nu}\right]^{1/2} \tag{18.8}$$

From Equations 18.7 and 18.8, the thermal conductivity, k, is given by:

$$k = Lc_p\rho\left(\frac{U_R}{V_v}\right)^3\left[\frac{3(1-v)}{1+v}\right]^{1/2} (J/s\cdot m\cdot K) \qquad (18.9)$$

where:

$$L \approx 5\times 10^{-11}\, m$$

Table 18.1

Comparison of Estimated Thermal Conductivities with Literature Values

Polymer	Thermal Conductivity, k (J/s · m · K)	
	Calculated	Literature
Poly(ethylene)	0.20	0.16-0.48
Poly(propylene)	0.20	0.17
Poly(isobutylene)	0.18	0.13
Poly(vinyl chloride)	0.18	0.17
Poly(vinyl fluoride)	0.16	----
Poly(chloro trifluoro ethylene)	0.12	0.14-0.25
Poly(vinyl acetate)	0.17	0.16
Poly(cis butadiene)	0.15	----
Poly(cis 1,4-isoprene)	0.15	0.13
Poly(styrene)	0.16	0.14
Poly(αmethyl styrene)	0.15	----
Polycarbonate	0.15	0.19
Poly(ethylene terephthalate)	0.16	0.22
Poly(butylene terephthalate)	0.17	----
Poly(3,3-bis chloromethyl oxacyclobutane)	0.19	----
Poly(methylene oxide)	0.22	0.16-0.42
Poly(vinyl methyl ether)	0.21	----
Poly(phenylene oxide)	0.16	----
Poly(hexamethylene adipamide)	0.23	0.23-0.43
Poly(acrylamide)	0.15	----
Poly(acrylonitrile)	0.20	----
Poly(methyl methacrylate)	0.23	0.19
Poly(cyclohexyl methacrylate)	0.21	----

The Rao-function as expressed in Equation 18.9 is utilized for calculating the thermal conductivity. Table 18.1 provides comparisons between calculated and literature values for the selected list of polymers.

18.7   MECHANISMS AND EQUATIONS OF DIFFUSION

Permeation of polymers by small molecules depends both on their solubility and their diffusivity. To understand the elements of permeation, the fundamentals of diffusion are first addressed.

Consider a volume represented by a cube using a Cartesian coordinate axes system as represented in Figure 18.8. The mass balance for component a diffusing through a control volume of this cube, i. e., Δx, Δy, and Δz can be obtained using the following considerations.

The net transfer of a by diffusion in each of the directions must be considered. In the x-direction, the mass transfer is given by:

$$\frac{\partial}{\partial x}\left(\frac{w_a}{A}\right)_x \Delta x \Delta y \Delta z \tag{18.10}$$

Similar expressions may be written for the y and z direction.

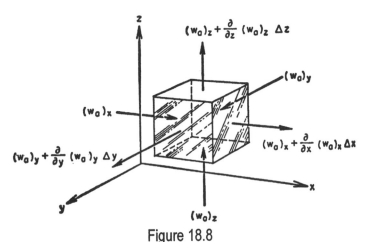

Figure 18.8
Representation of Volume Element Associated with the
Equation for Mass Diffusion

The rate of production or immobilization of a within the volume is:

$$Q_a \Delta x \Delta y \Delta z \qquad (18.11)$$

where:

Q_a = rate of production per unit volume

The rate of change of mass concentration of a within the volume is:

$$\frac{\partial c_a}{\partial t} \Delta x \Delta y \Delta z \qquad (18.12)$$

Dividing through by $\Delta x \Delta y \Delta z$, the following expression is obtained for the rate of change of mass concentration:

$$\frac{\partial c_a}{\partial t} = -\frac{\partial}{\partial x}\left(\frac{w_a}{A}\right)_x = \frac{\partial}{\partial y}\left(\frac{w_a}{A}\right)_y - \frac{\partial}{\partial z}\left(\frac{w_a}{A}\right)_z + Q_a \qquad (18.13)$$

Assuming the medium is isotropic and conforms to the usual experimental conditions of proportionality between the diffusion rate of the diffusing substance and the concentration gradient as given by:

$$\left(\frac{w_a}{A}\right)_x = -D\frac{\partial c_a}{\partial x} \qquad (18.14)$$

The rate equation becomes:

$$\frac{\partial c_a}{\partial t} = \frac{\partial}{\partial x}\left(D\frac{\partial c_a}{\partial x}\right) + \frac{\partial}{\partial y}\left(D\frac{\partial c_a}{\partial y}\right) + \frac{\partial}{\partial z}\left(D\frac{\partial c_a}{\partial z}\right) + Q_a \qquad (18.15)$$

or in vector notation:

$$\frac{\partial c_a}{\partial t} = dw\left(D\ Grad\ c_a\right) + Q_a \qquad (18.16)$$

Also, if D was independent of x, y, z, and c_a, and Q_a were zero:

$$\frac{\partial c_a}{\partial t} = D\left(\frac{\partial^2 c_a}{\partial x^2} + \frac{\partial c_a}{\partial y^2} + \frac{\partial c_a}{\partial x^2}\right) = D\nabla^2 c_a \qquad (18.17)$$

It will be noted that Equation 18.14 is the mass transfer which is analogous to the Fourier equation of heat conduction, Equation 18.2.

18.7.1 Steady-State Diffusion Through A Plane Membrane

Assuming that Equation 18.17 applies, this is directly analogous to that of heat conduction through a slab. Equation 18.17 reduces to:

$$\frac{d^2 c_a}{dx^2} = 0 \qquad (18.18)$$

for boundary conditions:

$$x = 0, \qquad\qquad c_a = c_{a1}$$
$$x = l, \qquad\qquad c_a = c_{a2}$$

the concentration profile and mass transfer rate would be given by:

$$c_a = \left(c_{a2} - c_{a1}\right)\frac{x}{l} + c_{a1} \qquad (18.19)$$

$$\frac{w_a}{A} = \frac{D}{l}\left(c_{a1} - c_{a2}\right) \qquad (18.20)$$

as illustrated in Figure 18.9.

The concentration of w_a at any distance in a semi-infinite slab starting from time $= 0$, from $x = 0$ to $x = l$, is:

$$w_a = \frac{D\left(c_{a1} - c_{a2}\right)}{A\left(x_0 - x_l\right)} = -\frac{D}{A}\frac{dc}{dx} \qquad (18.21)$$

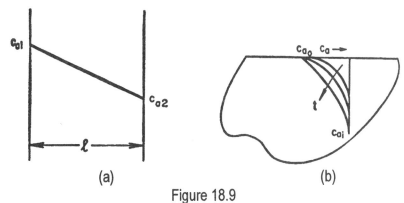

Figure 18.9
Concentration and Mass Transfer Profiles
a. Steady-state in a plane membrane
b. Transient distribution in a semi-infinite medium

Alternate expressions for the rate equation use: mass fraction; $c_a^* = c_a/\rho$; the molal concentration, n_a, the number of moles of a per unit volume; and the mole fraction, n_a^*, the ratio of the number of moles of a to the total number of moles of the solution mixture.

Equation 18.21 is simply the net flux through the material over a cross section of area A and is given by:

$$J = -D\frac{dc}{dx} \tag{18.22}$$

This equation is identical to that first proposed by Adolph Fick in 1855 on theoretical grounds for diffusion in solutions [S5]. In this equation, called Fick's first law, J is the flux, or quantity per second, passing through the area A under the action of a concentration gradient dc/dx. The factor D is the diffusion coefficient and has units of L^2/T such as ft^2/hr or cm^2/s.

For simple gases the interactions with polymers are weak, and the result is that the diffusion coefficient is dependent on temperature and independent of time and the concentration of the penetrant. In this case the penetrant molecules act as different sized probes. In general, diffusion is a thermally activated process and is given by an Arrhenius type equation.

$$D = D_o \exp(- E_D/RT) \qquad (18.23)$$

where D_o and E_D (activation energy) are constants for that particular polymer and gas.

All the known data on the diffusivity of gases in various polymers were collected by Stannett [S6].

The diffusivities of even simple gases in different polymers show large variations. The diffusion depends on both the type of gas and the structure of the polymer. The activation energy, E_D, is the most dominant parameter in the diffusion process, since it is the energy needed to enable a molecule to jump into another hole. Since larger holes are necessary for larger molecules, the activation energy must necessarily be larger, and the diffusivity will be smaller.

Crystallinity or crystalline regions obstruct the movement of the diffusing molecules and increase the average length of the paths they must travel.

Van Krevelen [S7] has listed general equations for diffusion derived from experimental data in the literature as follows.

- Rubbers (elastomers)

$$\log D_o = 10^{-3} E_D/R - 4.0 \pm 0.4 \qquad (18.24)$$

- Glassy Polymers

$$\log D_o = 10^{-3} E_D/R - 5.0 \pm 0.18 \qquad (18.25)$$

- All Types of Polymers

$$\log D(T) = \log D_o - \frac{435}{T} 10^{-3} E_D/R$$

$$= \log D(298) - 0.435 \frac{E_D}{R} \left(\frac{1}{T} - \frac{1}{298} \right) \qquad (18.26)$$

- Crystalline Polymers

$$D_{sc} = D_a \left(1 - x_c \right) \qquad (18.27)$$

These values of E_D/R show significant variation for the same polymer only in the cases of the smaller gas molecules, e.g. H_2. However, there is significant variation between polymers. For specific systems, Van Krevelen has tabulated these values (ref. S7, pg. 544-545).

18.8 SOLUBILITY

Here we consider only the solubility of simple materials in polymers and concentrate mainly on the solubility of gases. In the broader topic of solubility of a wide range of materials, particularly large molecules in polymers, such as exist in polymer blends, the reader is referred to Chapter 21 on Polymer Miscibility.

18.8.1 Gases

The solubility of a component, such as a gas, per unit volume of solvent (polymer) in equilibrium with a unit partial pressure is given by the well known Henry's Law. This law simply states that the mass of gas dissolved per unit volume is directly proportional to the pressure p.

$$w = kp \qquad (18.28)$$

The factor k is a proportionality constant and is often represented by S since it is the amount of substance (gas) per unit volume of polymer (solvent) in equilibrium with a unit partial pressure. The units are $cm^3(STP)/cm^3 \cdot bar$ or in S.I. units $10^{-5} m^3(STP)/m^3 \cdot Pa$.

The solubility of gases in polymers is low and this makes

quantification difficult especially when employing conventional techniques. Contemporary analytical techniques of isolation and measurement using gas chromatography, etc., have improved this situation. For organic vapors or moisture that are soluble in a polymer, simple weight measurement techniques are generally employed. Visible light scattering can also be used to determine the "cloudiness" point or point at which phase separation and/or deviation from Henry's Law occurs.

The nature of the gas is a determining factor of solubility, as might be imagined. If the solubility of oxygen and carbon dioxide, e.g., are normalized to nitrogen, it is observed that the solubility of these gases are, respectively, about 2 and 25 times higher than that of nitrogen.

The structure of the polymer also has a large influence on the solubility of a particular component. For polymers that contain groups that can strongly interact with the component or gas such as carbon dioxide, the solubility changes. This can readily be seen by comparing the estimated individual solubility parameters of the polymers in question.

The morphology of the polymer also influences solubility. The first consideration is whether or not the polymer is amorphous or crystalline and if crystalline, the extent of its crystallinity.

Fortunately, for a given simple gas, there is no great variation in solubility in different polymers, amorphous or crystalline as there is for diffusivity as was pointed out previously. For the smallest gas molecules dissolving in elastomers the heat of solution, ΔH_s is positive (endothermic) while for the larger gas molecules the reverse is true. The process is exothermic if the sorption energy evolved is greater than the energy needed to make a hole of molecular size in the polymer.

For simple gases the molecular interactions are small. The Lennard-Jones equation for the potential energy at high compression is:

$$\phi(r) = 4\varepsilon \left[\left(\frac{\sigma}{r}\right)^{12} - \left(\frac{\sigma}{r}\right)^{6} \right] \qquad (18.29)$$

where:

$\phi(r)$ = molecular interaction energy as a function of the
 separation distance
 r = separation distance
 ε = potential energy constant
 σ = collision diameter

Both ε and σ are the Lennard-Jones scaling factors. The division of ε by the Boltzmann constant, **k**, gives the Lennard-Jones temperature $\varepsilon/\mathbf{k}$ in Kelvin.

Van Krevelen has listed general equations for solubility derived from experimental data in the literature, using the above concept as follows.

- Rubbers (elastomers)

$$\log S(298) = -7.0 + 0.010\varepsilon/\mathbf{k} \pm 0.25 \qquad (18.30)$$

- Glassy Polymers

$$\log S(298) = -7.4 + 0.010\varepsilon/\mathbf{k} \pm 0.6 \qquad (18.31)$$

- All types of amorphous polymers

$$\log S_a(T) = \log S(298) - 0.435\frac{\Delta H_s}{R}\left(\frac{1}{T} - \frac{1}{298}\right) \qquad (18.32)$$

- Crystalline polymers

$$S(298) \approx S_a(298)(1 - x_c) \qquad (18.33)$$

where x_c is the degree of crystallinity.

Values of $\varepsilon/\mathbf{k}$ for various gases have been tabulated by Van Krevelen [S7].

18.9   PERMEATION OF POLYMERS

The category of permeation of greatest interest here is the

permeation of simple gases through polymers as occurs in coatings, packaging, and polymers used in sheets and composites. These applications are of broad interest and the permeation of gases and other materials are important considerations in the performance of the final product.

For simple gases a general relationship between the three main permeation properties P (permeability), S (solubility) and D (diffusivity) is closely approximated by the relationship:

$$P = S \cdot D \qquad\qquad (18.34)$$

This indicates that permeation is a sequential process, beginning with the solution of the gas at the surface of the polymer, followed by slow inward diffusion. All three physical quantities can be described by equations of the Van't Hoff-Arrhenius type:

$$S(T) = S_o \exp\left(-\Delta H_s / RT\right) \qquad\qquad (18.35)$$

$$D(T) = D_o \exp\left(-E_D / RT\right) \qquad\qquad (18.36)$$

$$P(T) = P_o \exp\left(-E_P / RT\right) \qquad\qquad (18.37)$$

where:

ΔH_s = molar heat of sorption
E_D = activation energy of diffusion
E_P = apparent activation energy of permeation
S_o, D_o, P_o = reference states derived from experimental measurements

Similar to diffusion and solubility equations, Van Krevelen has listed general equations for permeation derived from experimental data in the literature. These are as follows.

- Rubbers (elastomers)

$$\log P(298) - 10.1 - 0.46 x 10^{-3} \, E_P / R \pm 0.25 \qquad\qquad (18.38)$$

- Glassy Polymers

$$\log P(298) = -11.25 - 0.46x10^{-3}\ E_P/R \pm 0.25 \qquad (18.39)$$

- All Polymers

$$\log P(T) = \log P(298) - 0.435\frac{E_P}{R}\left(\frac{1}{T} - \frac{1}{298}\right) \quad (18.40)$$

- Crystalline Polymers

$$P_{sc} = P_a\left(1 - x_c\right)^2 \qquad (18.41)$$

These equations for rubbers and glassy polymers confirm that there is no great difference between rubbers and glassy polymers although the combined effect of both diffusion and solubility are operable as discussed previously. Values of E_P for simple gases have been tabulated by Van Krevelen (ref. S7, pg. 555).

18.10 COMPLEX PERMEATION

The permeation of simple gases assumes Fickian diffusion where the diffusion coefficient is independent of the concentration of the sorbed penetrant and the gas pressure is close to atmospheric.

When there are strong interactions between the penetrant and the polymer the adsorption isotherms graphically display this non-ideal behavior.

Here, we must distinguish between adsorption and sorption isotherms. Adsorption is the process that involves single layer or multilayer adsorption onto surfaces. There are five basic types as classified by Brunauer [S8].

In the case of sorption where a substance interacts with the polymer surface and the interior of the polymer as it penetrates, Rogers [S9] has proposed a classification of sorption isotherms.

- Type I - This is ideal behavior as described by Henry's Law and applies to simple gases around atmospheric pressure.

- Type II - This is the Langmuir isotherm. This results from the adsorption of gases at specific sites at higher pressures.
- Type III - This is the Flory-Huggins adsorption isotherm. Here the solubility coefficient increases continually with pressure. This is observed when the penetrant acts as a swelling agent for the polymer containing groups which "like" or interact with the penetrant but the structure of the penetrant is such that the penetrant groups do not interact sufficiently with the polymer for it to be a solvent.
- Type IV - This type is a typical two stage BET isotherm and may be considered as a combination of other types.

These various types are illustrated in Figure 18.10.

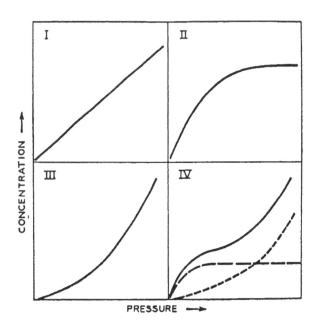

Figure 18.10
Typical Isotherm Plots of Adsorbed Concentration vs.
Ambient Vapor Pressure

18.11 AN ADDITIVE MOLAR FUNCTION FOR PERMEATION OF SIMPLE GASES

In 1986, Salame [S10] introduced a physical parameter which he coined the Permachor (π). It is given by:

$$P(298) = P^*(298)\exp(-s\pi) \tag{18.42}$$

or

$$\pi = -\frac{1}{s}\ln\frac{P(298)}{P^*(298)} = -\frac{2.3}{s}\log\frac{P(298)}{P^*(298)} \tag{18.43}$$

where:

$P(298)$ = permeability of an arbitrary simple gas in an arbitrary polymer

$P^*(298)$ = permeability of the same gas in a chosen standard polymer

s = a scaling factor

Nitrogen is normally used as the standard gas but in principle any other simple gas may be used since the permeabilites have a constant ratio determined by the collision diameter of the gas molecules.

Rubber (cis-isoprene) is used as the standard polymer since it is of defined composition and its permeability fits into the middle of the range for elastomers. Salame chose poly(vinylidene chloride with $log\,P = -17$ or an approximate spread of five orders of magnitude between zero (relatively high permeability) and 100 (no permeability) for the π scale. Polymers with a permeability greater than rubber would be represented by negative values. These concepts are illustrated in Figure 18.11.

For the number of characteristic groups per structural unit, Salame [S9] found that the relative permeability can be represented by:

$$N \times \pi = \Pi = \sum_i \left(N_i \cdot \Pi_i\right) \tag{18.44}$$

where:

Π = molar permachor

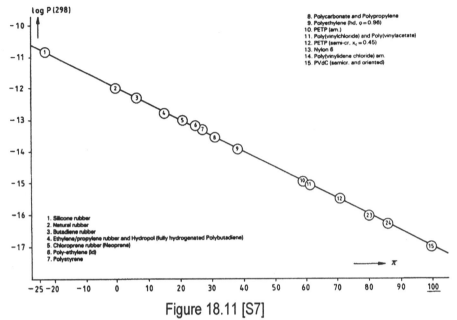

Figure 18.11 [S7]
Permeability of Various Polymers As A Function of the Permachor

Π_i = contribution for the group i

N_i = number of characteristic groups per structural unit

18.12 MOLCALC ESTIMATIONS OF PERMEABILITY

If the numerical value of π is known, the permeability at room temperature can be estimated from the following equation:

$$\log P(298) = \log P^*(298) - \frac{s \cdot \pi}{2.3} \qquad (18.45)$$

$$\mathrm{cm^3(STP \cdot cm)}/\mathrm{cm^2 \cdot s \cdot Pa}$$

In the case of nitrogen, $\log P^*(298) = -12$ and $s = 0.122$, so that Equation 18.45 reduces to:

$$\log P(298) = -12 - 0.053 \qquad (18.46)$$

Molcalc uses Equation 18.44 to calculate π followed by Equation 18.46 and then applies a correction factor based on the relative rate of permeation of nitrogen compared to the relative permeation of other simple gases listed in the gas permeation factor table.

The final permeation values shown in the table of Calculated Molecular Properties are valid only for amorphous polymers. For semi-crystalline and oriented polymer films a correction must be made. Salame [S9] recommended the following correction for semi-crystalline polymers:

$$\pi_{sc} \approx \pi_a - 18\ln_a - 41.5\log\left(1 - x_c\right) \qquad (18.47)$$

where:

a = volume fraction of the amorphous portion

x_c = degree of crystallinity

Table 18.2 provides comparisons between calculated and literature values for the selected list of polymers for permeation of the gas, as specified.

Calculated values for polymers should be considered approximate values only. This is because of the magnitude of the absolute values and the difficulty in estimating and comparing values for polymers whose morphologies may be quite different. In addition, only an average value is calculated here with the assumption that the morphology of the polymer whose permeability is being estimated, is in fact amorphous.

Table 18.2

Comparisons of Estimated Permeability with Literature Values

Polymer	log Permeability(P), $cm^3(STP) \cdot cm/cm^2 \cdot s \cdot Pa$	
	Calculated	Literature
Poly(ethylene)	-13.8 for N_2	-13.2 LDPE,-13.9HDPE
Poly(propylene)	-13.2 for N_2	-13.3, $\rho = 0.902$
Poly(isobutylene)	-13.3	----
Poly(vinyl chloride)	-15.1 for N_2	-15.1
Poly(vinyl fluoride)	-15.9 for N_2	-15.9
Poly(chloro trifluoro ethylene)	-14.5 for N_2	-14.3
Poly(vinyl acetate)	-13.6 for O_2	-14.0
Poly(cis butadiene)	-12.3 for N_2	-12.3
Poly(cis 1,4-isoprene)	-12.3 for N_2	-12.2
Poly(styrene)	-12.9 for O_2	-12.7
Poly(αmethyl styrene)	-13.5 for N_2	----
Polycarbonate	-13.7 for N_2	-13.7
Poly(ethylene terephthalate)	-13.0 for N_2	-14.9
Poly(butylene terephthalate)	-13.2 for N_2	----
Poly(3,3-bis chloromethyl oxacyclobutane)	-14.7 for N_2	----
Poly(methylene oxide)	-12.8 for CO_2	-12.9
Poly(vinyl methyl ether)	-14.5 for N_2	----
Poly(phenylene oxide)	-12.5 for N_2	-12.5
Poly(hexamethylene adipamide)	-14.2 for CO_2	-14.3
Poly(acrylamide)	-13.2	----
Poly(acrylonitrile)	-16.8 for O_2	----
Poly(methyl methacrylate)	-13.8 for N_2	-13.8
Poly(cyclohexyl methacrylate)	-15.7	----

18.13 ABBREVIATIONS AND NOTATION

c_a	mass concentration of component a
$c(\omega)$	contribution to the specific heat per frequency interval for lattice waves of that frequency
c_p	specific heat at constant pressure
c_v	specific heat capacity volume
D	diffusion coefficient
D_o	reference state constant for a particular polymer and gas
D_{sc}	diffusion
D_a	diffusion in amorphous polymers
E_P	apparent activation energy of permeation
J	joule
k	proportionality constant or thermal conductivity
$\mathbf{k}$	Boltzmann constant
$l(\omega)$	attenuation length for lattice waves
L	average mean free path
L	constant $\left(\approx 5 \times 10^{-11} \text{m} \right)$
P	pressure
P	permeability
$P(298)$	permeability of a given gas in a given polymer
$P^*(298)$	permeability of the same gas in a reference or standard polymer
P_o, S_o	reference states
q''	rate of heat flux
Q_a	rate of production of component a per unit
r	separation distance
S	substance/unit volume in equilibrium with unit partial pressure
S	solubility

S	scaling factor
T	temperature
U_R	Rao function
v	average molecular velocity
V	velocity of elastic waves
V_l	longitudinal sound velocity
V_v	volume of repeat structural unit
v	Poisson's ratio
W	watts
w	weight
w_a	mass of component a
Λ	constant of about 1
ρ	density
ΔH_S	heat of solution
$\phi(r)$	molecular interaction energy
ε	potential energy constant
σ	collision diameter
Π	molar Permachor
π	Permachor
π_{sc}	Permachor for semi crystalline polymers
π_a	Permachor for amorphous polymers
x_c	degree of crystallinity

18.14 SPECIFIC REFERENCES

1. Incropera, F.P. and Dewitt, D.P., *Fundamentals of Heat and Mass Transfer*, 2nd Ed., J. Wiley, New York, N. Y. 1985.
2. Debye, P., *Math. Vorlesungen Univ. Göttingen* **6**, 9(1914).
3. Kardos, A., *Forsch. Geb. Ingenieurw.* **5B**, 14(1934).
4. Sakiadis, B.C. and Coates, J., *A.I.Ch.E.J.*, **1**, 275(1955): 2, 88,(1956).
5. Fick, A., ÜberDiffusion, *Poggendorff's Analen*, **94**, 59(1855).

6. Stannett, V., Diffusion of Simple Gases, in *Diffusion in Polymers*, Crank, J. and Park, G.S., eds., Academic Press, London, 1968.

7. Van Krevelen, D.W. and Hoftyzer, P.J., *Properties of Polymers*, Elsevier, New York, N.Y., 1990.

8. Brunauer, S., *The Adsorption of Gases and Vapors*, Oxford University Press, London, 1945.

9. Rogers, C. E., *Polymer Permeability*, J. Comyn, (Ed.), Elsevier, London, 1985.

10. Salame, M., *Polym. Eng. and Sci.* 26, 1543(1986).

18.15 GENERAL REFERENCES

1. Van Krevelen, D.W. and Hoftyzer, P.J., *Properties of Polymers*, 2nd Ed. Elsevier, New York, N.Y., 1976.

2. Barrer, R.M., *Diffusion in and Through Solids*, Cambridge University Press, London, 1951.

3. Hopfenberg, H.B., (Ed.), *Permeability of Plastic Films and Coatings to Gases Vapours and Liquids*, Plenum Press, New York, 1974.

4. Vrentas, J.S., and Duda, J.L., Diffusion, in *Encyclopedia of Polymer Science and Engineering*, Vol 5, 2nd Ed., Wiley & Sons, New York, 1986.

5. Brandup, J. and Immergut, E.H., *Polymer Handbook*, 3rd Ed., J. Wiley & Sons., N. Y. 1989.

6. Mark, J.E., *Physical Properties of Polymers Handbook*, American Institute of Physics, Woodbury, N. Y., 1996.

7. Wylie, C.R., *Advance Engineering Mathematics*, 4th Ed., McGraw-Hill Book Co., N. Y., 1976.

8. Van Krevelen, D.W., *Properties of Polymers*, Elsevier, New York, N.Y., 1990.

Formation and Decomposition Properties
of Materials

19.1   INTRODUCTION

Many branches of chemistry are concerned with the formation of various products by carrying out appropriate reactions. From the understanding and control of these reactions comes a range of useful products, of which polymeric materials represent a single entity.

As we have seen, thermodynamics deals with equilibrium states and allows us to evaluate the likelihood that a given or proposed reaction may proceed. This does not say that a reaction will proceed but only that it is feasible for it to do so. In many cases it may prove impossible to find the right catalyst or conditions that favor desired products.

In order for a reaction that is feasible to proceed, a required amount of energy must first be added to the reactants. This energy is the activation energy which allows the reaction to proceed kinetically. This and other factors which control the kinetics determine how fast the reaction proceeds.

In some instances, a reaction will proceed through multiple paths or steps, each one of which may require activation energy to produce an activated complex that ultimately provides the desired reaction products. These concepts are illustrated in Figure 19.1.

19.2   THERMODYNAMICS RELATIVE TO FORMATION

From the gas laws in Chapter 2, one of the expressions for the change in Gibbs Free Energy (Equation 2.29) was:

551

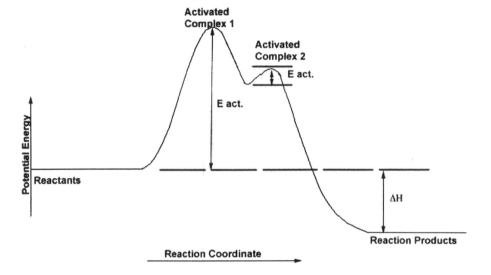

Figure 19.1
Activation Energy and Heat of Reaction in a System

$$dG = V.dP - SdT \qquad (19.1)$$

At constant temperature:

$$dG = VdP \qquad (19.2)$$

$$\text{or} \left(\frac{\partial G}{\partial P} \right)_T = V \qquad (19.3)$$

For an ideal gas, Equation 19.2 becomes:

$$dG = nRT \frac{dP}{P} \qquad (19.4)$$

For a pressure change from P_A to P_B :

$$\Delta G = G_B - G_A = nRT \int_{P_A}^{P_B} \frac{dP}{P} \qquad (19.5)$$

$$\Delta G = nRT \ln \frac{P_B}{P_A} \qquad (19.6)$$

The free energy of a gas is related to the standard free energy G^o, which is the free energy of one mole of gas at one atmosphere pressure.

We have:

$$G - G^o = RT \ln \frac{P_B}{P_A(=1)} = RT \ln P \qquad (19.7)$$

In this simple example, we can see that the free energy of a substance other than standard conditions is given by the equation:

$$G = G^o + RT \ln[P] \qquad (19.8)$$

$$= G^o + RT \ln K \qquad (19.9)$$

where:

R = gas constant
T = temperature
$[P]$ = concentration or more precisely, the activity

The discussion, so far, has been limited to a system containing only one chemical component. Additional moles of various components of the system, $n_1, n_2, \ldots n_i$, may be accounted for by adding further terms to Equation 19.1, which results in an equation given by:

$$dG = VdP - SdT + \sum_i \mu_i dn_i = \left(\frac{\partial G}{\partial n_i} \right)_{P,T,n_j} \qquad (19.10)$$

(cf. Chapter 21, Equation 21.77)

For the reaction:

$$A_2 + B_2 \Leftrightarrow 2AB \qquad (19.11)$$

and assuming 1.0 moles of A_2 and 1.0 moles of B_2 are added to a liter flask, the free energy of the system will be:

$$G = (1.0)G_{A_2} + (1.0)G_{B_2} \qquad (19.12)$$

After 0.1 mole of A_2 and 0.1 mole of B_2 have reacted to form 0.2 moles of AB, the free energy of the system will be:

$$G' = (0.9)G'_{A_2} + (0.9)G'_{B_2} + (0.2)G'_{AB} \qquad (19.13)$$

The free energy of this system can be followed as a function of the extent of the reaction, with the resulting curve depicted in Figure 19.2. The reaction, ζ, proceeds until the position of equilibrium is the position of minimum energy of the system represented by the arrow where $\left(\dfrac{\partial G}{\partial \zeta}\right) = 0$.

Since $dG = 0$ at equilibrium, the chemical potentials of the components in the phases (A, B, AB) must be equal.

$$\mu_i(A) = \mu_j(B) = \mu_k(AB) \qquad (19.14)$$

If x moles of A_2 and B_2 react to form $2x$ moles of AB, then:

$$G = (1-x)\left(G^o_{A_2} + RT\ln[A_2]\right) + (1-x)\left(G^o_{B_2} + RT\ln[B_2]\right)$$
$$+ (2x)\left(G^o_{AB} + RT\ln[AB]\right) \qquad (19.15)$$

Differentiating Equation 19.15 and setting $\dfrac{dG}{dx} = 0$, we obtain:

$$0 = -\left(G^o_{A_2} + RT\ln[A_2]\right) - \left(G^o_{B_2} + RT\ln[B_2]\right) + 2\left(G^o_{AB}\right) + RT\ln[AB]$$
$$(19.16)$$

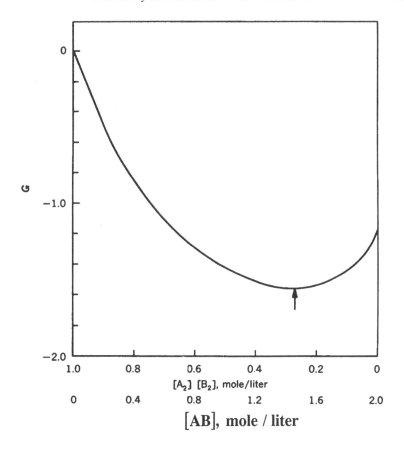

Figure 19.2
Change in the Total Free Energy , G, of a System During a Reaction

Rearranging Equation 19.16 gives:

$$\Delta G^o = 2G^o_{AB} - G^o_{A_2} - G^o_{B_2} = RT \ln[A_2] + RT \ln[B_2] - 2RT \ln[AB]$$
(19.17)

$$\Delta G^o = -RT \ln \frac{[AB]^2}{[A][B]} = -RT \ln K$$
(19.18)

In general terms, the extent of the reaction is given by:

$$\left(\frac{\partial G}{\partial \zeta}\right)_{T,P} = 2\mu_{AB} - \mu_A - \mu_B = \Delta G^o + RT \ln K \qquad (19.19)$$

From Chapters 2 and 22, the Gibbs free energy is:

$$G = H - TS \qquad (19.20)$$

and $\quad \Delta G = \Delta H - T\Delta S \qquad (19.21)$

Using the above relationship with the relationship in Equation 19.18, we have:

$$\Delta G^o = \Delta H^o - T\Delta S^o = -RT \ln K \qquad (19.21)$$

The free energy of a change then becomes a useful criterion for estimating the driving force or spontaneity of a reaction. The temperature dependence is expressed by the well known Van't Hoff equation which describes the temperature dependence of the equilibrium constant.

$$\ln K = -\frac{\Delta H^o}{RT} + \frac{\Delta S^o}{R} \qquad (19.22)$$

Analogous to the Van't Hoff equation for the reaction rate constant, the Arrhenius equation is given by:

$$\ln k = -\frac{E_{act}}{RT} + \ln A \qquad (19.24)$$

where:

E_{act} = activation energy

A = frequency factor or pre exponential constant

Figure 19.1 shows the relationship between ΔH^o and E_{act}.

19.3 FREE ENTHALPY OF REACTION FROM GROUP CONTRIBUTIONS

If a product, e.g. a polymer, is formed from its constituents, we have:

$$\sum_i n_i C_i \Leftrightarrow P \tag{19.25}$$

and the equilibrium of this reaction at some temperature is given by the equilibrium constant K_f as:

$$K_f = \frac{[P]}{\Pi [C]^n} \tag{19.26}$$

Since $\Delta G_f = -RT \ln K_f$, Equation 19.26 shows that the equilibrium is shifted to the right if $K_f > 1$ and to the left if $K_f < 1$ because a negative free energy always favors formation of a stable product. The relative value of ΔG_f^o is a quantitative measure of the stability of the products with respect to its constituents.

In general terms, the equilibrium at a given temperature is determined by the constant:

$$K_{eq} = \frac{\Pi [B]^{n_B}}{\Pi [A]^{n_A}} \tag{19.27}$$

For the reaction:

$$\sum n_A A \Leftrightarrow \sum n_B B \tag{19.28}$$

The change of free energy or enthalpy is:

$$\Delta G^o = \sum n_B \Delta G_{fB}^o - \sum n_A \Delta G_{fA}^o \tag{19.29}$$

This equation shows that if the free enthalpies of formation of the components of the reaction are known, it is possible to calculate the position of the equilibrium of this reaction.

For polymerization, degradation, and other reactions, it is of great practical importance to have available the numerical value of the free enthalpy of formation.

Theoretically, it is possible to calculate thermodynamic data by means of statistical mechanical methods but these methods are laborious and the spectroscopic data required are often lacking.

Thus in this instance as well as in previous examples in this book, the use of group contributions provides a powerful tool.

Van Krevelen and Chermin [S1,S2] calculated the free enthalpy of formation from group contributions with some corrections due to structural influences according to:

$$\Delta G_f^o = \sum_{\substack{Contributions\ of \\ Component\ groups}} + \sum_{\substack{Structural \\ corrections}} \qquad (19.30)$$

The group contributions are considered linear functions of temperature according to the following:

$$\Delta G_{f\,group}^o = A + BT \qquad (19.31)$$

where A has the dimension of heat of formation and B that of entropy of formation. According to Ulrich [S3]:

$$A \approx \Delta H_f^o(298) \qquad (19.32)$$

$$B \approx -\Delta S_f^o(298) \qquad (19.33)$$

Since $\Delta G = \Delta H - T\Delta S$, $\Delta G \approx A - BT$.

The dependence of free enthalpy on temperature is more complex than the simple relation given in Equation 19.31 but it is sufficiently accurate in the temperature interval from 300-600 K [S4].

19.4 MOLCALC ESTIMATIONS AND COMPARISONS TO LITERATURE VALUES FOR FREE ENTHALPY OF FORMATION

19.4.1 General

Molcalc estimates the free enthalpy of formation at 298K using the group contributions in accordance with Equation 19.31 as pointed out previously. At equilibrium, this equation becomes:

$$\Delta G_f^o = 0 = A + BT \qquad (19.34)$$

where:

$$A = \quad \Delta H_f^o(298)$$
$$B = \quad -\Delta S_f^o(298)$$

The values of A and B must be calculated using the group contributions for both the monomer and polymer and then subtracted to arrive at the enthalpy of formation from their component groups in their standard states as Equation 19.29 states.

The values of A and B can then be used to estimate the temperature below which polymerization is thermodynamically feasible which is given by:

$$T = -\frac{A}{B} \qquad (19.35)$$

The value of ΔG_f^o is calculated from the Gibbs equation for free energy at any temperature specified and the total free energy change can then be estimated from the numerical difference between the product and the reactants. Several examples are given in Chapter 5. The degree of negativity of ΔG_f^o gives an indication of the extent of the driving force.

Table 19.1 provides comparisons between calculated and literature values for the selected list of polymers for the molar free enthalpy of polymerization.

19.4.2 Factors Influencing Calculations

It will be noted that Table 19.1 contains listings for the polymers prepared from gaseous or liquid monomers and the final state of liquid, condensed, or gaseous (imaginary) polymers. In view of the paucity of data for comparisons, the agreement in all but two cases appears to be adequate.

Table 19.1

Comparison of Estimated Enthalpy of Formation with Literature Values

Polymer	Enthalpy of Polymerization, ΔH_f°, kJ/ mol ,@298K	
	Calculated	Literature
Poly(ethylene)	-90 gg	-93.5 gg
Poly(propylene)	-85.7 gg	-86.5gg
Poly(isobutylene)	-75 gg	-72 gg
Poly(vinyl chloride)	-85.5 gg	-71 lc
Poly(vinyl fluoride)	-85.7 gg	----
Poly(chloro trifluoro ethylene)	-60 gg	----
Poly(vinyl acetate)	-85.7 gg	-89.5 ls
Poly(cis butadiene)	-72 gg	-73 gg
Poly(cis 1,4-isoprene)	-72 gg	-70.5 gg
Poly(styrene)	-85.7 gg	-74.5 gg
Poly(αmethyl styrene)	-75 gg	-35 lc
Polycarbonate	----	----
Poly(ethylene terephthalate)	----	----
Poly(butylene terephthalate)	----	----
Poly(3,3-bis chloromethyl oxacyclobutane)	----	----
Poly(methylene oxide)	-24 gg	----
Poly(vinyl methyl ether)	-85 gg	----
Poly(phenylene oxide)	----	----
Poly(hexamethylene adipamide)	----	----
Poly(acrylamide)	-85.7 gg	-81.5 gg
Poly(acrylonitrile)	-85.7 gg	-76.5 lc
Poly(methyl methacrylate)	-75 gg	-56.5 lc@347.5K
Poly(cyclohexyl methacrylate)	----	----

gg=gas-gas, lc=liquid-condensed, ls=liquid-solid

However, if the reactants and products are not in the ideal or gaseous state, corrections must be made [S5] for the greatest accuracy. These corrections have not been incorporated into this version of Matprop©.

It will also be observed that only those reactions resulting from additional polymerizations are included. Data for condensation reactions have not been estimated since the necessary group contributions for other necessary components such as gases and water which might be reactant or products (H_2O elimination during condensation) are also not included in this software version. For the interested reader, these details can be found in Van Krevelen and Hoftyzer [S4].

19.5 STABILITY/DECOMPOSTION

19.5.1 General

The stability and/or decomposition of materials has been a subject of much study over time. Before the advent of widespread use of polymers, metals were replaced with ceramic materials for those applications requiring either temperature or chemical stability. As polymers began to replace these materials in applications where feasible, the question of thermal stability assumed larger dimensions.

In many instances, the materials scientist is concerned with the serviceability of a material with respect to one or more properties. Thermal and environmental stability are two key concerns in many different types of applications ranging from engineering thermoplastics to automobile tires to adhesives and coatings.

Quite often changes in physical properties as a result of dynamic or static temperature exposure are accompanied by weight loss of the polymer. Even if it is known that a material experiences weight loss over a period of time, it is difficult to choose a reference point in comparing the thermal stability of different materials since stability may be affected by a number of complex and interrelated phenomena.

The above factors affecting stability may range from the shape of the material or surface area to the structure of the material itself (primary controlling variable) to the morphology of the bulk phase. For industrially prepared polymers, the addition of additives such as mold release agents may adversely influence thermal stability while the addition antioxidants and UV absorbers normally added to polymers enhances the stability (thermal and otherwise).

These considerations lead to the approach that a useful way of measuring the stability of polymers is to consider the stability from the opposite extreme, i.e., to measure polymer decomposition or degradation. In this way, sample to sample variations of the same polymer and comparisons with different polymers can be made reliably.

It should be recalled from Chapter 2 that molecular weight, structure, and morphology, which prescribe many of the materials physical properties also influence polymer decomposition, the kinetic and thermodynamic parameters of the degradation process, and the role of the structural repeat unit in the decomposition process (monomer elimination, chain scission, cross-linking).

If we define the general thermal stability of materials as their ability to maintain properties over a wide range of environmental conditions, we can appreciate the differences in behavior between metals, ceramics, and polymers. Many ceramics are usable to very high temperatures under reductive or oxidative atmospheres and fail often due to thermal stresses which cause spalling and cracking. Special ceramics are required for ultra high temperatures, of course.

Metals will often fail at elevated temperatures due to oxidation or reduction, or embrittlement caused by diffusion of impurities to grain boundaries. At very high temperatures melting and/or loss of physical properties occurs and suitable alloys or noble metals must be chosen for these types of applications.

The following characteristics must be displayed by polymers that are considered stable.

- Must possess a high melting or softening temperature
- Must be capable of exhibiting high resistance to spontaneous thermolytic decomposition
- Must be relatively unaffected by chemical reagents or radiation

Since real materials never fulfill all these requirements, the major routes or some suitable combination must be used to produce stable materials. These are:

- The use of crystalline materials
- The use of crosslinked materials
- The use of rigid polymers such as polyaromatics or heterocyclics
- The use of composites or filled materials
- The use of materials containing suitable additives to provide stability for the application

The general stability of polymers can thus be represented by a triangle with crystallinity, cross-linking, and chain stiffening representing the apexes, with the interior of the triangle being the property map that each major category provides. Each axis and its corresponding interior property map can be modified by blending with other polymers or using appropriate fillers to increase the modulus and/or modify specific property requirements.

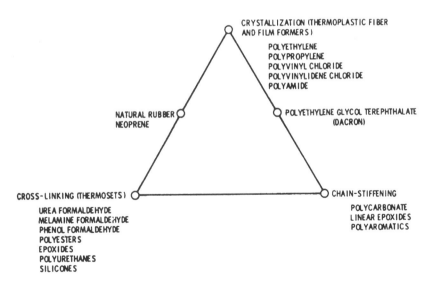

Figure 19.3 [S10]
Polymer Structure Influence on Thermal Stability

The property map that these categories represent is illustrated in Figure 19.3 [S10].

19.5.2 Thermal Stability

Thermal degradation begins for many polymers by breaking of the weakest bond. We might expect that because of this, there would be a direct relationship to bond energies shown in Table 19.2. A negative sign on these values represents the loss of energy by the system as the bond is formed between a mole of the two atoms involved in each case.

In the simple examples of the enthalpies of formation of HCl, HBr, and HI, the respective bond energies show that HCl is

Table 19.2
Bond Energies of a number of Bond Types for the Gaseous State

Bond	Energy (kJ/mole)
H-H	435.8
C-C	347.5
C=C	606.4
C$\equiv$C	828.0
O-O	138.8
O=O	489.3
O-H	455.8
C-O	330.4
C=O	723.5
C$\equiv$N	878.2
H-F	562.9
H-Cl	431.6
H-Br	365.9
H-I	298.6
N-N	160.6
N$\equiv$N	940.9
N-H	390.6
Cl-Cl	242.6
Br-Br	192.4
I-I	146.4

the most stable of the three. Other bonds such as $C \equiv N$ require more heat energy to break the bond than the O-O bond for example. From the previous section, however, we have seen that the change in the heat content is only a partial indication of the course of a reaction.

As the temperature is increased, and thermal dissociation occurs, the scission probability for all kinds of bonds increases. Although bond breakage is more likely to occur at weak bonds which are expected to break with a higher probability than other stronger bonds, the situation is considerably more complex even

Table 19.3

Bond Dissociation Energies, E_D (temperature 25° C , For Some Typical Bonds Encountered in Polymers

bond	E_D (kJ/mol)	reference compound
C=O	729	ketones
C−O	331	$H_5C_2\!\!-\!\!\!\overset{\downarrow}{}\!\!O-C_2H_5$
C≡C	838	$H-C\overset{\downarrow}{\equiv}C-H$
C=C	524	$H_2C\overset{\downarrow}{=}CH_2$
C−C	406	$F_3C\overset{\downarrow}{-}CF_3$
C−C	373	$C_6H_5\overset{\downarrow}{-}CH_3$
C−C	335	$H_3C-\overset{\overset{\displaystyle CH_3}{\displaystyle \mid}}{\underset{\underset{\displaystyle CH_3}{\displaystyle \mid}}{C}}\overset{\downarrow}{} CH_3$
C−H	507	$H\overset{\downarrow}{\cdots}C\equiv C-H$
C−H	432	$H\overset{\downarrow}{-}CF_3$
C−H	411−427	prim. aliphatics
C−H	394	sec. aliphatics
C−H	373	tert. aliphatics
C−H	325	$C_6H_5CH_2\overset{\downarrow}{} H$

though bond breakage in polymers is purely a stochastic process.

Table 19.3 lists bond energies for a number of selected bond values and shows that the energies for the same type of bond can vary appreciably, depending upon the structure. For example, the bond dissociation energy for a carbon-carbon bond varies from 335kJ/mol to 406 kJ/mole for a fluorine containing hydrocarbon compound.

Bond breakage is accompanied by chemical changes occurring during thermal treatment resulting from chemical bonds in the main and side chains rupturing. These changes are evidenced by lower molecular weights and the evolution of gaseous products. Typical intermolecular reactions are cyclizations and eliminations. Intermolecular cross-linking can occur in the case of linear polymers based on polyolefins. This process which is more significant the higher the temperature, is referred to as depolymerization. From the kinetic point of view, depolymerization proceeds as a chain reaction.

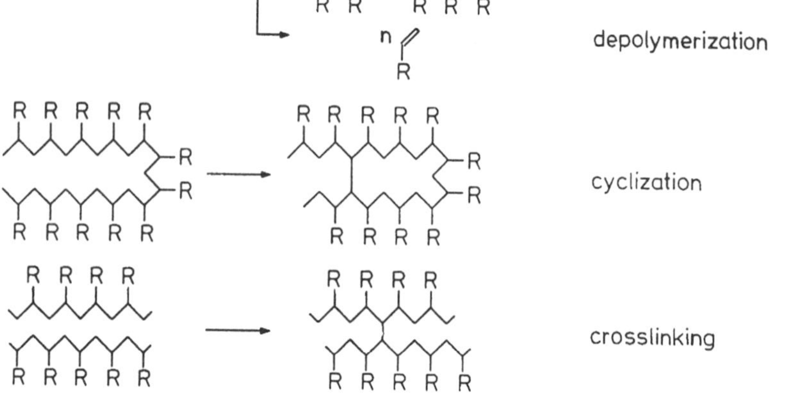

Figure 19.4 [S11]
Typical Reactions Occurring During Thermal Degradation of Polymers

The strong temperature dependence of the rate of depolymerization is due mainly to an increase in the rate of initiation and also to some extent due to an increase of the kinetic chain length with increasing temperature.

Random thermal decomposition can often be described as a first-order reaction if the decomposition products are volatile. The kinetic chain length $\left(k_{cl}\right)$ which is the ratio of the probability of propagation to the probability of termination plus transfer can vary tremendously. For example, in the case of polyethylene, the k_{cl} is zero while for poly(methyl methacrylate) it becomes nearly 200 [S6-S9]. Apart from depolymerization, other processes, such as radical transfer reactions, can also proceed as chain reactions at elevated temperatures. Figure 19.4 shows the general form of reactions that may occur [S11].

19.6 MOLCALC ESTIMATIONS AND COMPARISONS TO LITERATURE VALUES FOR THERMAL DECOMPOSITION

The initial temperature of decomposition is of great interest in many applications because of its obvious relationship to the ultimate use temperature or serviceable temperature. This temperature is generally determined from thermogravimetric analysis (TGA) measurements. In practice this temperature prove to be an elusive quantity because of gentle changes in the slope of the weight loss curve depending upon the sensitivity used in the measurement.

Because of these issues, the temperature of half decomposition, $T_{d,1/2}$, often proves to be a more reliable and useful temperature to compare the performance between materials. This is the temperature at which the loss of weight during pyrolysis, using a constant heating rate, reaches 50% of its final value. While there is good correlation with absolute values of $T_{d,1/2}$ in the literature, this temperature comparison is of even greater value in comparing the anticipated effects of changes made to polymer structures and experimentally using this temperature to compare changes made to the bulk by subtracting impurities or adding antioxidants.

Van Krevelen and Hoftyzer [S4] found that there was a relationship between the chemical structure of polymers and their characteristic temperature of decomposition, $T_{d,1/2}$. This relationship was based on additive properties of the product $M \cdot T_{d,1/2}$ and is given by:

$$T_{d,1/2} = Y_{d,1/2} = \sum_i N_i Y_{d,1/2,i} \Big/ M \qquad (19.36)$$

where:

    $T_{d,1/2}$ = temperature of half decomposition

    M = molecular weight of a structural unit or group

    $Y_{d,1/2,i}$ = group contribution

This relationship is used by Molcalc to estimate $T_{d,1/2}$. Table 19.3 provides comparisons between calculated and literature values for the selected list of polymers.

Table 19.3
Comparison of Estimated Temperature of Half Decomposition
with Literature Values

Polymer	Temperature of Half Decomposition K, $T_{d,1/2}$	
	Calculated	Literature
Poly(ethylene)	677	677
Poly(propylene)	665	660
Poly(isobutylene)	622	621
Poly(vinyl chloride)	544	543
Poly(vinyl fluoride)	662	663
Poly(chloro trifluoro ethylene)	650	653
Poly(vinyl acetate)	544	542
Poly(*cis* butadiene)	680	680
Poly(*cis* 1,4-isoprene)	595	596
Poly(styrene)	634	637
Poly(αmethyl styrene)	563	559
Polycarbonate	602	----
Poly(ethylene terephthalate)	721	723
Poly(butylene terephthalate)	716	----
Poly(3,3-bis chloromethyl oxacyclobutane)	542	----
Poly(methylene oxide)	583	----
Poly(vinyl methyl ether)	620	----
Poly(phenylene oxide)	753	753
Poly(hexamethylene adipamide)	694	693
Poly(acrylamide)	----	----
Poly(acrylonitrile)	726	723
Poly(methyl methacrylate)	608	610
Poly(cyclohexyl methacrylate)	594	----

19.7 ABBREVIATIONS AND NOTATION

A	frequency factor
C_i	constituent components

E_{act}	energy of activation
G	Gibbs free energy
G_i^o	standard free energy of a pure component, i
ΔG^o	standard free enthalpy change of reaction
ΔG_f^o	standard free enthalpy of formation
H	enthalpy
ΔH^o	standard enthalpy of reaction
K	equilibrium constant
K	Kelvin
P	product of formation
P	pressure
$[P]$	concentration
R	gas constant
S	entropy
T	temperature
$T_{d,1/2}$	temperature of half decomposition
V	volume
μ	chemical potential

19.8   SPECIFIC REFERENCES

1. Van Krevelen, D.W. and Chermin, H.A.G., *Chem. Eng. Sci.* **1**, 66(1951).
2. Van Krevelen, D.W. and Chermin, H.A.G., *Chem. Eng. Sci.* **1**, 238(1952).
3. Ulrich, H. *Chemische Thermodynamik*, Dresden, Leipzig, 1930.
4. Van Krevelen, D.W. and Hoftyzer, P.J., *Properties of Polymers*, Elsevier, New York, N.Y., 1990.
5. Dainton, F.S. and Ivin, K.J., *Trans. Faraday Soc.* **46**, 331(1950).
6. Van Krevelen, D.W., Van Heerden, C., and Huntjens, F.J., *Fuel* **30**, 253(1951).
7. Reich, L., *Makromol. Chem.* **105**, 223(1967).
8. Reich, L. and Levi, D.W., *Makromol. Chem.* **66**, 102(1963).
9. Broido, A.J., *Polymer Sci.* A2, 7,1761(1969).

10. Conley, R.T., *Thermal Stability of Polymers*, Vol.1, Marcel Dekker, Inc, New York, 1970.

11. Schnabel, W., *Polymer Degradation, Principles and Practical Applications*, Akademie-Verlag, Berlin, 1981.

19.9   GENERAL REFERENCES

1. Van Krevelen, D.W. and Hoftyzer, P.J., *Properties of Polymers*, 2nd Ed. Elsevier, New York, N.Y., 1976.

2. Grassie, N. and Scott, G., *Polymer Degradation & Stabilisation*, Cambridge University Press, New York, 1985.

3. Van Krevelen, D.W., *Properties of Polymers*, Elsevier, New York, N.Y., 1990.

CHAPTER 20

Processing of Thermoplastics

20.1 INTRODUCTION

Polymeric materials possess a useful combination of chemical, mechanical, thermal, and electrical properties which makes them an extremely versatile class of materials available for a wide spectrum of applications. Their end-uses can be classified in several categories including general purpose plastics, engineering plastics, fibers, elastomers, adhesives and coating. The plastic deformable state in thermoplastic polymers is achieved at elevated temperatures and in thermoset polymers before being chemically set which allows them to be shaped into a wide variety of finished products, some of great geometrical complexity. Polymeric materials possess a very high specific strength, can be fabricated into complex shapes at relatively low temperatures, and are suitable for mass production at relatively low cost using appropriate processing techniques. There has been a rapid and dynamic growth of polymeric materials in the last 40 to 50 years which accounts for our contemporary time being referred to as the "Plastics Age", from a materials point of view.

The polymer industry is a diverse one and includes, (a) production of polymers, (b) compounding of polymers, and (c) fabrication/conversion into useful products. In the production based industries low molecular weight raw materials, commonly known as monomers, are synthesized to produce polymers-copolymers of desired molecular weight chosen for end-applications. Compounding industries in turn mix these materials with several different kinds of additives such as stabilizers, pig-

573

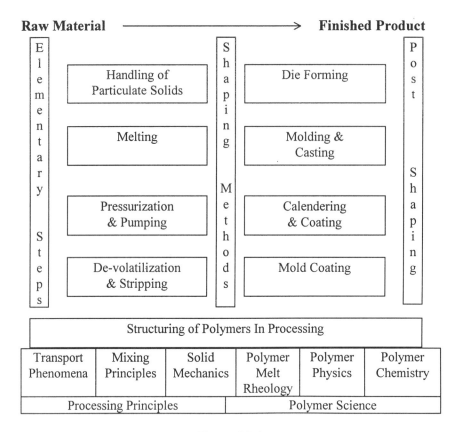

Figure 20.1
A Structural Breakdown of Polymer Processing

ments, fillers and reinforcing agents. Fabrication based industries use polymer processing operations to convert the virgin or compounded polymers into final products. Polymer processing of thermoplastics makes the basis of this chapter. A breakdown of polymer processing operations is given in Figure 20.1.

Polymer processing is basically an engineering specialty concerned with the operations carried out on polymeric materials or systems to increase their utility. In these operations the thermoplastic polymers must flow to be shaped into final form and then solidified while still in the given shape. The thermoplastics are heated above their softening temperatures which are governed by their glass transition $\left(T_g\right)$ in amorphous polymers angd melting point $\left(T_m\right)$ in crystalline polymers.

The polymer processing industry has to be closely associated with the machine/equipment design, product design, mold or die design and plant design. The roles of these are closely linked to the actual processing of polymers. Polymer processing requires a good understanding of the flow analysis in the processing operations. The number of processing operations is large and only the important ones are discussed here with an emphasis on recent developments.

20.2   CLASSIFICATION OF PROCESSING
           OPERATIONS

The processing operations for thermoplastics results in forming specific shapes and thus are often referred to as forming operations. In some cases there may be a greater emphasis on mixing rather than forming. Forming operations produce a wide variety of products of diverse shapes and sizes such as pen holders, buckets, refrigerator liners, thin films for packaging, fibers, car bumpers, bottles, television cabinets, gears, and overhead water tanks. The most important and versatile operations are extrusion and molding. Extrusion produces continuous profiles of constant cross-section to make products such as pipes, films, fibers, sheets, wire coatings, and window frame sections. Molding is used for making products with all three dimensions varying and include a variety of operations such as injection molding, blow molding, thermoforming, and rotomolding. Some operations use a combination of two techniques such as blow molding which can be either extrusion based or injection molding based. Similarly thermoforming utilizes extrusion prior to the actual forming operation. Other important related areas to processing are the machining and welding of plastics. These are of considerable importance since many applications require post processing operations but will not be discussed here.

The forming operations can be further subdivided into three categories which are as follows:

1. One dimensional operations: In these only the thickness of the product is of importance such as in a coating operation.
2. Two dimensional operations: In two dimensional operations both forming the cross-section and controlling the thickness are important. An example of this is extrusion of a profile or a pipe.
3. Three dimensional operations: In these all three dimensions can vary such as in the molding of a car bumper or the blowing of a bottle or the forming of a refrigerator liner. One and two dimensional processes are generally continuous in nature while the three dimensional process is carried out batchwise.

20.3   EXTRUSION PROCESSES

20.3.1 General

Extrusion is one of the most important forming techniques in polymer processing and more thermoplastics are converted into useful products by this process than any other method. The process of extrusion is a continuous one in which a polymeric fluid is forced through a shaped die by means of pressure to form products such as films, rods, pipe, coated wire or any other similar product with a long length and constant cross-section. In some cases the extrusion process may be discontinuous when a regulated amount of a fluid is forced into a mold as in the case of injection molding and blow molding. In the earliest form of extrusion a ram-driven plunger type machine was used to transport the polymeric liquid. However, in the modern process helical screws are used to transport the polymeric fluid through the barrel of the machine which has come to be known as an extruder. The most widely used machine is the single screw extruder containing one screw housed inside a barrel. Twin-screw extruders containing two inter-meshing screw housed in a single barrel are also used especially where a higher level of mixing is required.

An extruder is basically a screw pump capable of performing several operations simultaneously. A melt extruder is one in which the feed is a fluid, generally a molten polymer and it primarily provides a pumping/metering action. A plasticating extruder is one in which the feed is solid polymer granules or powder with melting, compression, and pumping taking place sequentially. The goal of an extruder is to produce a homogeneous molten material at a given flow rate, pressure and temperature for the next operation in the process line such as passing through a die or feeding into molding operation. An additional important feature of an extruder is its ability to intermittently mix the feed. Hence a polymer along with other additives or polymers may form the feed to give a homogenous molten mixture. In some cases the extruder may be used simply as a mixing/blending device which is generally the case with twin screw extruders. A major difference between extrusion and injection molding is that the extruder processes plastics at a lower pressure and operates continuously.

A cross section of the most widely used plasticating single screw extruder is given in Figure 20.2 which consists of a barrel and a screw as its main components. In the region of the hopper the barrel is cooled to allow a free flow of the feed to the barrel. The remainder of the barrel is encircled with electric resistance heaters which are generally split into three to five zones. Each zone is controlled individually at temperatures to give the most optimum melting and mixing of the polymer system.

The versatility of the extruder lies in the designing of the screw which may be specialized with respect to the function to be

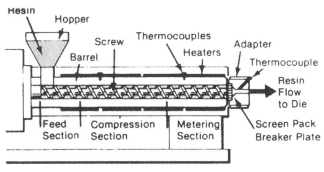

Figure 20.2
Cross Section of a Plasticating Single Screw Extruder

achieved such as mixing, metering or removal of volatile solvent. The design of the screw also varies with the type of polymer that is handled. The barrel diameter for industrial extruders ranges from 5 to 25 cm and the choice of size depends on the output capacity required. The length to diameter ratio of the screw varies from 15:1 to 35:1 with the criteria that a higher ratio is used when greater degree of homogenization is required such as in the case of polymer blends and filled polymer systems. A plasticating extruder screw is divided into three zones namely feed, compression and metering zones. In the feed zone the polymer granules are heated coupled with some melting. The screw depth is constant and the length of this zone is such as to ensure a correct rate of forward feed, neither starving nor overfed. In the compression zone complete melting takes place thus increasing the bulk density and entrapped air is expelled towards the feed zone. The depth of the screw decreases to adapt to the change in the bulk density and to build up pressure in the polymeric fluid. The length of this zone varies from polymer to polymer depending upon the melting behavior of polymer. Polymers which melt sharply require a short compression zone and those which melt gradually required a long compression zone. In view of this, there are three popular types of screws referred to as the nylon screw, PE screw, and PVC screw with increasing lengths of compression zones in this order. Lastly, the metering zone has a constant screw depth of sufficient length to homogenize the polymer and stabilize the flow.

The screw terminates in the die zone which shapes the final product. The die zone consists of a screen pack which contains a sieve pack supported by a breaker plate. The three functions of this are to sieve out extraneous material, to provide a resistance to flow so that pressure builds up in the screw, and to remove the turning memory from the melt since polymers are generally viscoelastic in nature. The equations for flow rate in the screw and the die are called screw characteristic and die characteristic. These coupled with a flow analysis show that a higher level of pressure build up results in more mixing flow in the screw at the expense of overall output from the extruder. In the extreme case when the exit of the extruder is blocked it simply becomes a mixer. Therefore, a suitable combination of level mixing and output can be achieved. Mixing can also be increased by lengthening the screw which primarily would result in lengthening of the metering zone where most of the homogenization occurs.

The heat necessary for softening or melting of a polymer is supplied by conduction from an external source such as electric band heaters and/or from the heat generated by mechanical action of the screw on the polymer due to frictional heating. Of the two, the latter is superior since it gives minimal temperature gradients. In fact, extrusion of soft polymers like LDPE or rubbery polymers can be carried out without supplying any external heat and this is referred to as operating under adiabatic conditions.

20.3.2 Special Cases

In some cases specialized design of screws can be used such as those with varying pitch or with reverse pitch. In the case when volatiles may be given off, the screw is designed to reduce the pressure midway in the barrel where a vent is provided. After this point, second compression and metering sections are provided. Such extruders are called vented extruders. In some cases the opening (vent) in the middle can be used to feed additives which are liable to degrade if added right at the beginning.

A very special type of screw is the recycle screw in which the main screw is made hollow towards the metering section and contains an additional stationary screw. As the polymeric fluid passes along the main channel some of it enters the inner screw through a drainage hole, travels in the opposite direction and then re-enters the main stream. Such screws can provide energy efficiency and less waste.

20.3.3 Twin Screw Extruders

Twin screw extruders contain two screws inside one barrel and these can be either co-rotating or counter-rotating, i.e. either rotating in the same direction or in opposite directions respectively. Furthermore, the two screws can be either meshing or non-meshing. The non-meshing type is not a true twin-screw extruder and can be taken as simple double screws in a barrel. In the meshing type there is another subdivision into conjugated and non-conjugated machines, depending on whether the meshing flights fully occupy the channels in the machine area. Twin screw extruders act as positive displacement pumps with little dependence on friction hence they are specially chosen for heat sensitive materials. Counter-rotating machines, if conjugated, may have no passage at all for materials to move around the screws and the material must move axially towards the die end. Similarly, co-rotating machines will have no passage around each screw and only a small and tortuous one around both, also leading to positive axial flow. Hence, the length to diameter (L/D) ratio is unimportant for propulsion and there is no need for a long metering zone. The length must be of course sufficient for proper melting of the polymer. Due to positive pumping action the rate of feed is not critical in maintaining output pressure. The conveying is not by drag flow and this allows good control of shear rate and hence temperature.

20.3.4 Extrusion Based Products
   20.3.4.1 Extrusion Dies
   The proper design of an extrusion die is critical and also difficult. Principles have been established but the die design and construction requires considerable experience. To design a die properly one needs to maintain linear flow in the melt and avoid dead spots. Thus it is necessary to establish converging flows from one cross-section to another which streamline the flow while not producing abrupt changes. In some cases the die temperature is kept different from the metering zone temperature in order to alter the rheological properties of the polymer for obtaining desired flow in the die. Improper die designs and die temperatures can lead to undesirable levels of die-swell and to melt fracture which mars the product appearance.

   Some typical products and the extrusion techniques used for fabrication are:

     1. Profile extrusion which is the direct manufacture of products from the extruder die. Examples of this

include pipes, window frames, rainwater gutters and downspouts, sealing strips for windows and doors, curtain rails, garden hose, corrugated sheets, etc.

2. Downstream operations on profiles which require an additional secondary step such as pressure or vacuum sizing and stretching of an extruded sheet to make it thinner.

3. Cross-head extrusion requires the melt to turn $90°$ before emerging from its die such as in wire coating.

4. Tubular film blowing is carried out by extruding a cylindrical tube of material which normally points upwards. The tube is inflated with air while still in molten state and is simultaneously drawn upward. The air inside is contained as a large bubble by a pair of collapsing rollers. This process is capable of making very thin films of uniform thickness.

5. Films and sheets can be produced as direct profiles. Special precautions have to be taken in the die design as a circular output from the extruder has to be spread over a wide area to form the film.

6. Synthetic fibers constitute a special class of profile extrusion whereby very thin filaments are formed. This technique is often referred to as spinning. The extruded filaments are also drawn to several times the extruded length to produce much lower diameters than the extrusion diameters.

20.3.4.2 Co-extrusion

Simultaneous extrusion of two or more polymers such that their melts are in contact with each other to maintain phase integrity is termed co-extrusion. In this forming process the polymer melt is supplied to the die through separate extruders. In the advanced world of packaging co-extrusion finds enormous applications. Sometimes a product is designed to achieve differential barrier properties in packaging materials and sometime the sole purpose is the economy of the product. The product in turn can be designed for two or more layers. Two layers are sufficient when the polymers used are compatible. When polymers are incompatible, adhesive layers are used which are also polymeric in nature. Such products can be of three, five or seven layers and are designed for special properties in specific applications such as packaging of milk, fruit juices and similar food items. Typical co-extruded products are multilayered films for packaging, sheets for thermoforming and parisons for blow molded containers. Obviously, the application requires consideration of the permeation rate of the specific gas(es) of interest (e.g., O_2, CO_2, H_2O, etc.) so that the proper combination of polymers can be selected to provide the desired barrier properties.

20.4   INJECTION MOLDING

Injection molding has been a means of converting thermoplastics into finished products for over 40 years. It is one of the most common molding operations in the plastics industry. It is used to create finished articles which range from paper clips to automobile front-end assemblies. In essence, injection molding is a process in which a solid thermoplastic material is heated until it reaches a state of fluidity, then transferred under pressure and in a short time injected into an enclosed hollow space or mold cavity. It is then cooled in the mold until it again reaches a solid state, conforming in shape to the mold-cavity. An injection molding machine consists of two main sections, the injection unit and the clamp unit which houses the mold. A typical injection molding machine is represented by a schematic diagram in Figure 20.3

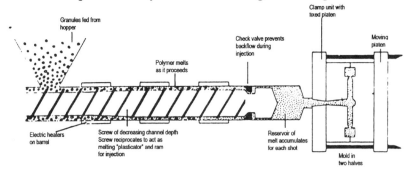

Figure 20.3
Main Features of a Reciprocating Screw Injection Molding Machine

The simplest molding machine is the plunger type, in which plastic is pushed forward by a plunger through a heated region. In more common use is the reciprocating screw type machine which is required for the processing of specific engineering articles. In this system the screw function is principally to melt and mix the feed material. For injection, the entire screw moves forward as a plunger, thereby pushing the polymer melt into the cold mold. A specific valve prevents back flow. Currently many variations of injection molding units are available commercially. Machines are available in which the injection unit can be vertically placed in reference to the clamp unit, for there can be two injection units for double color moldings. In all these machines the injection unit has the same basic components. It has a hopper for feeding the material and a barrel to which the materials are fed. There are heater zones on the barrel which ranges from 3-5 zones depending on the capacity of machine and the rate of plasticized polymer required. The nozzle also varies depending upon the type of plastic to be processed. A stiff-sluggish flow plastic like ABS requires a general purpose nozzle but a more fluid polymer melt like that of nylon or polyesters requires a special type of nozzle to avoid dripping the molten polymer.

Injection molding machines utilize different types of clamping units such as a hydraulic clamping unit or a toggle type mechanical clamping unit. Variations in toggle systems are also available from single toggle systems to double toggle systems. The clamping units of these machines have provisions for mold mounting which can vary from vertical to horizontal mold mounting. These variations are designed according to the specific product formulations. For example, small general purpose articles might require a horizontal injection unit with either a vertical or horizontal clamping unit but large size articles would require only a vertical clamping unit as the projected area of component is large. Various types of injection molds are required depending upon the nature of the article. They could be designed for semi-automatic to automatic production. These molds are available from single cavity to multiple cavity which are common place today. Newer mold design technologies have appeared for faster and safer production, which also produce less waste of material by using hot runner mold design technology.

A typical molding cycle has the following operations.

1. The mold is closed. At this stage it is of course empty. A shot of melt is ready in the injection unit.
2. Injection occurs. The valve opens and the screw, acting as a plunger forces the melt through the nozzle into the mold.
3. The "hold-on" stage is when pressure is maintained during the early stages of cooling to counteract contraction. Once freezing commences the pressure can be reduced.
4. The valve closes and the screw rotation starts. Pressure develops against the closed-off nozzle and the screw moves backwards to accumulate a fresh shot of melt in front.
5. Meanwhile, the part in the mold has continued to cool; and when ready, the press and the mold open and the molding is ejected.
6. The mold closes again and the cycle repeats.

The injection molding cycle is depicted in the chart shown in Figure 20.4. It can be seen that a large proportion of the cycle time is taken by cooling of the component. As a consequence, cooling rates are an important concern in the economics of injection molding. It is also observed that the final dimensions and quality of the molded parts are intimately related to the packing and holding stages of the injection molding. For example, part defects such as warpage, sink marks, uneven shrinkage and excess stress may result from the packing and holding processes. Thus an understanding of flow behavior during these stages becomes essential for the production of warp free parts.

Extensive applications of this forming process has triggered development in various fields like product design, mold design, mold filling, and machine design. With the help of these studies better products are produced which have good mechanical strength

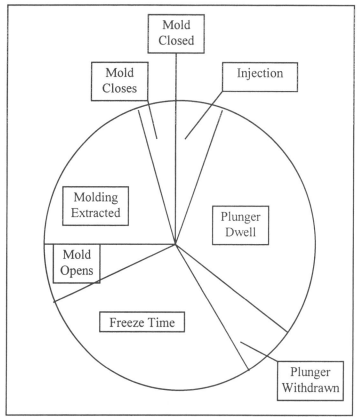

Figure 20.4
Pie Chart Showing the Cycle of Operations for the
Production of Injection Moldings

and appearance, i.e. no visible splash, sink, or shrinkage marks.

20.5   BLOW MOLDING

20.5.1 General

The process of blow molding is an established technique for manufacturing hollow plastic products such as bottles and other containers. It can also be used for the production of toys, automobile parts and several other engineering components. Containers with capacity as high as 220 liters are being made for the handling of materials by replacing large metal drums with the added advantage that plastic can also store hazardous chemicals. The basic principle of blow molding is similar to that used in the production of glass bottles although the processes are different. A semi-molten tube is first formed which is clamped between two halves of a split mold and then inflated with air to adapt to the shape of the mold. The mold surfaces are cooled hence producing a frozen product which is still under pressure. The product is recovered by opening the mold.

There are two major types of blow molding, namely, extrusion blow molding and injection blow molding. The former was the main technique initially in use but in recent times the latter one has emerged as a major method specially for PET bottles. In extrusion blow molding a semi-molten tube called a parison, is extruded which is ready, hot, and soft for feeding into the mold. In injection blow molding a tube closed at one end is molded by injection molding. This is called the preform. The preform is then reheated and fed to the blow molding. Although any thermoplastic can be blow molded, those with higher melt strength are generally extrusion blow molded while those with lower melt strength are injection blow molded.

The parisons of lower melt strength polymers are likely to sag by their own weight in an extrusion based process. Bottles and similar packaging containers constitute the largest product group made by blow molding. In addition large blow molded products include such items as drums for chemicals, including specialty products for acids, coolant and expansion tanks, traffic light housings and petrol tanks. As is obvious, this is an excellent technique for making goods that have a re-entrant curve, such as a narrow neck, and thus cannot be made by injection molding. It is perceived that the number of products made by blow molding will increase in time to come. The sudden spurt of PET bottles in recent times and big HDPE drums for chemical storage are examples of this. These two processes are described below.

20.5.2 Extrusion Blow Molding

The principle of the process is shown in Figure 20.5. The extrusion process may be continuous, in which case it will be nec-

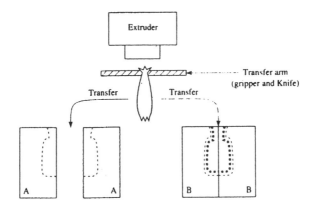

a. Parison transfer system for continuous extrusion blow molding

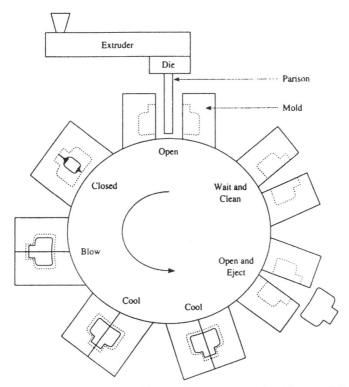

b. Rotating mold system used in continuous extrusion blow molding

Figure 20.5
Illustrations of Transfer and Rotating Extrusion Blow Molding Processes

essary either for the parison to be cut off and moved to the mold or the mold to move away carrying the parison. Alternately the extrusion can be intermittent, and in this case the mold can remain beneath the extrusion point. The former arrangement results in a faster production rate, hence it is more popular. The parison can sag due to gravity thus resulting in thick-walled bottoms and thin walled tops. Also the wall thickness varies with the diameter of the blow up sections. In order to offset the sag and the thickness variation during extrusion, the parison thickness can be varied by the use of a parison variator. Generally molds have a bottom blow arrangement in which the parison descends onto the blow pin. This method allows no delay between mold closure and blowing and hence gives a shorter molding cycle. One end of the parsion is pinched off by the mold thus forming the closed bottom end of the bottle. This leads to a pinch mark and can be a source of weakness in the finished product if the welding between the two contacting surfaces is not proper.

The matching of the molds should be excellent otherwise a mold separating line appears and if the product is thin walled this can become a source of failure. The blowing is generally done by the expansion of the parison with air. Evenness of wall thickness of the blown product is important. In practice optimum conditions must be developed so that rupture of the parison during blowing or entrapment of air between the parison and the mold does not occur. This is the art portion of the processing.

A glossy surface appearance is more desirable and is usually obtained from a slightly roughened mold surface. The mold interior is sandblasted to provide an appropriately rough surface based on the viscosity and other properties of the polymer and thus the actual mold surface is not reproduced. The gloss comes when the original extruded parison expanded by air comes in contact with cold mold and forms a molten glaze by immediate freezing at the surface. A matte finish can be obtained by heating the mold and allowing the polymer to flow at the surface before solidifying.

The rigidity of a blown product depends upon the thickness of the walls as well as on the flexural modulus of the material. Achieving the highest rigidity with the thinnest wall is desirable from economic considerations. The design of the product can incorporate ribs which would enhance the buckling strength of the container. For crystalline polymers, crystallization would increase the strength and would allow the thickness to be reduced. A bottle of HDPE can be of thinner wall than LDPE simply on strength criteria.

20.5.3 Injection Blow Molding

Injection blow molding is a relatively recent process and is most widely used for making bottles of polyethylene terephthalate (PET) the majority of which are used for carbonated soft drinks. Plastic bottles allows packaging in larger bottles of less weight with less chance of breakage. In this process the injection molding is used to make a test-tube like preform. This preform is generally shorter than the length of the final product. It is reheated to just above softening temperature and stretch blown to its final shape. The stretch blowing is accomplished by pushing down the blow pin which stretches the preform downwards and simultaneous blowing to give radial expansion. This process develops biaxial orientation in the product which can also reduce gas permeation, depending upon the specific polymer.

In the case of crystalline polymers such as PET, the mold is kept cold to quickly cool the polymer to prevent crystallization thus producing a transparent bottle. An additional advantage of injection blow molding is that it does not produce a weld line and can be easily employed for polymers with low melt strengths. For example, if PET was used for extrusion blow molding the parison would deform very rapidly.

20.6 THERMOFORMING

20.6.1 General

Thermoforming is a process in which a thermoplastic sheet is heated above its softening temperature and then deformed by stretching into a mold where it cools to produce the final product. Thermoforming is further subdivided, on the basis of the shaping force into three main categories, vacuum forming, pressure forming (blow forming) and mechanical forming (matched mold forming). In some cases a combination of two of these can be used for improving the performance of the operation and producing a better product. Commodity polymers that are generally thermoformed include polystyrene, cellulose acetate, cellulose acetate butyrate, polyvinyl chloride, polystyrene, ABS, polymethyl methacrylate, HDPE and PP. The bulk of thermoforming is done with extruded sheets, although cast, calendered or laminated sheets can also be formed. Thermoforming requires much less pressures for molding than injection molding. Hence longer articles can be easily produced coupled with the low cost of the mold. However, thermoforming is not able to make solid intricate shapes that can be made by injection molding. Recently, thermoformed products using engineering thermoplastics such as polycarbonate (PC) and PET have been introduced.

20.6.2 Vacuum Forming

Vacuum forming is the most widely used technique in which a hot softened sheet is drawn into a mold by applying vacuum on one side for the deformation of the sheet. A typical vacuum forming operation is shown in Figure 20.6(a) and 20.6(b). A male or a female mold can be used to make the articles. The female mold uses a cavity for the main formation while the male mold employs a projection. The variation in thickness of the final product depends upon the level of drawing and hence on the kind of mold used. In some instances, mechanical aids are used to assist the forming process especially when deep drawn or articles of complicated shapes are produced. Other variations are employed for improvement of thickness distribution.

20.6.3 Pressure Forming

For this forming operation, a pressure higher than atmospheric pressure is applied to the softened polymer sheet. This driving force deforms the sheet into the mold shape in a manner similar to vacuum forming. In some cases pressure may be used to assist vacuum forming.

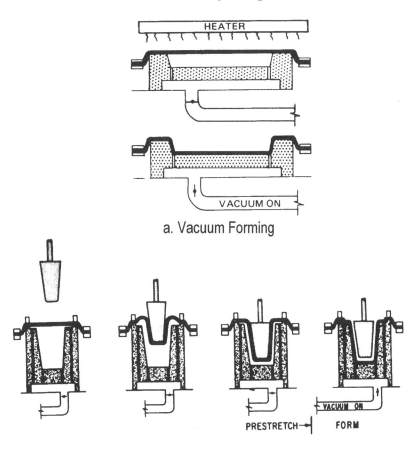

a. Vacuum Forming

b. Plug Assisted Vacuum Forming

Figure 20.6 [G11]
Schematic View of Two Thermoforming Processes

20.6.4 Matched Mold Forming

In this operation a softened sheet is deformed into the shape of the article by the use of matching male and female molds. However, this technique is generally not used in isolation but as a means to assist vacuum or pressure forming. Matched "mold assist" is specially suitable for articles requiring deep drawing and for articles requiring a special surface finish on both sides of the product.

20.6.5 Thermoformed Products

There are three main classes of products that are made by thermoforming as listed below:

1. Thin walled containers: This include cups, containers for packed juices, packs for chocolates, and other such items. For these, the forming is run in line with a sheeting extruder with an accurate control of sheet thickness to minimize scrap.

2. Large technical moldings: For these applications, thermo-moldings are much larger and the products generally require greater strength. The materials used can be quite complex such as co-extruded sheets, composites and laminates in addition to simple polymers. The applications include items like refrigerator liners, freezer liners, bath tubs, and garage doors. The product dimensions can be as much as 3 m wide and 10 m long coupled with deep drawn moldings. The control of extruded sheet is important in achieving good quality control.

3. Skin and Blister Packing: Skin and blister packing is currently used in display settings of a wide variety of household goods as tools and toiletry items. The skin packs are those where a flexible polymer film is tightly drawn on to the goods placed on a cardboard base. Blister packs are preformed foils which roughly conform to the shape of the article and are sealed to the base after the article is put inside.

In general, thermoforming has come a long way to become a major processing operation in the plastics industry. With manufacturers thinking of innovative articles made from polymers, thermoforming has a very bright future. The use of co-extruded sheets is currently prevalent such that the outside layers are of virgin material while the middle layer is of recycled material. Such flexibility of recycling does not exist in injection molding.

20.7   ROTATIONAL MOLDING

Rotational molding or rotomolding is a process which can

make hollow articles of any variety without seams. Blow molding is somewhat limited in the shape of the product that it can make, whereas rotomolding can make any shape, even those with no openings. The technique consists of putting powdered polymer between a closed mold. The mold is then placed in an oven and rotated in two planes around two axes to give a tumbling action. This process softens and sinters the polymer granules into a continuous article, basically by forming a coating on the inside surface of the mold. The mold, while still being rotated, is then taken to another chamber for cooling and rotation is continued until the product solidifies. The mold is then opened and the product removed. This process also requires much lower pressure and gives a uniform thickness with no residual stresses. A typical rotomolding operation is represented in Figure 20.7.

Rotomolding generally requires long cycle times and hence polymers used should have relatively good heat stability as they are likely to degrade. In some cases nitrogen purging can minimize this problem. Rotomolding can be used to make a wide range of hollow products such as footballs, marine buoys, surf boards, bins, containers and petrol tanks. One of the largest developing appli-

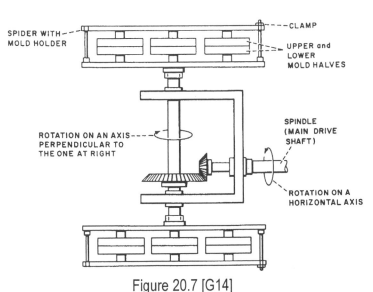

Figure 20.7 [G14]
Schematic Presentation of a Typical Mechanical System Used to Obtain
Biaxial Mold Rotation in Two Perpendicular Planes

cations of rotomoldings is the manufacture of overhead water tanks and other storage containers. With the need for such items the rotomolding industry has expanded considerably in recent times. The machinery and mold costs are relatively low and the capital requirements for equipment are cheap by comparison with other molding processes. Its main drawback is its slow output rate. In some special cases a two step molding can be made giving a different surface on the inside and outside of the product. The main innovativeness in rotomolding is to develop the most efficient way of heating the polymer and then having it disperse uniformly by the rotational arrangements.

20.8   CALENDERING

Calendering is a continuous processing operation utilizing a pair of drive rolls which convert a mass of softened material into a sheet of uniform thickness. The molten material is dragged through the narrow region between two co-rotating rollers as shown in Figure 20.8. Film or sheets are produced by calendering at high speeds with uniform thickness controlled to close tolerances. Thermoplastics can be calendered at speeds up to 100 m/min. with widths up to 2 m. Films as thin as 0.05 mm or less can be produced with thickness variations of less than 5%. Calenders may also be used to produce a specific surface finish as a result of the interaction between the roll surface and fluid in the region of separation. When thermoplastic materials are calendered, the main process occurring in the fluid is laminar viscous flow. An analysis of the calendering process involves developing a relationship between:
   1. Performance variables such as sheet thickness
   2. Operating variables such as roll diameter, roll speed, gap between rolls, and the film withdrawal rate

20.9   COATING OPERATIONS

The main operation in this category is the coating of a polymer liquid onto a moving sheet referred to as the web. The web may be paper, paperboard, cellulose film, plastic film, textile fabrics or metal foils. The polymer liquid can be a melt, a solution or an

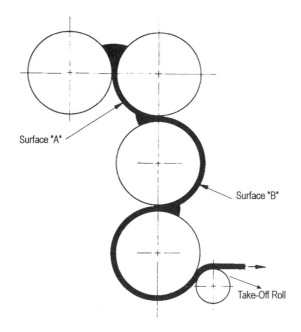

Figure 20.8
Schematic Diagram of Four-Roll Inverted L Calender

emulsion. In the case of a melt, the coating is simply cooled while in the other two cases the carrier solvent has to be evaporated after coating. Figure 20.9 shows various kinds of coating operations. In a roll coater, the lower roll picks up liquid from bath and delivers it to a second roll, or directly to the moving web. The web is squeezed between two rolls and the amount of coating applied depends upon fluid properties such as viscosity, surface and interfacial tension, and the spacing between the rolls. In a kiss coater the web is run over the roll without any backup roll on the other side.

The contact on the roll, the tension on the web and the fluid properties control the thickness. In a reverse coater, the roll that delivers the fluid to the web moves in the opposite direction to the web. A blade coater normally follows a kiss coater which can directly control the amount of fluid supplied to the moving web from a pond of coating fluid. Fluid properties, the shape and the positioning of the blade determine the coating thickness. Withdrawal

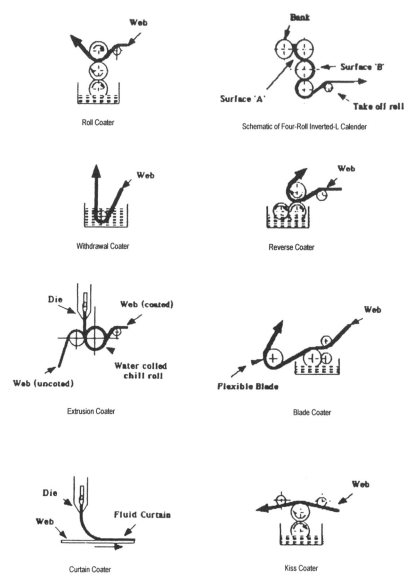

Figure 20.9
Illustration of a Number of Coating Processes

coating is employed when it is undesirable to contact the coating with another surface. The coating thickness depends upon the fluid properties. All these operations generally use polymer solutions or emulsions as the fluids. Sometimes a coating of a polymer melt is applied directly from the extruder with the web and the extrudate passing between rolls. Alternately the extrudate may be directly laid on the web which is called curtain coating. In both these cases, the coating thickness can be precisely controlled by the extrusion process.

20.10 MIXING

20.10.1 General

Most finished plastic products are not pure polymers but are mixtures of a polymer with a variety of additives which may include one or more of the following: stabilizers, antioxidants, plasticizers, pigments, fillers, reinforcing agents, and foaming agents, as well as others. The choice and need of additives varies with the base polymer and the end applications. The required additives must be mixed with the polymer prior to the forming operation either as a post polymerization step or a pre-processing step. The level of additives can vary from small amounts such as that of stabilizers to large amounts such as that of fillers. Some of these may be compatible with the polymer and others may not. In all cases an intimate mixture of all the components must be made before processing. In some cases two or more polymers may be blended to form the base polymer systems. The component polymers may be miscible, compatible, or incompatible. In some cases two polymers can be dispersed intimately with the help of compatibilizers. Blending of polymers has acquired a very special significance special with respect to mixing. The mixing processes thus form a very important part in thermoplastic polymer processing industry. Some of the common mixing methods are listed below.

20.10.2 Batch Mixers

Batch mixers are versatile units in which operating conditions can be varied during the mixing cycle and additives can be added at an optimal time. They provide good temperature control and they are generally available in various sizes to adapt to the processing operation which they feed. These can be subdivided into three categories consisting of particulate solids mixers, extrusion liquid mixers, and intensive liquid mixers. As is obvious the classification is based on the nature of the primary mixing mechanism.

Particulate solid mixers are also referred to as blenders and involve a random distributive mixing mechanism. In practice, the mixers generally consist of three types; tumbling, agitating ribbon, and fluidized bed type. The tumbling type is the simplest and least expensive but cannot handle difficult mixtures due to a tendency of segregation. Ribbon blenders consist of a moving spiral element which induce connective motion and are good for sticky mixtures. Fluidized bed mixers provide fast mixing but cannot handle sticky mixtures.

Liquid mixers involve laminar mixing and bring about an increase in the interfacial area between the components and the distribution of interfacial elements throughout the mixer volume. In the low viscosity range an impeller and high speed mixer are used. Medium viscosity ranges use double blade units such as the sigma blade mixer. This consists of a rectangular trough curved at the bottom to form two half-cylinders. The two blades revolve towards each other at different speeds to induce mixing by axial as well as tangential motion. High viscosity mixtures are mixed on a roll-mill or high intensity internal mixers such as the Banbury type. Both types impart extensive mixing at high shear rates with accompanying heat generation.

20.10.3 Continuous Mixers

Single as well as twin screw extruders are excellent examples of continuous mixers. When they are used for mixing purposes alone the design of the extruder, especially that of the screw is modified to increase the level of mixing. In many cases a properly designed extruder can provide the mixing action as well as supply material for the forming operation. Major modifications in extruders led to the development of mixers such as the Buss Ko-Kneader and the Transfer mixer. The Banbury mixer has been transformed into the Farrel continuous mixer. The two roll-mill can be converted into a continuous mixer by feeding raw material on one side and continuously stripping product on the other side.

20.11 NEW PROCESSING TECHNOLOGIES

20.11.1 General

The need for a newer technology arises when a processor encounters a limit to a forming process or when he needs to replace conventional applications with products made out of plastic raw materials or newly developed polymer blends. Development has taken place at almost every front of polymer technology, whether it be material, machine, or product design.

In material technology emphasis is given in the discovery of newer systems with better properties by combination of raw materials, development, modification of polymers, and recycling technology of thermoplastics.

In the field of product design, mold/machine design, extensive use of microprocessors and computers has been made. Robotics have been introduced for thermoplastic production and automation. Product design and mold design is coupled with production by CAD/CAM and CAE.

When a new application is uncovered, a lengthy, time consuming development process is shortened by the use of computers, which assists in product design, development, and manufacture of the part. CAD/CAM and CAE are avenues to the required production tools. Computer analysis also provides information to a processor about the molding parameters and other machine settings necessary for warp-free plastic production.

It has been firmly established that the forming machine plays an important role in high quality plastic part production and control must be exercised over the total forming process, e.g., in injection molding, hold and packing forms an important stage to eliminate defects from moldings.

Molding parameters can be best controlled by the use of microprocessors and computers using new technologies that have emerged as a result of specific applications. All of these aspects constitute processing technology.

20.11.2 Some Examples of Recent Developments

1. Doubled walled blow moldings to replace large size injection moldings: Since installation of injection molding equipment is costlier than blow molding equipment and as injection molding processes require larger size machines and intricate molds in comparison to the ones used in blow molding, replacement is ongoing. In recent years, blow molding processes have been modified so that double walled blow molded articles can be obtained. These have the strength and performance capabilities to replace large size injection moldings.

2. Open mold technology for replacement of thermoforming: Conventional thermoforming requires a preform in the form of an extruded sheet which after softening is thermoformed into profiles. There are, however, limitations to this forming process as very large size articles are difficult to form and very accurate heating of the sheet is required. Also deep drawing of such sheets is not possible because of the softening limitation. However, in this proposed technology, a combination of compression molding and extrusion is used to form large size articles.

   In this forming technique, a sheet is extruded by the backward movement of an extruder between the two halves of the forming machine, and while the sheet is still soft, the platens are closed to give the shape to the preform. Upon cooling the product is ejected. Through this newer technology, products which were difficult to form via thermoforming can be produced.

There are various other technologies available which are either combinations of two technologies like in example 2 or a modification of the original process as in example 1. The sole purpose is to achieve products which were not possible or difficult to produce, previously.

20.12 GENERAL REFERENCES

1. Beck,R.D., *Plastic Product Design*, 2nd ed, Van Nostrand Reinhold, New York,1980.
2. Brydson J.A., *Plastics Materials*, 4th ed., Butterworth Scientific, London, 1982.
3. Cheremisinoff, N.P., *Polymer Mixing and Extrusion Technology*, Marcel Dekker, New York, 1987.
4. Marqolis, J.M., ed., *Engineering Thermoplastics, Properties and Applications*, ed., Marcel Dekker, New York, 1985.
5. Mashelkar, R.A., Majumdar, A.S., Kamal, M.R., ed., *Transport Phenomena in Polymeric Systems*, Ellis Horwood, Halsted Press (distributors), New York, 1989.
6. McKelvey, J. A., *Polymer Processing*, John Wiley, New York, 1962.
7. Middleman, S., *Fundamentals of Polymer Processing*, McGraw Hill, New York, 1977.
8. Morton-Jones, D.H., and Ellis, J.W. *Polymer Products: Design, Materials, and Processing*, Chapman and Hill, 1986.
9. Powell, P.C., *Engineering with Polymers*, Chapman and Hall, 1986.
10. Rauwendaal, C., *Polymer Extrusion*, Hanser, New York, 1986.
11. Rosato, Donald, and Rosato, Dominick, Plastics Processing Data Handbook, Van Nostrand Reinhold, New York, 1990.
12. Tadmor, Z., and Gogos, C.E., *Principles of Polymer Processing*, Wiley Interscience, New York, 1979.
13. Throne, J. L., *Technology of Thermoforming,* Hanser, New York, 1996.
14. Zimmerman, A.B., Stabenau, W.H., Molding Equipment, Rotational, in *The Encyclopedia of Plastics Technology*, H.R. Simmonds, ed., Reinhold Publishing, New York, 1964.

CHAPTER 21

Miscibility and Estimation of Polymer Blends Miscibility

21.1 INTRODUCTION AND BACKGROUND

The mixing of materials, whether they be liquids, solids, or gases has intrigued and fascinated people for centuries. This interest and study of why certain materials mix, others that do not, and still others that mix but do not stay mixed, was the result of very specific practical needs and applications. An effort to understand the interaction of materials and the underlying science that allows them to mix continues to the present day in such diverse fields as surfactants, ceramics, metallurgy, and polymer science.

About four thousand years ago (2000BCE), in many instances it was becoming impractical to barter. Using stones as currency for common everyday exchanges also became impractical. Although there were a number of possibilities, metal became the obvious choice but it too had to possess certain qualities that allowed it to be easily manufactured, and identified while providing the necessary durability and portability required. It was at this point that bronze was selected as the material of choice for currency.

Basically, bronze consists of a mixture or alloy of copper and tin. The people in those primitive times moved to this form of exchange because it provided them a unique set of properties that fulfilled their needs. They probably found that the properties of this material exceeded, in combination, those that any other single material available to them at that time could provide. Thus it was very early that mankind took advantage of "enhanced" properties available through alloying without fully understanding the scientific ramifications.

21.1.1 Metals, Ceramics, and Polymers

In the examination of the structure of metals, ceramics, and polymers in Chapter 1, we have seen large differences between the structure of these materials. These differences account for remarkable variations in many if not most of their properties.

In this section we will explore the differences and similarities behind the fundamentals that cause them to mix and not mix. That is, we will look at many but not all of the factors that permit the formation of homogeneous or single phase mixtures. Metals and polymers will be compared since many considerations concerning metals are also applicable to ceramics. Then we will concentrate more heavily on the practical aspects of polymer miscibility.

21.2   TERMINOLOGY AND BASIC DEFINITIONS

The term alloy, applied to polymers, has been borrowed from metallurgy. The term alloy, in metallurgy, defines a homogeneous mixture of two or more kinds of atoms in the solid state.

For all three fields, i.e., metals, ceramics, and polymers, we are concerned with phase diagrams and whether or not there are one or more phases at a particular temperature at a specific component composition. A phase is defined as a macroscopically homogeneous body of matter and, thermodynamically, this means when this situation exists, the materials are miscible.

In the case of polymers, the term compatibility has been used often synonymously with miscibility. This is imprecise terminology and has led to confusion in the literature. Compatibility denotes that there is an interaction between the components in the blend or mixture and that they do not separate, at least over the time scale of interest. Miscibility implies "thermodynamic miscibility" which means there can be only one phase and that the free energy of mixing, $\Delta G_m \approx \Delta H_m \leq 0$. The exact mathematical relationships that define miscibility are presented in a subsequent section. However, the underlying concept is that there are degrees of compatibility and polymer blends that are miscible are of course compatible but blends that are compatible may not be thermodynamically miscible. This is not the case for metals and ceramics where a homogeneous mixture is known as a solid solution. For polymers, compatibility is often used to describe whether a desired or beneficial result is achieved when two or more materials are mixed or blended. In many cases miscibility in polymer blends is neither a requirement or desirable.

Polymer miscibility is best understood in thermodynamic terms which are related to the interaction of forces between two or more polymers. These aspects will be addressed in detail in subsequent sections.

21.3   METALS

From some standpoints consideration of metals represents a simplified system compared with a polymer system. The principles of the entropy of mixing are applicable to both systems. We first consider a simple system in order to understand the principles.

Consider a closed system of two gases A and B at temperature T_1 and pressure P_1 that are initially separated by a partition as illustrated in Figure 21.1

From our previous relationships developed in Chapter 2,

$$dE = dQ - dW \tag{21.1}$$

If we remove the partition and maintain constant temperature and pressure,

$$dE = d\overset{0}{Q} - d\overset{0}{W} \tag{21.2}$$

$$dE = 0 \tag{21.3}$$

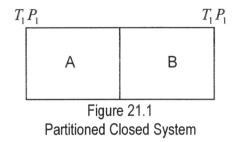

Figure 21.1
Partitioned Closed System

From the Helmholtz free energy:

$$dF = dE - TdS \qquad (21.4)$$

Therefore,

$$dF = -TdS \qquad (21.5)$$

A decrease in free energy can only mean that dS must be positive. The entropy of the system has increased and the system has become more disordered. Also recall from our previous relationships that this is a spontaneous and irreversible reaction.

To understand why systems seek a random distribution, such as the one in this example, we refer to the relationship first expressed by Boltzmann.

$$S = k \ln P \qquad (21.6)$$

where P is the probability of the state. A change in entropy of mixing from state 1 to state 2 is:

$$\Delta S = k \ln \frac{P_2}{P_1} \qquad (21.7)$$

where:

P_2 = probability of mixed state
P_1 = probability of unmixed state
k = Boltzmann constant (1.38×10^{-16} ergs/K)

If we have an empty closed system and two types of molecules or atoms A and B, with volumes V_A and V_B respectively, and we introduced an atom A into the system, the probability of finding it in V is $\dfrac{V_A}{V}$. If a second atom of A is introduced, the probability of finding it is:

$$(V_A/V) \cdot (V_A/V) \tag{21.8}$$

Therefore, the probability of finding all A atoms in V_A and all B atoms in V_B is:

$$P_1 = (V_A/V)^{n_A} \cdot (V_B/V)^{n_B} \tag{21.9}$$

21.3.1 Homogeneous Mixtures and Mixing

An extension of the above is that a homogeneous mixture is one with a probability ratio that does not vary sufficiently from the mean to be detectable, i.e., P_2 will approach 1.

$$\therefore \Delta S = k \ln \frac{1}{P_1} = -k \ln P_1 \tag{21.10}$$

$$\Delta S = -k \ln (V_A/V)^{n_A} \cdot (V_B/V)^{n_B} \tag{21.11}$$

$$\Delta S = -kn_A \ln(V_A/V) - n_B k \ln(V_B/V) \tag{21.12}$$

$$n_A/n = V_A/V = c \tag{21.13}$$

$$(n_B/n) = V_B/V = 1 - c \tag{21.14}$$

Since $$k = R/n \tag{21.15}$$

$$\Delta S_m = -R[c \ln c + (1-c) \ln(1-c)] \tag{21.16}$$

Metal crystals are never perfect and contain defects similar to polymer crystals. One of the most important kind of defects in a metal is called a vacancy. A vacancy can be viewed very simply as an absence of an atom and was originally postulated to explain diffusion.

Since gas atoms are small and relatively more elastic, and have much lower melting points, they have much greater mobility and their movement within a metallic lattice is not difficult to visualize. However, it is well known that copper and nickel will diffuse into each other if placed in intimate contact and heated to high temperatures.

Thus we can think about vacancies in metals being similar to free volume in polymers. In case of amorphous polymers the calculated WLF volume under equilibrium conditions at T_g is ≈ 0.025 cm^3 as an iso free volume state [S1] or alternatively 0.113 cm^3 using the Simha-Boyer method of calculation which depends on the difference of the cubical coefficients of expansion between the glass and the liquid ($\Delta \alpha T_g = K_1$) [S2], as discussed in Chapter 8. The free volume is specified at T_g and remains constant at all values of $T < T_g$. Above T_g the free volume is dependent upon the absolute value of the polymers' liquid coefficient of cubical expansion ($\alpha_L T_g = K_2$). These are functions of molecular weight, the degree of cross linking, and presence of sub group transitions.

The creation of a vacancy in a metal and its movement can be visualized as illustrated in Figure 21.2. In Figure 21.2(c), any of the surrounding atoms can move into the vacancy and the process repeated to move the vacancy through the crystal. In Figure 21.2(a) and (b), additional vacancies can be created by repetition of these steps. When a vacancy has been created by movement of atoms within the crystal to the surface, it is called a Schottky defect [S3].

If we represent the number of vacancies as n_v and the number of atoms available as n_o, then the total number will be:

$$n = n_o + n_v \tag{21.17}$$

and
$$c = \frac{n_v}{n_o + n_v} \tag{21.18}$$

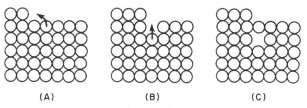

Figure 21.2
Creation of Vacancies in a Metal

(A)　　　　(B)　　　　(C)

$$1 - c = \frac{n_o}{n_o + n_v} \tag{21.19}$$

This allows a recasting of the entropy of mixing as:

$$S_m = \Delta S = k\left[\left(n_o + n_v\right)\ln\left(n_o + n_v\right) - n_v \ln n_v - n_o \ln n_o\right] \tag{21.20}$$

The free energy in this case is:

$$F_v = n_v E - T\Delta S_v \tag{21.21}$$

where E is the energy associated with forming a vacancy.
At equilibrium conditions of the metal crystal,

$$\frac{\partial F_v}{\partial n_v} = 0 = E + Tk\left[\ln\left(n_v / \left(n_o + n_v\right)\right)\right] \tag{21.22}$$

or

$$\frac{n_v}{n_o + n_v} = e^{-E/kT} \tag{21.23}$$

Since $n_v \ll n_o$, $n_o + n_v \approx n_o$,

$$\frac{n_v}{n_o} = e^{-E/kT} \tag{21.24}$$

and $\qquad e^{-E/kT} = e^{-EN/kNT} = e^{-E_{,1}/RT} \tag{21.25}$

where N is Avogadro's number and E_A is the energy of activation.

Using copper as an example, the experimental value for the activation energy in copper is $\approx 20,000$ cal/mole of vacancies. At $300\ °K$ ($\approx$ room temperature), we have:

$$\frac{n_v}{n_o} = 4.45 \times 10^{-15} \tag{21.26}$$

At $1350\ °K$, six degrees below the melting point, this becomes:

$$\frac{n_v}{n_o} \approx 10^{-3} \tag{21.27}$$

or about one vacancy for every 1000 atoms. Thus for each temperature there is a unique equilibrium number of vacancies (n_e) which affects the free energy of Equation 21.21 as illustrated in Figure 21.3.

Employing a simple hard ball model and using copper whose structure is face centered cubic, we estimate the volume occupied by one mole of copper atoms and the corresponding volume of vacancies, assuming each vacancy is roughly the size of the copper atom.

At room temperature for copper, from Equation 21.26; $\frac{n_v}{n_o} \approx 4.45 \times 10^{-15}$ for 1 mole of atoms.

$$n_v = 4.45 \times 10^{-15} \times 6.023 \times 10^{23} = 2.68 \times 10^9\ \frac{\text{atom vacancies}}{\text{mole of atoms}} \tag{21.28}$$

for copper $r = 1.28 \times 10^{-8}$ cm and the packing is :

$$\text{P.F.(Packing Fraction)} = \frac{\text{volume of spheres (atoms)}}{\text{volume of unit cells}} \tag{21.29}$$

$$= 0.74 \text{ for face centered cubic}$$

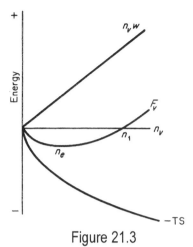

Figure 21.3
Free Energy as a Function of the Number of Vacancies, n_v, in a Crystal at High Temperature

∴ The volume of vacancies or free volume/mole is:

$$V = 2.68 \times 10^9 \frac{\text{atom vacancies}}{\text{mole of atoms}} \times 4/3 \times \pi (1.28 \times 10^{-8})^3 \frac{\text{cm}^3}{\text{atom}} \times 0.74$$

(21.30)

$$= 2 \times 10^{-14} \frac{\text{cm}^3}{\text{mol} \times 66.5 \text{g/mol}} = 3 \times 10^{-16} \text{cm}^3/\text{g}$$

(21.31)

For a polymer, say polystyrene (PS), that is at its T_g or slightly below, we have for the WLF free volume:

$$2.5 \times 10^{-2} \text{cm}^3/\text{g} = \text{specific volume}$$

(21.32)

and therefore PS contains roughly 10^{14} times as much free volume based on this comparison to a metal "hard ball" model.

Although no hard and fast conclusions can be drawn, it is fair to say that on the basis of the difference in free volume alone between a metal and a polymer in this simple model example, we would expect differences in the behavior and properties between these two types of materials. This expectation is based upon this single property resulting from structural differences.

21.3.2 Interstitial and Substitutional Atoms

An example of an interstitial site, in the case of a face centered cubic crystal, has been given previously (cf. Chapter 1, Structural Considerations /Why Materials Differ). Recall that an interstitial atom is one that occupies a place in a crystal that normally would be unoccupied. Such places are the holes or interstitial positions that occur between the atoms lying on the normal lattice sites. Extensive interstitial solid solutions occur only if the solute atom has an apparent diameter smaller than 0.59 that of the solvent.

A substitutional atom is one that is of a proper type that will allow it to be directly substituted for another type so that this atom can act as a solute atom and enter the crystal to take a position normally occupied by solvent atoms. Both of these cases are illustrated in Figure 1.20, Chapter 1.

The miscibility of substitutional atoms with solvent atoms to form alloys is governed by a set of rules that have been formulated from experimental observations and are known as the Hume-Rothery Rules.

Hume-Rothery Rules

1. Solute to solvent ratio: $\dfrac{\text{Solute }(r)}{\text{Solvent }(R)} \geq 0.85$
2. Materials must have the same crystal structure.
3. If an added atom or ion has a valence different from that of the host ion (ceramics), substitution is limited.
4. The atoms or ions must be close in the electromotive series, i.e., in the same group or close in the periodic Table. This determines whether or not there will be metallic or ionic bonding.

As an example of the solubility of interstitials, we consider the solubility of carbon in body centered cubic (BCC) iron.

$$r^c = 0.914 \qquad\qquad \frac{r^c}{R^{Fe}} = 0.78$$
$$R^{Fe} = 1.17$$

Therefore, we can conclude that the solubility of carbon in BCC iron is low and produces high lattice strain.

Each interstitial atom increases the entropy of the crystal in the same manner as does each vacancy. This intrinsic entropy increment is due to the fact that an interstitial atom affects the normal modes of lattice vibration. The solute atom distorts the orderly array of iron atoms and this makes the thermal vibrations of the crystal more random or irregular. The total entropy contribution from this source is $n_c S_c$ where n_c is the number of carbon atoms.

Accounting for this intrinsic entropy, the ratio of vacancies to atoms in the crystal is similar to Equation 21.24 and reduces to:

$$\frac{n_v}{n_o} = Be^{-E_f/RT} \tag{21.33}$$

where B is a constant, E_f is the energy required to introduce a mole of vacancies into the lattice, R is the universal gas constant, and T is the temperature in Kelvin, as before.

Analogous to Equation 21.16, the entropy of mixing of carbon and iron atoms to form a solid solution is:

$$S_m = k\left[\left(n_{Fe} + n_c\right)\ln\left(n_{Fe} + n_c\right) - n_c \ln n_c - n_{Fe} \ln n_{Fe}\right] \tag{21.34}$$

The situation is quite opposite for the substitutional solid solution of silver in copper. In this case, we have:

$$r^{Ag} = 1.34 \qquad\qquad \frac{r^{Ag}}{R^{Cu}} = 0.87 \tag{21.35}$$
$$R^{Cu} = 1.17$$

This fits the Hume-Rothery rules and these two elements are miscible and have different phase compositions depending upon the concentrations of copper and silver.

21.3.3 Ideal and Non-Ideal Solutions

This consideration is of great importance in the diffusion of metals and also offers some stark contrasts from polymers. Since diffusion and self diffusion also bear directly on miscibility, we consider these concepts here.

21.3.3.1 Ideal

In an ideal solution, there is no tendency for a mixture of A and B atoms or molecules to exhibit preference for one another. They will therefore not attract one another nor cluster. Such a solution is said to be ideal and the free energy of mixing is expressed as:

$$G = G_A^\circ N_A + G_B^\circ N_B - T\Delta S_m \qquad (21.36)$$

$$\Delta S_m = -R\left[N_A \ln N_A + N_B \ln N_B\right] \qquad (21.37)$$

$\left(G_A^\circ N_A + G_B^\circ N_B\right)$ may be considered to be the free energy of one total mole of the two components if the components are not mixed or do not mix.

where:

$\quad G_A^\circ$ = free energy /mole of pure A

$\quad G_B^\circ$ = free energy/mole of pure B

$\quad T$ = absolute temperature

$\quad \Delta S_m$ = entropy of mixing, as before

$$N_A = \frac{n_A}{n_A + n_B}$$

and $\qquad\qquad\qquad\qquad\qquad\qquad\qquad\qquad$ (21.38)

$$N_B = \frac{n_B}{n_A + n_B}$$

$$\therefore G = N_A\left(G_A^\circ + RT\ln N_A\right) + N_B\left(G_B^\circ + RT\ln N_B\right) \quad (21.39)$$

where the quantities in parentheses are the respective partial molal free energies of components A and B and are customarily denoted by $\overline{G}_A$ and $\overline{G}_B$.

The increase in free energy when one mole of A or one mole of B is dissolved in a very large quantity of the solution is:

$$\Delta\overline{G}_A = \overline{G}_A - \overline{G}_A^{\circ} = RT\ln N_A \qquad (21.40)$$

$$\Delta\overline{G}_B = \overline{G}_B - G_B^{\circ} = RT\ln N_B \qquad (21.41)$$

21.3.3.2 Non-Ideal

In general, it is not expected that two atomic forms, chosen at random, will show no preference either for their own or their opposites and observed results show that this is indeed the case. This means most solutions are non-ideal.

For non-ideal solutions, a quantity, known as the activity, defines this departure from ideality. Thus $\Delta\overline{G}$ is given by:

$$\Delta\overline{G}_A = RT\ln a_A \qquad (21.42)$$

$$\Delta\overline{G}_B = RT\ln a_B \qquad (21.43)$$

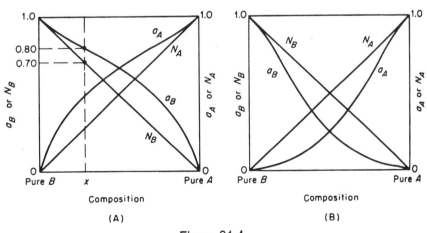

Figure 21.4
Variation of the Activities with Concentration
(A) Positive Deviation and (B) Negative Deviation

Referring to Figure 21.4

$$\Delta \overline{G}_B = RT \ln 0.7 = -0.356 RT \text{ (Ideal)} \qquad (21.44)$$

$$\Delta \overline{G}_B = RT \ln 0.8 = -0.223 RT \text{ (Non - Ideal)} \qquad (21.45)$$

$$\left(\Delta \overline{G}_B\right)_{Ideal} > \left(\Delta \overline{G}_B\right)_{Non-Ideal} \qquad (21.46)$$

∴ There is a smaller decrease in free energy for positive deviation, i.e., the attraction between likes is greater than between unlikes. A similar analogy exists in the case of polymers where an attraction between unlikes results in a larger decrease in free energy as a result of the formation of a solution (negative deviation).

21.3.4 One Component Systems

In a one component system (a pure substance) of a given phase, the thermodynamic state is determined uniquely if any two of its thermodynamic variables or properties are known. Solutions, however, have additional degrees of freedom and it is necessary to specify temperature, pressure, and composition, in general.

From the previous thermodynamic relationships, under conditions of constant temperature and pressure for a reversible reactions:

$$dG = VdP - SdT \qquad (21.47)$$

and

$$dH = TdS + VdP \qquad (21.48)$$

$$dH = \frac{Tdq}{T}^{\nearrow 0} + VdP \qquad (21.49)$$

$$\therefore dH = dQ \qquad (21.50)$$

For allotropic transformations, i.e., changes in crystalline structures, say from body centered to face centered, we have:

$$G = H - TS \tag{21.51}$$

$$\text{but} \quad dH = q = \int C_p dT \tag{21.52}$$

$$H = H_o + \int_0^T C_p dT \tag{21.53}$$

$$\text{and} \quad dS = \frac{dq}{T} = \frac{C_P}{T} dT \tag{21.54}$$

$$S = S_o + \int_0^T \frac{C_p dT}{T} = \int_0^T \frac{C_p dT}{T} \tag{21.55}$$

$$G^\alpha = H_o^\alpha + \int_0^T C_p dT - T \int_0^T \frac{C_p^\alpha dT}{T} \tag{21.56}$$

$$G^\beta = H_o^\beta + \int_0^T C_p dT - T \int_0^T \frac{C_p^\beta dT}{T} \tag{21.57}$$

$$\Delta G = G^\beta - G^\alpha = H_o^\beta - H_o^\alpha + \int_0^T C_p^\beta dT - \int_0^T C_p^\alpha dT \tag{21.58}$$

$$- T \left(- \int_0^T \frac{C_p^\beta dT}{T} + \int_0^T \frac{C_p^\alpha dT}{T} \right)$$

Thus, the change in free energy is due to the differences in heat capacities and the stable phase at lower temperatures is the one having the lower internal energy $(H = E + PV)$, i.e., the one that is capable of resulting in a tighter binding of the component atoms.

21.3.5 Two Component Systems

Metallurgical systems often exist in more than one phase,

depending upon temperature and composition. Let us now consider a two component system, specifically, a binary alloy. The phase that surrounds the other phase is the continuous phase while the one that is surrounded is the discontinuous phase.

For a two component system, for example, from Equation 21.38:

$$N_A = \frac{n_A}{n_A + n_B}$$

$$N_B = \frac{n_B}{n_A + n_B}$$

$$N_A + N_B = 1 \tag{21.59}$$

and the change in the Helmholtz free energy is:

$$dF = \left| \frac{\partial F}{\partial n_A} \times dn_A + \frac{\partial F}{\partial n_B} \times dn_B \right|_{T,P} \tag{21.60}$$

and

$$\left| \frac{dF}{dn_A} = \frac{\partial F}{\partial n_A} \right|_{T,P,n_B} \tag{21.61}$$

$$dF = \overline{F}_A dn_A + \overline{F}_B dn_B \tag{21.62}$$

and the total free energy is:

$$F_{TTL} = n_A \overline{F}_A + n_B \overline{F}_B \tag{21.63}$$

$$F^\alpha = n_A^\alpha \overline{F}_A^\alpha + n_B^\alpha \overline{F}_B^\alpha \tag{21.64}$$

$$F^\beta = n_A^\beta \overline{F}_A^\beta + n_B^\beta \overline{F}_B^\beta \tag{21.65}$$

Let dn_A transfer α to β then,

$$dF = dF^{\alpha} + dF^{\beta} \tag{21.66}$$

$$dF = \overline{F}_A \left(-dn_A \right) + \overline{F}_A^{\beta} \left(dn_A \right) \tag{21.67}$$

and finally $\quad dF = \left(\overline{F}_A^{\beta} - \overline{F}_A^{\alpha} \right) dn_A \tag{21.68}$

If phases are in equilibrium, i.e., $dF = 0$:

$$dF = \left(\overline{F}_A^{\beta} - \overline{F}_A^{\alpha} \right) dn_A = 0 \tag{21.69}$$

$$\overline{F}_A^{\alpha} = \overline{F}_A^{\beta} \tag{21.70}$$

and similarly it can be shown that:

$$\overline{F}_B^{\alpha} = \overline{F}_B^{\beta} \tag{21.71}$$

From Equation 21.47, the change in Gibbs free energy when the number of moles of various components vary, we have:

$$dG = Vdp - SdT + \left(\frac{\partial G}{\partial n_i} \right)_{T,P,n_j} \tag{21.73}$$

or $\quad dG = VdP - SdT + \sum_i \mu_i dn_i \tag{21.74}$

and the chemical potential is: $\quad \mu_i = \left(\frac{\partial G}{\partial n_i} \right)_{T,P,n_j} \tag{21.75}$

This is sometimes called the fundamental equation of chemical thermodynamics. The quantity μ_i may be thought of as the increase in the free energy of the system when one mole of component i is added to an infinitely large quantity of the mixture so that it does not significantly change the overall composition.

Chemical potential is an intensive property and can be regarded as providing the force which drives systems to equilibrium.

For alloying of metals, the miscibility is determined by the entropy of mixing and the partial molal free energies which in turn are affected by atomic or molecular preferences of one atom for another which is defined by the activities of each atom in a system.

21.4   POLYMER MISCIBILITY

21.4.1 Introduction

New types of alloys have appeared in recent years as opposed to the conventional alloys, discussed in the previous sections that have been around for several thousand years. These, of course, are *polymer alloys* which are achieved by blending two or more polymers that have a modified interface and/or morphology.

The growth rate of polymer blends and alloys has been exceeding the polymer industry growth rate by a factor of about five due to the ever increasing cost of development of new materials. Employing polymer blending commercially, viable products can be achieved with either unique properties or with lower costs or some optimal combination of both.

The consumption of high value added polymer blends is about 1.75 million tons per year. In the automotive industry alone, plastics consumes about 2.5 million tons and blends of polyphenylene-ether or polycarbonate are being used for body panels, bumpers, instrument panes, etc. In general, the move is towards more complex systems. This, in itself, necessitates a better understanding of the fundamentals of polymer blending and polymer miscibility.

Interestingly, Thomas Hancock in England, who had long been attempting to improve the properties of natural rubber and was awarded an English patent for vulcanization over Charles Goodyear, was probably the first to gain extra performance by blending. He mixed natural rubber with gutta percha and obtained a mixture easily applied for waterproofing cloth [S4].

In general the following can be identified as the driving forces for polymer blending:

1. Extending engineering resin performance by diluting it with a low cost polymer.
2. Controlling or tailoring properties without the invention or creation of completely new polymers.
3. Preparing a high performance blend from synergistically interacting polymers.
4. Tailoring the properties of the blend to the customers specifications by adjusting the composition.
5. Recycling of industrial and/or municipal scrap.
6. Decreasing the time and costs to develop polymer(s) with desirable properties.
7. Understanding the underlying fundamentals that control miscibility in order to provide both property and quality control.

Polymer blending can be accomplished by a number of techniques.

1. Mechanical mixing followed by melt blending.
2. Dissolution in a co-solvent followed by film casting or spray drying.
3. Latex blending (e.g., SAN + AB → ABS).
4. Use of monomer(s) as solvents for another blend component then polymerization as in interpenetrating networks (co-continuous networks) or high impact poly styrene (HIPS), manufacture.
5. Reactive blending.

21.4.2 Polymer/Polymer Miscibility

Polymer-polymer miscibility often means different things to different people. As was mentioned previously, the term compatibility is often used synonymously with miscibility. In general, most polymer pairs are immiscible, although more are miscible than was recognized a short time ago. Nevertheless, there are a limited number of systems which are miscible in a thermodynamic sense, i.e., when $\Delta G_m \cong \Delta H_m \leq 0$. Thermodynamic miscibility of polymer blends simply means that there is only one phase present. Compatibility is often used to characterize a beneficial result when two or more materials are mixed. The term compatibility is used to denote a mixture or polymer blend which is homogeneous to the eye (a necessary but not sufficient condition), remains homogeneous over the time scale and conditions of use, and has enhanced or desirable properties.

The distinction is made here between miscibility and compatibility. Polymers that are miscible, in the thermodynamic sense, are compatible but polymers that are compatible are not miscible. Miscibility in polymer blends is neither a requirement nor is it necessarily desirable. However, the interaction of blend components is desirable and the interfacial surface properties and their miscibilities are thermodynamically interrelated. Two or more polymers, constituting a blend, are compatible if they exhibit two phases on a microscopic level but interact with each other in a manner that provides useful properties and in many cases enhances one or more properties. This implies that there are degrees of compatibility. In many instances, it is desirable to have two phases present since this morphology can improve properties as long as polymer interactions and phase sizes can be controlled.

Miscibility or regions of miscibility generally occur due to the following: 1) appropriate and intimate mixing that maximizes surface interactions, 2) chemical reactions which result in chemical bond formation, 3) favorable group intermolecular interactions, and 4) influence of molecular weight of the species involved.

Ultimately, thermodynamics is the key to understanding the behavior and properties of polymer alloys or blends. Although miscibility defines the performance of the blend product, due to the low value of the self diffusion coefficient for macromolecules, equilibrium thermodynamic conditions are often difficult to achieve. Miscibility theory and the factors which influence miscibility have been comprehensively reviewed [S5-S9].

Although polymer blending can be accomplished by any of the techniques enumerated above, more often than not blends are

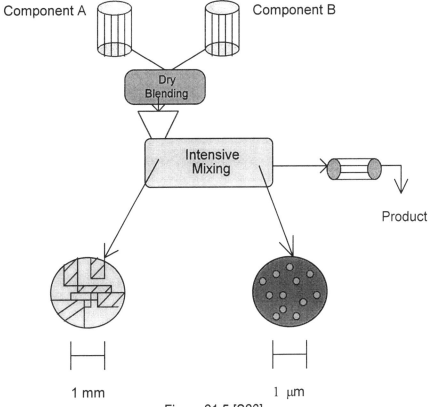

Figure 21.5 [S26]
Effects of Processing on Polymer Blending

prepared by melt blending/mixing in an extruder. Whether or not this is passive blending or active blending, the end result is that the component particle size must be reduced so that there is intimate mixing that allow particle surfaces and the molecular groups to freely interact. Thus, a blending and intensive mixing operation resulting in hoped for reduction of domain size and optimized interaction at the molecular level is illustrated in Figure 21.5.

By passive blending, as used here, is meant a mixing, albeit intensive, to maximize particle and group interactions of two components each of which contain groups or chemical moieties that would be expected to interact favorably. On the other hand, active blending is the blending of components which will not form a compatible or miscible blend unless groups are introduced onto one of the components that will be miscible with the other component(s). Active blending is also used to chemically react two incompatible components so that they cannot later phase separate. An example of active or reactive blending is when polyolefins are *maleated* (reacted with maleic anhydride) in order to introduce polar groups into the backbone of the polyolefin chain. These groups can and often further react with adjacent polymer chains which results in grafts whereby these chains are chemically linked.

Criteria used to establish polymer miscibility are: film clarity, a display of properties which are at least a weighted average of the components, and evidence of a single T_g [S6], and a homogenous morphology on the molecular scale.

21.4.3 General Principles of Thermodynamics and Phase Equilibria

There is no single theory that is able to predict details of the phase equilibria of commercially attractive systems. The most difficult parameter, for a theory to take into account, is molecular polydispersity which for many systems is difficult to determine experimentally. Also, commercial systems can seldom be represented by a single chemical species.

Polymer-polymer miscibility is determined by a delicate balance of enthalpic and entropic forces, significantly smaller than those observed in small molecule liquid mixtures (solutions) or in the metal alloys discussed previously. In particular the combinatorial entropy term, which we have seen previously, is responsible for miscibility of alloys and solvents in polymer solutions, is negligibly small in high molecular weight polymer blends. If this indeed be the case, then we must conclude that polymer miscibility is primarily due to the negative heat of mixing, i.e., to specific interactions between two or more types of macromolecules.

We refer to the thermodynamic concepts discussed in the above sections, and once again begin with the expression for the Gibbs free energy. From Equation 21.51, $G = H - TS$.

The change in free energy is:

$$\Delta G_{mix} = \Delta H_{mix} - T\Delta S_{mix} \tag{21.76}$$

or

$$dG = VdP - SdT + \sum_i \mu_i dn_i = \left(\frac{\partial G}{\partial n_i}\right)_{P,T,n_i} \tag{21.77}$$

The free energy, enthalpy, entropy, and chemical potential of mixing are respectively defined as a difference between the mixture and pure state.

$$\Delta F_m = F - F_o; \quad F = G, H, S, \mu_i \tag{21.78}$$

The conditions of miscibility in a binary system are described in terms of binodal:

$$\Delta\mu_i^1 = \Delta\mu_j^2 \tag{21.79}$$

where the superscripts are the phases and μ_i and μ_j are the chemical potentials of the components and in terms of spinodal:

$$D = \left(\frac{\partial^2 \Delta G_m}{\partial x_2^2}\right)_{P,T} = 0 \tag{21.80}$$

and the critical point (Critical Solution Temperature, CST) is:

$$D' = \text{CST} = \left(\frac{\partial^3 \Delta \partial G_m}{\partial x_2^3}\right)_{P,T} = 0 \tag{21.81}$$

where x_2 is the mole fraction of component 2. These quantities are illustrated in Figure 21.6.

The upper part of Figure 21.6 shows the isothermal and isobaric variation of ΔG_m with x_2, whereas the lower part represents the isobaric phase diagram for binary mixtures with the upper critical solution temperature, UCST. The points of contact

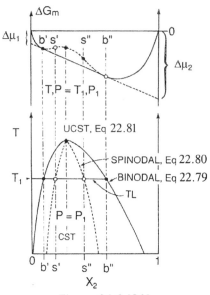

Figure 21.6 [S9]
Dependence of Chemical Potentials and Critical Solution Temperature
(CST) On Mole Fraction $\left(X_2 \right)$

Upper part: Gibbs free energy of mixing as a function of concentration in a binary liquid system at constant T and P.

Lower part: phase diagram at $P = P_1$. Points b, s, and CST represent binodal, spinodal, and critical solution temperature.

define the limiting conditions for the miscible system, $x_2 < b'$ or $x_2 > b''$. The inflection points, s' and s'', define the spinodal conditions, i.e., for $s' < x_2 > s''$ the system is phase separated. With the concentration range between the b and s curves the system is metastable.

In polymer blends other types of phase diagrams illustrated in Figure 21.7 are also observed as a function of temperature. The LCST is more frequently encountered than the UCST. The free energy of mixing will be very small owing to the small number of moles of each polymer in the blend as a result of their large molecular weights. Although the sign of the combinatorial entropy favors mixing, it is usually too small to result in the necessary negative free energy because the heat of mixing is generally slight-

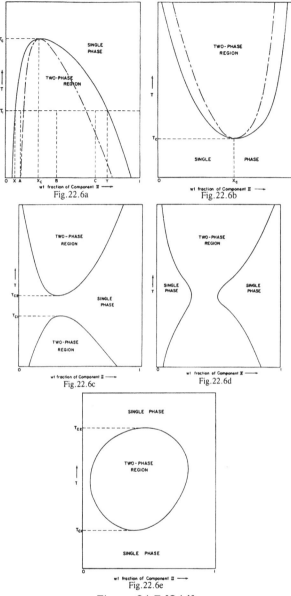

Fig.22.6a

Fig.22.6b

Fig.22.6c

Fig.22.6d

Fig.22.6e

Figure 21.7 [S14]

Phase Diagrams for Various Mixtures as a Function of Temperature

(a) A mixture with a UCST: _ binodal; _ _ _spinodal.
(b) A mixture with an LCST: _ binodal; _ _ _ spinodal.
(c) A mixture with an LCST above a UCST.
(d) A mixture with a tendency toward greater solubility at intermediate temperatures.
(e) A mixture with a UCST above an LCST (closed phase diagram).

627

ly positive at least for nonpolar systems which are monodisperse.

The Flory-Huggins solution theory [S7] provides an approximation for the free energy of mixing of polymers A and B. The free energy of mixing is given by:

$$\Delta G_{mix} = RTV\left[\frac{\phi \ln \phi_A}{\widetilde{V}_A} + \frac{(1-\phi_A)\ln(1-\phi_A)}{\widetilde{V}_B} + \widetilde{\chi}_{AB}\phi_A(1-\phi_A)\right] \quad (21.82)$$

where: V = total volume of the two polymers
ϕ_A = volume fraction of A
$\widetilde{V}_A$ = molar volume of polymer A
$\widetilde{V}_B$ = molar volume of polymer B
$\widetilde{\chi}_{AB}$ = is an interaction parameter of the two polymers related to the heat of mixing

Scott [S10] and Tompa [S11] were the first to apply the Flory-Huggins theory of polymer solutions to mixtures of polymers, with or without added solvent.

If Equation 21.82 is divided by a reference volume, V_r, which is taken as a volume which is close to the molar volume of the smallest polymer repeat unit, the Gibbs free energy of mixing becomes:

$$\Delta G_{mix} = \frac{RTV}{V_r}\left[\left(\frac{\phi_A}{x_B}\right)\ln \phi_A + \left(\frac{\phi_B}{x_B}\right)\ln \phi_B + \chi_{AB}\phi_A\phi_B\right] \quad (21.83)$$

This is the result obtained by Scott, where:

x_A and x_B are the degrees of polymerization of polymers A and B, respectively.

Scott determined the critical conditions by letting:

$$\left|\frac{\partial^2 \Delta G_{mix}}{\partial \phi_A^2}\right|_{T,P} = \left|\frac{\partial^3 \Delta G_{mix}}{\partial \phi^3}\right|_{T,P} = 0 \quad (21.84)$$

and obtained:

$$\left(\chi_{AB}\right)_{cr} = \frac{1}{2}\left[\frac{1}{x_A^{1/2}} + \frac{1}{x_B^{1/2}}\right]^2 \tag{21.85}$$

$$\left(\phi_A\right)_{cr} = \frac{x_B^{1/2}}{x_A^{1/2} + x_B^{1/2}} \tag{21.86}$$

$$\left(\phi_B\right)_{cr} = \frac{x_A^{1/2}}{x_A^{1/2} + x_B^{1/2}} \tag{21.87}$$

The polymer interaction parameter between molecules of comparable size can often be written in terms of Hildebrand solubility parameters [S12, S13].

$$\chi_{AB} = \frac{V_r}{RT}\left(\delta_A - \delta_B\right)^2 \tag{21.88}$$

and converting V_r to the molecular weight:

$$\chi_{AB} = \frac{M_r}{\rho RT}\left(\delta_A - \delta_B\right)^2 \tag{21.89}$$

From Equation 21.85 above, the limiting case for a reference DP of 1 is:

$$\left(\chi_{AB}\right)_{cr} \approx 2 \tag{21.90}$$

when χ_{AB} becomes $\left(\chi_{AB}\right)_{cr}$, M_r becomes M_{cr} for a nonpolar case and:

$$M_{cr} \approx \frac{2\rho RT}{\left(\delta_A - \delta_B\right)^2} \tag{21.91}$$

$$\left(\delta_A - \delta_B\right)^2 \approx \frac{2\rho RT}{M_{cr}} \tag{21.92}$$

Substituting in Equation 21.88:

$$\tilde{\chi}_{AB} \approx \frac{2V_r \rho}{M_{cr}}$$ (21.93)

For two polymers having the same molecular weight M and density ρ, $\tilde{\chi}_{AB}$ can be replaced with the equivalent parameter in Equation 21.82 and rearranging, ΔG_{mix} becomes:

$$\Delta G_{mix} \approx \frac{\rho V_r RT}{M_{cr}} \left\{ \frac{M_{cr}}{M} \left[\phi_A \ln \phi_A + \left(1 - \phi_A\right) \ln\left(1 - \phi_A\right)\right] + 2\phi_A\left(1 - \phi_A\right) \right\}$$ (21.94)

which is the result obtained by Paul [S5].

The first two terms arise from the combinatorial entropy of mixing and each is inversely related to the size or molecular weight of that component and the third term, χ_{AB}, is a polymer interaction term related to the heat of mixing as stated above.

From Equation 21.91, it can be seen that when the absolute difference of the solubility parameters is 1.0 $\left(\text{cal}/\text{cm}^3\right)^{1/2}$, $M_{cr} <$ 1200. When this difference is 0.1 $\left(\text{cal}/\text{cm}^3\right)^{1/2}$, M_{cr} rises to about 120,000.

Table 21.1
Values for $\left(\chi_{AB}\right)_{cr}$ when $x_A = x_B$

$x_A = x_B$	$\left(\chi_{AB}\right)_{cr}$	$\left|\delta_A - \delta_B\right|_{cr}$
50	0.04	0.49
100	0.02	0.35
200	0.01	0.25
300	0.0067	0.20
500	0.0040	0.15
1000	0.0020	0.11
2000	0.0010	0.077
3000	0.00067	0.063
5000	0.00040	0.049

The molar reference volume V_r is conveniently taken to be 100 cm^3/mol which means for $M_{cr} = 120,000$, a DP of ≈ 1200. It is instructive to examine how $\left(\chi_{AB}\right)_{cr}$, the DP, and $\left|\delta_A - \delta_B\right|_{cr}$ are interrelated. Krause [S14] has summarized these results which are shown in Table 21.1.

We now use Equation 21.94 to explore the interrelationship of combinatorial entropy and the heat of mixing terms.

The variation of ΔG as a function of volume fraction is illustrated in Figure 21.8 for various ratios of M_{cr}/M.

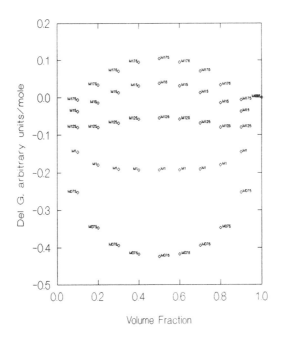

Figure 21.8[S26]
Free Energy of Mixing of Polymers *A* and *B* for Molecular Weights as Multiples of the Critical Molecular Weight
M075 = $M_{cr} \times 0.75 / M_w$: M1 = $M_{cr} \times 1/M_w$: M125 = $M_{cr} \times 1.25 / M_w$:
M1.5 = $M_{cr} \times 1.5 / M_w$: M175 = $M_{cr} \times 1.75 / M_w$

The cases for various polymer molecular weights relative to M_{cr}, show that the free energy of mixing tends to more positive values as the molecular weight is increased, as expected, from the previous discussion. Although it is never positive when $M_{cr} = 1.25$, there is a node of slightly lower energy that occurs at 0.2 and 0.8 volume fraction which means that intermediate compositions would tend to phase separate in order to achieve the lower free energies.

From Equation 21.80, thermodynamic stability of a single phase for a blend of two components, A and B, exists only when:

$$\left(\partial^2 \Delta G_{mix} / \partial \phi_B^{\ 2} \right)_{T,P} > 0 \qquad (21.95)$$

It will be noted that blend composition has a major effect on miscibility and when $M_{cr} \leq M/1.25$, stable one phase mixtures are indicated.

In an attempt to further clarify these issues, a 3-D surface plot is shown in Figure 21.9 where the volume fraction ϕ, the combinatorial entropy, ΔS, and the enthalpy, ΔH, have been plotted for the various ratios of M_{cr}/M discussed above. This provides an appreciation of the relative contributions of the entropy and enthalpy terms to the heat of mixing. Figure 21.9(a) shows that the enthalpy term is always positive while the entropy term is always negative with the surface representing the interrelationship of the three variables. It is instructive to note that the surface is skewed as a result of the effects of the M_{cr}/M ratios shown in Figure 21.9(b). The total effect of the enthalpy and entropy terms then depend on the concentrations of A and B, i.e., the volume fractions (ϕ) of each component. Figure 21.9b also shows the average change in arbitrary units for ΔG_{mix} for comparison.

21.5   MISCIBILITY OF METALS VS POLYMERS

The previous discussions brings us to the point where we can attempt to draw some important distinctions for each class of materials. For both classes, materials will always attempt to lower

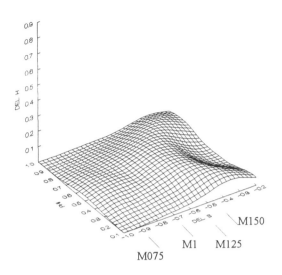

21.9a. Surface Plot of Enthalpy, DEL H (ΔH) ,Volume Fraction, PHI (ϕ) , and Entropy, DEL S, (ΔS)

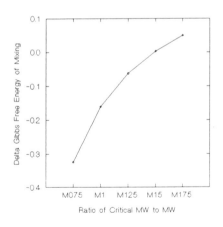

21.9 b. Average (ΔG_{mix}) vs. Various Ratios of M_{cr} to M_w

Figure 21.9 [S26]

Relations Between Combinatorial Entropy, Enthalpy, and Volume Fraction for Various Ratios of Critical Molecular Weight M_{cr} to Molecular Weight M_w For the Free Energy of Mixing

M075 = $M_{cr} \times 0.75 / M_w$: M1 = $M_{cr} \times 1/M_w$: M125 = $M_{cr} \times 1.25 / M_w$:

M1.5 = $M_{cr} \times 1.5 / M_w$: M175 = $M_{cr} \times 1.75 / M_w$

their free energy state and achieve the lowest possible thermodynamic form. In the case of metals, for alloys that conform to the Hume-Rothery rules, the driving force is the entropy of mixing since the total number of molecules in a given volume is large. The enthalpy of mixing term turns out to be a minor contributor and in fact for allotropic, reversible phase transformations we have seen that (Equation 21.52):

$$dH = q = \int C_p dT$$

For high polymers, the situation is somewhat reversed. For this class of materials, as we have seen, the entropy of mixing is also a driving force but tends to positive values as M becomes greater than M_{cr}. For non polar polymers, the enthalpy of mixing is always positive which must be balanced by the entropy term for a favorable negative ΔG_{mix} to be obtained. The combinatorial entropy values are small and for two polymers of equal and high DP's ($x_A = x_B$), although the entropy is negative, it may not be sufficiently negative to counter the positive enthalpy. Because of the total number of molecules in a given volume is much smaller compared with metal, the enthalpy term $\tilde{\chi}_{AB}\phi_A\left(1-\phi_A\right)$ or χ_{AB} becomes the significant factor in determining miscibility, particularly when $x_A = x_B$ and $M_{cr} > M$.

The interrelationships of ΔH, ΔS, and ϕ show the importance of the composition of the mixture on miscibility. For polar polymers a negative value of χ ensures mixing since a negative enthalpy (exothermic) mixing results.

21.6   ESTIMATION OF POLYMER MISCIBILITY

The origin of forces between molecules, precluding covalent bond formation, arise from van der Waals forces. These forces consist of Keesom orientation forces, Debye inductive forces, London dispersion forces, and hydrogen bonding which is itself the most important contributor to acid-base or electrostatic forces.

The fact that two materials have the same solubility parameter does not ensure that there is an equal balance of London, Keesom, and hydrogen bonding forces. This accounts for the case where a polymer will not dissolve in a solvent although the two have similar solubility parameters.

For these reasons, a number of investigators have decomposed the classical Hildebrand solubility parameter into several terms, as Hildebrand [S12] himself did using dispersive and polar. In 1966, Crowley et al.[S15, S16] showed that solubility parameter exceptions could be accounted for by using a group of solvents with the same solubility parameter and dipole moment but with different hydrogen bonding affinities.

Hansen [S17-S21] and Hansen and Skaarup [S21] assumed that the cohesive energy arises from dispersive, permanent dipole-dipole interactions, and hydrogen bonding forces. This laid the foundation for resolving the solubility parameter into its component parts.

These concepts and those presented in Chapter 9 serve the basis for understanding the principles that are used to estimate polymer-polymer miscibility using the miscibility parameter (MP) model proposed by David and Sincock [S23]. It is this model upon which MPCalc is based.

The MP model is based on using the individually calculated solubility parameters. Considering a blend of two polymers, each polymer may be a homopolymer, a copolymer, or ternary system. Theoretically, more complex systems can be estimated but we concern ourselves with more realistic, industrially important systems.

Recall that each individual solubility parameter (cf. Chapter 9) is a vector quantity and therefore has both direction and magnitude. The vectors for each polymer are represented in Figure 21.10. These vectors always give the direction and distance from a common point, i.e., the origin.

According to Hildebrand [S12], the enthalpy of mixing per unit volume is;

$$\Delta h_m = \phi_1 \phi_2 \left(\delta_1 - \delta_2 \right)^2 \qquad (21.96)$$

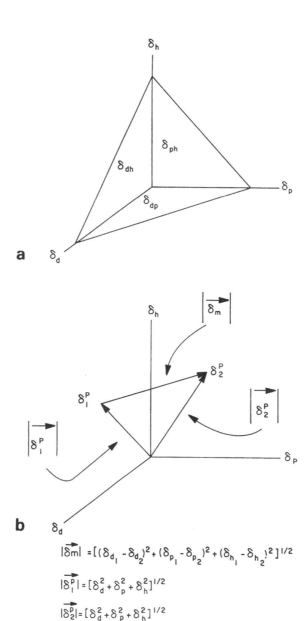

$$|\overrightarrow{\delta m}| = [(\delta_{d_1} - \delta_{d_2})^2 + (\delta_{p_1} - \delta_{p_2})^2 + (\delta_{h_1} - \delta_{h_2})^2]^{1/2}$$

$$|\overrightarrow{\delta_1^P}| = [\delta_d^2 + \delta_p^2 + \delta_h^2]^{1/2}$$

$$|\overrightarrow{\delta_2^P}| = [\delta_d^2 + \delta_p^2 + \delta_h^2]^{1/2}$$

Figure 21.10 [S23]
Three Dimensional Solubility Parameter Coordinate Axis System
(a) Solubility parameter coordinate axis system
(b) Illustration of solubility parameter vector quantities

636

Comparisons of solubility parameters necessitates examination of the contributions to the total solubility parameter.

$$\delta^2 = \delta_d^2 + \delta_p^2 + \delta_h^2 \qquad (21.97)$$

where d, p, and h refer to the dispersive, polar, and hydrogen bonding components, as before. The overriding reason for providing the capability for calculating the individual solubility parameter contributions is that all three quantities may not be used in subsequent calculations to estimate miscibilities. In some cases, only two of the three are used.

As an example we use a binary blend of two copolymers: For monomers 1 and 2 of polymer A:

$$\left(\delta_1^{poly}\right)^2 = \delta_d^2 + \delta_p^2 + \delta_h^2 \qquad (21.98)$$

$$\left(\delta_2^{poly}\right)^2 = \delta_d^2 + \delta_p^2 + \delta_h^2 \qquad (21.99)$$

For the miscibility estimation, the weight fraction ϕ' is used since this does not introduce error which in any way changes the interpretation of results. Additionally, for industrial applications, densities are often not known accurately. For polymer A, we have:

$$\delta_A^{poly} = \delta_1^{poly}\phi_1' + \delta_2^{poly}\phi_2' \qquad (21.100)$$

Similarly for a second polymer B:

$$\delta_B^{poly} = \delta_3^{poly}\phi_3' + \delta_4^{poly}\phi_4' \qquad (21.101)$$

These values are then used to calculate a miscibility parameter which is defined as:

$$MP = \left(\delta_A^{poly} - \delta_B^{poly}\right)^2 \qquad (21.102)$$

The miscibility parameter is related to the interaction parameter as shown in Equation 21.88, i.e.,

$$\chi_{AB} = \frac{V_r}{RT} \left(\delta_A - \delta_B \right)^2.$$

For any polymer blend system, there are four possible MP's: MP_{dph}, MP_{dp}, MP_{ph}, and MP_{dh}. When blend components have significant dispersive, polar, and hydrogen bonding SP's, the total MP_{dph} most often can be used to predict actual behavior. in other cases appropriate combinations can be used to predict actual behavior, examples of which are illustrated in the following tables.

21.7   General Guidelines For Interpretation of MP Results

1. When blend components have significant dispersive, polar, and hydrogen bonding solubility parameters, the total (MP_{dph}) most .often predicts actual behavior. In these cases the minimum in the MP_{dph} is usually associated with one or two other MP's in a plot of MP versus composition.

2. The miscibility/compatibility of polymer blends with groups that show strong polar and particularly hydrogen bonding tendencies can often be determined based on the "donor-acceptor" criterion. In other words, one can eliminate all hydrogen bonding MP combinations if one or other of the components in the blend are incapable of hydrogen bonding.

3. In the event of strong intramolecular hydrogen bond formation, the remaining interactions, i.e., dispersive and polar have been used successfully to predict miscibility and miscibility/compatibility ranges.

4. The window of miscibility/compatibility has often been found to be delineated by the MP value of approximately 0.1 as supported by experimental data. It is expected that the type and magnitude of blend component interactions will raise or lower this value.

A key point of the MP model is that an absolute level of MP is not set as the criteria for miscibility/compatibility. In place of an absolute level of MP, used to judge potential miscibility, MP values are examined for changes and blend compositions where the MP's are monotonically decreasing and display a minimum. The minima of these curves represent blend compositions where miscibility/compatibility is indicated consistent with the above guidelines.

The use of a common database (MATPROP$^©$) for calculating the solubility parameters that determine the MP's minimizes comparison errors between different materials. Relying on trends which indicate a minimum further minimize error since they decrease the necessity of highly accurate solubility parameters that are normally necessary because of the required small solubility differences illustrated previously. Thus, miscible/compatible blend compositions that are not precise may be indicated. Nevertheless, the estimated composition ranges serve as useful guides and starting points for investigating potentially compatible or miscible blend compositions.

The decomposition of solubility parameters into combinations other than the total $\left(\delta_{dph}\right)$ simply allows examination for molecular interactions which may be favorable to or promote mixing. The MP model, therefore allows separate examination of polar and hydrogen bonding interaction which is not possible when a *lumped* or total solubility parameter is utilized. These molecular interactions may range from weak to strong and thus may be indicative only of compatibility.

The following section addresses the obvious questions how well does this approach work, i.e., is it useful and, if so, can we account for this on the basis of theory and the simple MP model?

21.8 APPLICATION RESULTS OF THE MISCIBILITY PARAMETER (MP) MODEL

David and Sincock [S23] reported a number of examples to illustrate this approach using blends that were expected to provide a variety of molecular interactions, from weak to strong. The PVC-SAN blend which had been reported by Coleman et al. [S24] to have relatively weak molecular interaction which we will use as an example to estimate miscibility and compare the window of miscibility from the MP model to experimental results and the results of Coleman et al.

The results of using MPCalc to estimate the window(s) of miscibility of the PVC-SAN blend are shown graphically in Figure 21.11 where the various MP's are plotted as a function of the % SAN.

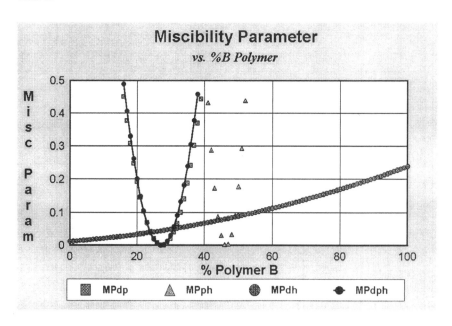

Figure 21.11
Typical Plot of MP vs. Weight % of a Component
of a Mixture Showing a Window of Miscibility for
MP_{dp} and MP_{dph}
Blends of PVC and SAN
MP vs. %AN in SAN

Table 21.2
Data for Polymer Repeat Units Used in Estimating Polymer Miscibility

Polymer Repeat Unit	Abbrev.	Solubility Parameters		
		δ_d	δ_p	δ_h
αMethyl Styrene	αMS	8.56	0.459	0
α-methylstyrene-ran-4-(2-hydroxy ethyl α-methylstyrene	αMSI	8.311	1.637	5.590
α-methylstyrene-ran-4-(hydroxy α-methylstyrene	αMSIII	9.030	2.323	6.60
Acrylonitrile	AN	8.51	12.00	3.65
Aliphatic Polyesters	AP	-	-	-
Aromatic Polyamides	APAS	-	-	-
Butadiene	BD	8.01	0	0
Butylene Terephthalate	BT	8.747	2.756	4.373
Cyclohexane Dimethylene Succinate	CDMS	8.020	2.389	4.084
Cyclohexyl Methacrylate	CHMi	8.493	1.582	3.324
Chlorinate Poly(ethylene)	CPE	-	-	-
Ethylene	E	8.021	0	0
Maleic Anhydride	MA	9.038	13.962	5.479
Methacrlonitrile	MAN	8.038	8.42	3.058
Methyl Methacrylate	MMA	8.075	2.767	4.396
Methylene Phenylene Oxide	MPO	9.395	1.252	2.068
Polyamide(Nylon6)	N6	8.447	4.599	3.42
Propylene	P	7.665	0	0
2,2Propane bis(4-phenyl carbonate)	PC	8.769	1.545	3.358
Phenoxy	PH	-	-	-
Phenylene Oxide	PO	9.099	1.708	2.457
Styrene	S	8.88	0.55	0
Vinyl Alcohol	VA	7.62	6.80	11.54
Vinyl Acetate	VAC	7.853	3.317	4.813
Vinyl Butyral	VB	7.72	2.90	3.26
Vinyl Chloride	VC	8.65	5.95	1.45
Vinyl Methyl Ether	VME	7.382	3.394	3.527

An examination of the solubility parameter contributions (Table 21.2) for styrene (S), acrylonitrile (AN) and vinyl chloride (VC), shows that styrene has no hydrogen bonding component while AN and VC have large polar contributions, relative to their hydrogen bonding components.

It would therefore be likely that we can eliminate the MP_{ph}

and MP_{dh} terms because of the styrene. The dispersive and polar forces would be expected to be the driving forces in blend miscibility. However, there may be hydrogen bonding forces between the AN part of SAN and the hydrogen bonding component of VC. The minima, observed at about 26% AN, or approximately from 21%-33% for MP_{dp} and MP_{dph}, are consistent with these expectations and above guidelines. Coleman et al. [S24] showed a minimum which occurred at approximately 10%-11% AN in SAN while Kim et al. [S25] reported miscibility of PVC with SAN copolymers that contained 11.5%-26% AN based on experimental results.

Comparisons of estimated miscibility/compatibility using MPCalc with literature values for a number of blends representative of the different types that might be encountered in practice have recently been reported by David [S26]. These data are listed in

Table 21.3

Comparison of Miscibility Parameter Model Estimations of Polymer-Polymer Miscibility with Experimental Results

Polymer Blend	MP	Miscibility Window Estimated From MP	Miscibility Window Determined from Other Models(M)/Exper.Results(E)	Ref [S]
PPO-PMPO	dph	Msc.(miscible)	Msc(E)	14 p89
PS-PPO	dph	I(immiscible)	Msc.(E)	14 p76
PVC-PVAC/PVC	dph	0-39%VAC	conditionally compatible @R.T.(E)	14 p67
PBDAN-PS	dp,dph	11-15%AN	30-34%AN(M)	24
PVC-BDAN	dp,dph	32-41%AN	23-45%AN(E) 35-42%AN(M)	24 27
PVC-PαMS	dph	28-32%AN	32%AN(E)	25
PVC-SAN	dph	21-33%AN	11-26%AN(E)	25 28
PH-AP	ph	$CH_X/COO = 3.5$	$CH_X/COO = 3.5$(E)	29
APAS-AP	dph	vol.frac $CH_2 = 0.80$	vol frac- $CH_2 = 0.81$ (E)	30
PEO-PEVAC	dph	59-100%VAC	56-100%VAC(E)	31
PSAN-PC	ph	23-28%AN	18-40%AN(E) @25%AN, max. toughness(E) I	32 33 34

continued

Table 21.3 (cont.)

Polymer Blend	MP	Miscibility Window Estimated From MP	Miscibility Window Determined from Other Models(M)/Exper.Results(E)	Ref [S]
PαMSAN-PC	ph	27-32%AN	30-50%AN(E)	34
PSMAN-PC	ph	34-42%MAN	32-63%MAN(M)	35
PVB-PN6/PP	dph	0-20%PP@18%OH	80%N6/20%PP/5%PVB@18%OH(E)	36
PVC-PMMA	dph	I	i-I; s-Msc. upto 60% s-PMMA(E)	37
PVC-PS	dph,dp	I	I(E)	38, 39
PVC-PBT	dph	Msc.	Msc.(E)	39
PVC-PCPE	dph	>48% Cl in PE	>42% Cl in PE(E)	39
PVC-PE	dph	I	I(E)	39
PVC-PE	dph	I	I(E)	39
PVC-PBD	dph	I	I(E)	39, 40
PVME-PS/PMMA	dph	60% styrene	60% styrene(E)	41
PαMS-PαMSI/PVME	dph	3-36% PαMSI	>4% PαMSI(E)	42
PαMS-PαMSIII/PVME	dph	1-20%PαMSIII	>4%PαMSIII(E)	42
PαMS-PVME	dph,dp	borderline I	I(E)	42
PS-PVME	dph	Msc.	Misc(E)	43
PS/PVME-PVME	dp	0-40%PS	χ_{23}' negative@35-65%PS(E)	43
PSAN-PMMA/PCHMA	dph	@0%CHMA 0-6%AN, I 7-16%AN, Msc. @20%CHMA 5-15%%AN,Msc. @60%CHMA 3-13%AN, Msc. @80%CHMA 5-15%AN, Msc @100%CHMA 1-11%AN, Msc.	0-6%AN,I 10-30% AN, Msc. 7-25% AN, Msc(E) 4-20%AN, Msc.(E) 0-16%AN, Msc.(E) 0-15%AN, Msc.	44
PVC-PEVAC-45	dph	I	Semi compatible(E)	45
PVC-PEVAC-65	dph	I	Semi compatible(E)	45
PAN-PC	dph,ph	I	I(E)	46
PCHMA-PC	dph,ph	Msc.	Msc(E)	46
PS-PC	dph,ph	I	I(E)	46
PSAN-PCDMS/PC	ph	@25%AN,1-17%PCDMS	<5%PCDMS(E)	47
PVC-ABS	dph	I	I, two T_g's but property improvement	48

643

Table 21.3. The results show that good comparisons between estimated blend miscibilities using MPCalc with the experimental results from this study also compare very well using the MP_{dh}. This indicates that the hydrogen bonding component of the polyester largely determines miscibility.

A separate study carried out by Ziska [S49] et al. examined the miscibility of PVC with a number of different polyesters. A comparison of miscibility results from MPCalc with the experimental results from this study also compare very well using the MP_{dh}. This indicates that the hydrogen bonding component of the polyester largely determines miscibility.

It is likely, that at the outset of a study of the miscibilities of PVC and polyester, without any experimental data, the use of MP_{dh} would not be obvious. However, by comparing the various MP combinations with a few initial experimental results and using

Table 21.4
Miscibility Data for Poly(vinyl chloride)-Polyester Blends

Repeat Unit	Solubility Parameters			MP's				MP Est	Exp Res
	δ_d	δ_p	δ_h	*dph*	*dp*	*ph*	*dh*		
ε-Caprolactone	7.959	2.241	3.956	2.052	4.974	2.488	0.014	M	M
β-Caprolactone	7.905	4.165	5.393	0.026	2.445	0.476	0.638	I	I
α-methyl-α-ethyl-β-Propiolactone	7.704	2.211	3.938	2.775	6.156	2.570	0.014	M	M
β-Propiolactone	7.746	1.927	3.668	3.290	6.334	3.923	0.040	M	M
Butylene Adipate	7.948	2.649	4.301	1.395	4.499	1.151	0.071	M	M
Butylene Terephthalate	8.747	2.756	4.373	0.192	1.763	0.912	1.017	I	M
Cyclohexane Dimethylene Succinate	8.020	2.389	4.084	1.656	4.539	1.940	0.053	M	M
Dipropylene Succinate	7.774	2.897	4.497	1.350	0.851	0.60	0.044	M	M
Ethylene Adipate	7.931	3.239	4.755	0.641	3.73	0.14	0.23	I	I
Ethylene Orthophthalate	8.378	3.190	4.705	0.215	2.353	0.193	0.702	I	I
Ethylene Succinate	7.905	4.165	5.393	0.026	2.45	0.48	0.64	I	I
Hexamethylene Sebecate	7.974	1.714	3.459	3.025	5.488	5.125	0.006	M	M
Pivalolactone	7.647	2.62	4.277	2.112	5.834	1.223	0	M	M

the comparison results as guidelines, much time could be saved in extending favorable regions of miscibility/compatibility estimated from MPCalc to favorable miscibilities that might be experimentally expected. An example of the results of such a study is shown in Table 21.4.

MPCalc can also be used to obtain results in constructing phase or miscibility diagrams. An example of one such system is the poly(styrene-co-acrylonitrile) and poly(styrene-co-maleic anhydride) blend. In blend *A* the % MA in styrene was varied and for each composition of blend *A*, the window of miscibility was determined for blend *B* as a function of % AN in SAN. Table 21.5 shows the results obtained.

These data plotted in Figure 21.12 are almost identical with the experimental results reported by Hall et al. [S50]. The cross hatched area shows the estimated window of miscibility for this system.

Polymer-plasticizer miscibility can also be estimated using MPCalc [S51]. The same technique as before is used except in this case because of the large molecules-small molecules interactions, the contribution of the combinatorial term will be increased thus, the interaction parameter is given by:

Table 21.5
Miscibility of the Blend Poly(styrene-co-acrylonitrile) and
Poly(styrene-co-maleic anhydride)

Composition of Maleic Anhydride-Polystyrene		Miscibility Window as a Function of % Acrylonitrile in Styrene-Acrylonitrile
%MA	%S	%AN
5	95	2-12
10	90	8-19
20	80	21-33
30	70	36-47
40	60	50-60
50	50	63-74
60	40	Immiscible

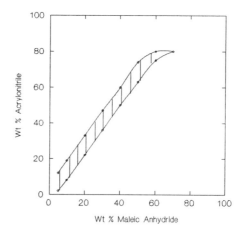

Figure 21.12 [S26]
Phase or Miscibility Diagram for the Poly(styrene-co-Acrylonitrile) and
Poly(Styrene-co-Maleic Anhydride) Blend

$$\chi = \beta + \frac{V_s}{RT}\left(\delta_1 - \delta_2\right)^2 \qquad (21.103)$$

In this case β is a lattice constant, usually 0.35 ± 0.1 (normally assumed to be zero when dealing exclusively with high polymers), V_s is the molar volume of the solvent (plasticizer) [S52].

For this case, a copolymer, poly(vinyl butyral), which precipitated from a plasticizer solvent, the temperature of precipitation, θ temperature, the solvent volume V_s, and the χ_{cr} define the theoretical MP_{cr} as:

$$\frac{\left(\chi_{cr} - 0.35\right)RT}{V_s} = MP_{cr} \qquad (21.104)$$

For seventeen different systems investigated MP_{cr} was found to be ≈ 0.30.

For a series of plasticizers that provided a range of compatabilities with PVB, the MP_{dph}'s were calculated. Small amounts of PVB were dissolved in plasticizers. These solutions were then cooled until the polymer precipitated and the temperature at which this occurred was found to correlate with the MP_{dph} (Figure 21.13). This presumably occurs because of the strong molecular interactions of the polymer with itself via the hydroxyl groups. MP's for systems with strong interactions between the polymer and the plasticizer do not precipitate. One such system is PVC. In this case the calculated MP was correlated with temperature of solution (Figure 21.14).

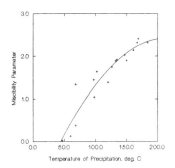

Figure 21.13 [S51]
Miscibility Parameter MP_{dph} of Poly(Vinyl Butyral)/Plasticizer vs.
Temperature of Precipitation (T_p)

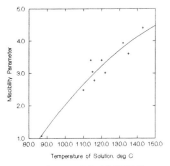

Figure 21.14 [S51]
Miscibility Parameter MP_{dph} of Poly(Vinyl Chloride) vs. Temperature of
Solution $\left(T_s \right)$

Thus, a few initial calculations of MP and determination of their corresponding solution or precipitation temperatures can serve to establish a relationship with MP that can be used to estimate miscibilities/compatibilities with a variety of plasticizers.

These results indicate that the simple MP model can be effectively used to estimate compatibility/miscibility in a wide range of applications. Thus it appears that the answer to the first part of the question raised previously is affirmative in that this technique is a useful one in guiding the selection of initial blend compositions in blending studies and in estimating polymer-polymer interactions for a variety of systems.

21.9   MECHANISMS OF THE MISCIBILITY PARAMETER MODEL

From the previous discussions of mixing theory, we have seen that the entropy term is always negative and the enthalpy term is always positive in the equation for the free energy of mixing (Equations 21.83 and 21.94). Obviously, the absolute contribution of each of these terms determines whether or not ΔG_{mix} will be negative. Also, from Equation 21.94 and Figure 21.8 we have seen that the likelihood that ΔG_{mix} will be negative will be much greater if $M < M_{cr} \times 1.25$ or the equivalent $M_{cr} > M/1.25$ and compositions are used that constitute smaller volume fractions of one component mixed with the other. Of course if the interaction parameter, χ_{AB}, is zero or negative, mixing is ensured since ΔG_{mix} will be negative.

However, Equation 21.102 shows that the MP can never be negative since it is based on the difference of the solubility parameters of the two polymers raised to the 2nd power. In spite of this the results show that the MP model works reasonable well. The explanation appears to involve three considerations.

1. In a majority of industrially prepared polymers, the polydispersity is never 1, but more likely, 2 or greater. Consequently, it is likely that there are present molecular weight species that fulfill the condition of

$M < M_{cr} \times 1.25$. Blends of two or more polymers having some components that meet this criteria would therefore be miscible. The miscibility of these components with each other would then in turn promote miscibility/compatibility with other molecular weight species not meeting these criteria due to sufficiently *matched* solubility parameters.

2. In some cases at least, neglecting for the moment the effects of molecular weight, differences between solubility parameters are sufficiently small that the magnitude of the entropy term ensures a negative ΔG_{mix} at the blend composition examined.

3. By providing a technique to examine various MP combinations, molecular interactions can be identified that lead to correctly estimating blend miscibilities-compatibilities. A simple example of this to illustrate the point occurs in the case of polyvinyl alcohol (PVA) or poly(vinyl butyral) (PVB) that also contains residual PVA groups which have a large tendency to hydrogen bond. Thus, a polymer like PVB containing residual PVA groups will intra and inter molecularly hydrogen bond and the miscibility of this polymer with other blends will depend largely (but not entirely) on the match of the MP_{dp} with other components in the blend.

   Similar arguments can be made for other groups in different applications.

21.10 OTHER APPROACHES FOR DETERMINING POLYMER BLENDS MISCIBILITY

Coleman et al. [S24] used a similar approach that also employed the Flory-Huggins free energy of mixing to which they added an additional term, ΔG_H, to account for the presence of favorable intermolecular interactions.

This approach used solubility parameters that were calculated based upon non-hydrogen bonded solubility parameters derived from hypothetical analogs of unassociated molecules. The matching of these solubility parameters with other components then determines the level of χ_{cr} and potential miscibility.

The authors, ten Brinke, Karasz, and MacKnight [S53] presented a miscibility model based on mean field theory that explains miscibility based on repulsion between the different covalently bonded groups of a copolymer rather than specific group interactions. These investigators have demonstrated that miscibility in a system can occur if the bonded monomer groups of a copolymer A prefer the constituent groups of a second polymer or copolymer B over those of A.

This approach was applied recently to a mixture of two copolymers that differed only in copolymer composition, i.e., $A_x B_{(1-x)} / A_y B_{(1-y)}$ and therefore requires one dimensionless interaction parameter χ_{AB}. Although mean field theory is mainly directed to systems where specific interactions are absent, i.e., equivalence of lattice sites, if appropriate temperature and composition dependencies are taken into account, this approach can generally treat such systems.

Mean field theory has shown that for this type of blend, the blend interaction parameter is given by:

$$\chi_{blend} = \chi_{cr} = (x - y)^2 \chi_{A,B} \qquad (21.105)$$

where x and y denote the copolymer compositions expressed in volume fractions. From Equation 21.85, we have:

$$\chi_{A,B} = \frac{1}{2}(x-y)^{-2}\left[\frac{1}{N_1^{1/2}} + \frac{1}{N_1^{1/2}}\right]^2 \qquad (21.106)$$

From the experimental data of critical composition of the two copolymers, the interaction parameter $\chi_{A,B}$ can be determined. The magnitude of the interaction parameter determines miscibility. The window of miscibility determined experimentally in this study, [S54] matched results from the MP model.

21.11 ESTIMATION OF METALS MISCIBILITY

Although we have considered principles that account for the miscibility (alloying) of metals previously, there are additional factors that are worthwhile considering. Identical principles related to the free energy of mixing apply to metals as well as polymers. Thus Equations 21.16, 21.76, and 21.88, apply here. These equations are listed here for ease of reference:

$$\Delta S_m = -R\left[c\ln c + (1-c)\ln(1-c)\right] \tag{21.16}$$

$$\Delta G_{mix} = \Delta H_{mix} - T\Delta S_{mix} \tag{21.76}$$

$$\chi_{AB} = \frac{V_r}{RT}\left(\delta_A - \delta_B\right)^2 \tag{21.88}$$

and remembering that the following equation also applies.

$$\Delta H \approx \chi_{AB}\phi_A\phi_B \tag{21.107}$$

The solubility parameter concept has been extended to metallic solutions although it is not as well developed and therefore not cited or used as extensively for metals as it is for polymers. In this instance, it is not clear how to interpret the concepts of separate dispersive, polar, and hydrogen bonding components for metals. The total solubility parameters for a variety of metals have been reported by Hildebrand and Scott [S12] and are listed in Table 9.2 in Chapter 9, Cohesive Properties and Solubility Parameter Concepts.

Since metals have relatively small expansion coefficients and δ is a measure of cohesive energy, the solubility parameter is relatively insensitive to temperature. Solubility parameters decrease with increasing temperature and show discontinuities at phase transitions.

Table 21.6
Miscibility Parameters for Solid Solutions of Face Centered Cubic
Transition Metals

System	MP	Solubility Limits, Mole%
Ir-Pt @ 298 °K	324	Miscible
Ir-γFe 298 °K	484	Miscible
Ir-Cu 298 °K	1024	0.5% Ir
Ir-Au 298 °K	2116	2.8% Ir
Ir-Ag 298 °K	3249	Immiscible
Rh-Ni298 °K	25	Miscible
Rh-P t298 °K	64	Miscible
Rh-γFe 298 °K	144	Miscible
Rh-Pd 298 °K	729	Miscible
Rh-Ag 298 °K	2209	< 0.1% Ag
βCo-Ni 298 °K	4	Miscible
βCo-Pt 1000 °K	9	Miscible
βCo-γFe 1000 °K	100	Miscible
βCo-Pd 298 °K	576	Miscible
βCo-γMn 298 °K	961	Miscible(?)
βCo-Ag 1000 °K	1936	0.001% Co
Ni-Pt 298 °K	9	Miscible
Ni-γFe 298 °K	49	Miscible
Ni-Cu 298 °K	289	Miscible
Ni-Pd 298 °K	484	Miscible
Ni-Au 298 °K	961	Miscible above 1120 °K
Ni-Ag 298 °K	177241	Immiscible
Pt-γFe 1000 °K	49	Miscible
Pt-Cu 1000 °K	256	Miscible
Pt-Pd 298 °K	361	Miscible
Pt-Au 1000 °K	841	Miscible above 1430 °K
γFe-Ag 1000 °K	1156	Immiscible

The application of χ_{cr} does not apply to metals since the concept of degree of polymerization does not apply. However, from the experimental data, we apply the concept of miscibility parameter to metallic solid solutions of the same crystalline structures. The data in Table 21.6 [S12] indicates that for a general guide to metal miscibility the MP must be on the order of ≤900.

This approach should be viewed simply as an additional criterion to the Hume-Rothery rules (see section 21.3.2) for determining mutual compatibility. As an example iron and magnesium do not form alloys and their solubility parameters provide an MP of ≈ 4500.

Solubility parameters of metals are particularly useful for estimating the overall solubility parameters of metallo-organics or metal carboxylates. These solubility parameters can then be used to estimate miscibilities of these compounds with other compounds, oligomers, or polymers in order to help understand and guide experimental work.

It should be remembered that the Hume-Rothery rules may not permit two metals that have differences in size, valency, or crystal structure to be miscible even though they have similar solubility parameters.

21.12 ABBREVIATIONS AND NOTATION

a_i	activity coefficient of component i
B	constant
c	concentration
C_P	heat capacity at constant pressure
E	internal energy of a system
E_f	energy required to introduce a mole of vacancies
F	Helmholtz free energy
$\overline{F}_i$	partial molal free energy of component i
F_i^o	free energy per mole of pure component
G	Gibbs free energy
ΔG_{mix}	Gibbs free energy of mixing
G_i^o	free energy per mole of pure component i
$\Delta \overline{G}_i$	partial molal free energy of component i

H	enthalpy
k	Boltzmann constant
K	temperature in Kelvin
M	molecular weight
M_{cr}	critical molecular weight
MP	miscibility parameter
n_i	number of atoms or sites of type i
N	Avogadro's number
P	probability
P	pressure
Q	heat absorbed by the system
r	radius of solute
R	radius of solvent
R	universal gas constant
S	entropy of a system
V	volume
V_i	molar volume of polymer i
V_r	reference molar volume element close to the polymer group repeat unit
V_S	solvent volume
x_i	degree of polymerization of polymer i
$\left(\chi_{AB}\right)_{cr}$	critical interaction parameter
χ_{AB}	polymer interaction parameter
$\tilde{\chi}_{AB}$	χ_{AB}/V_r
δ_i	solubility parameter of component i
μ_i	chemical potential of component i
μ_i^1	chemical potential of component i in phase 1
ϕ_i	volume fraction of component i

21.13 SPECIFIC REFERENCES

1. Williams, M.L., Landel, R.F., and Ferry, J.D., *J. Am. Chem. Soc.*, **77**, 3701 (1955).

2. Simha, R., Boyer, R.F., *J. Chem. Phys.,*37,1003(1962).
3. Wagner, C. and Schottky, W., *Z. Phys. Chem.*, **b11**, 163(1931).
4. Hancock, T., *Engl. Pat.* No. 11147, 1846.
5. Paul, D.R. and Newman, S., (Eds) *Polymer Blends*, Vol. I and II, Academic Press, London, 1978.
6. MacKnight, W.J., Karasz, F.E, and Fried, J.R., in *Polymer Blends*, (Eds. D.R. Paul and S. Newman), Vol. I, Ch.5, p 185, Academic Press, London 1978.
7. Flory, P.J., *Principles of Polymer Chemistry*, Cornell Univ. Press, Ithaca, New York, N.Y., 1953.
8. MacKnight, and Karasz, F.E., in *Comprehensive Polymer Science*. Sir G. Allen and J.C. Bevington, (Eds), Vol 7, Ch. 4, p111, Pergamon Press, 1989.
9. Utraki, L.A., *Polymer Alloys and Blends*, Hanser Publishers, New York, 1990.
10. Scott, R.L., *J. Chem. Phys.* **17**, 279(1944).
11. Tompa, H., *Trans. Faraday Soc.*, **45**, 1142(1949).
12. Hildebrand, J.H., and Scott, R.L., *The Solubility of Nonelectrolytes*, 3rd ed. Van Nostrand-Reinhold, Princeton, New Jersey, 1950: reprinted, Dover, New York 1964.
13. Hildebrand, J.H., and Scott, R.L., *Regular Solutions*, Prentice-Hall, Englewood Cliffs, New Jersey, 1962.
14. Krause, S.J., in *Polymer Blends*, (Eds D.R. Paul and S. Newman), Vol. I Ch.2, p.15, Academic Press, London, 1978.
15. Crowley, J.D., Teague, G.S., and Lowe, J.W., *J. Paint Technol.* **38**(496), 296(1966).
16. Crowley, J.D., Teague, G.S., and Lowe, J.W., *J. Paint Technol.* **39**(504), 19(1967).
17. Hansen, C.M., *Ind. Eng. Chem. Prod. Res. Dev.*, **8**, 2(1969).
18. Hansen, C.M., *Doctoral Dissertation*, Technical University of Denmark, Danish Technical Press, Copenhagen, 1967.
19. Hansen, C.M., *Faerg Lack*, 14(1), 18; (2) 23(1968).
20. Hansen, C.M., *J. Paint Technol.*, **39**(505), 104(1967).
21. Hansen, C.M., *J. Paint Technol.*, **39**(511), 505(1967).
22. Hansen, C.M., and Skaarup, K., *J. Paint Technol.*, **39,** 511(1967).
23. David, D.J., and Sincock, T.F., *Polymer*, **33**, 4505(1992).

24. Coleman, M.M., Serman, C.J., Bhazwagar, D.E., and Painter, P.C., *Polymer*, **31**, 1187(1990).
25. Kim, J.H., Barlow, J.W., and Paul, D.R., *J. Polym. Sci., Polym. Phys. Edn.*, **27**, 2211(1989).
26. David, D.J., *Advances in Polymer Technol.*, **15**, 326(1996).
27. Huang, J.C. and Wang, M.S., *Adv. Poly. Technol.*, **9**,293(1989).
28. Zakrzewski, G.A., *Polymer*, **14**, 347(1973).
29. Harris, J.E., Goh, S.H., Paul, D.R., and Barlow, J.W., *J. Appl. Polym. Sci.*, **27**, 839(1982).
30. Ellis, T.S., *Polym. Eng. Sci.*, **30**, 998(1990).
31. Cimmino, S., Maruscelli, E., Saviano, M., and Silvestre, C., *Polymer*, **32**, 1461(1991).
32. Keitz, J.D., Barlow, J.W., and Paul, D.R., *J. Appl. Polym. Sci.*, **29**, 3131(1984).
33. Janarthanan, V. Stein, R.S., and Garett, P.D., *J. Polym. Sci., Polym. Phys. Edn.*,**31**,1995(1993).
34. Silvestri, R., Rink, M., and Pavan, A., *Macromolecules*, **22**, 1402(1989).
35. Private Communication, 1993.
36. Misra, A., and David, D.J., Development of Blends/Alloys of Poly(vinyl butyral), *Private Communication*, 1992-1993.
37. Schurer, J.W., de Boer, A., and Challa, G., *Polymer*, **16**, 201(1975).
38. Rudin, A., Bloembergen, S., and Cork, S., *J. Vinyl Tech.*, **7**, 103(1955).
39. Huang, J.C., Chang, D.C., and Deanin, R.D., *Adv. in PolymTech.*, **12**, 81-90(1993).
40. Gong K. and Liu, B., *SPE ANTEC* 1225(1986).
41. Chien, Y.Y., Pearce, E.M., and Kwei, T.K., *Macromolecules*, **21**, 1616(1988).
42. Cowie, J.M. and Reilly, A.A.N., *Polymer*, **23**, 4814(1992).
43. Kwei, T.K., Nishi, T., and Roberts, R.F., *Macromolecules*, **7**, 667(1974).
44. Nishimoto, M., Keskkula, H., and Paul, D.R., *Macromolecules*, **23**, 3633-3639(1990).
45. Shur, Y.J., and Ranby, B. *J. Appl. Polym. Sci.*, **19**, 2337(1975).

46. Nishimoto, M., Keskkula, H., and Paul, D.R., *Polymer*, **32**, 1274(1991).

47. Shah, V.S., Keitz, J.D., Paul, D.R., and Barlow, J.W., *J. Appl. Polym. Sci.*, **32**, 3863(1986).

48. Maiti, S.N., Saroop, U.K., and Misra, A., *Polym. Eng. Sci.*, **32**, 27(1992).

49. Ziska, J.J., Barlow, J.W., and Paul, D.R. *Polymer*, **22**, 918(1981).

50. Hall, W., Kruse, R., Mendelson, R.A., and Trementozzi, Q.A., in "The Effects of Hostile Environments on Coatings and Plastics", (D.P. Garner and G.A. Stahl, Eds.), *ACS Symposium Series*, **229**, 49-64(1983).

51. David, D.J., Rotstein, N.A., and Sincock, T.F., *Polymer Bull.*, **33**, 725(1994).

52. Bristow, G.M., *Trans. Faraday Soc.*, **54**, 1731(1958).

53. ten Brinke, G., Karasz, F.E., and MacKnight, W.J., *Macromolecules*, **16**, 1827-1832(1983).

54. Zhou, Z.M., David, D.J., MacKnight, W.J., and Karasz, F.E., *Tr. J. of Chemistry*, **21**, 229-238(1997).

21.14 GENERAL REFERENCES

1. Reed-Hill, R.E., *Physical Metallurgy Principles*, 2nd Ed., D. Van Nostrand Company, New York, N.Y., 1973.

2. Kingery, W.D., Bowen, H.K., Uhlmann, D.R., *Introduction to Ceramics*, 2nd Ed., John Wiley & Sons, New York, N.Y., 1976.

3. DiBenedetto, A.T., *The Structure and Properties of Materials*, McGraw-Hill, Inc., New York, N.Y., 1967.

4. Coleman, M., Graf, J.F., and Painter, P.C., *Specific Interactions and the Miscibility of Polymer Blends*, Technomic Publishing Co., Lancaster, PA, 1991.

5. Van Krevelen, D.W., *Properties of Polymers*, Elsevier, New York, N.Y., 1990.

6. Seymour, R.B., Carraher, C.E. Jr., *Polymer Chemistry*, Marcel Dekker, Inc., New York, N.Y., 1988.

CHAPTER 22

Composites: Concepts and Properties

22.1 INTRODUCTION

In a general sense, a composite material is one which consists of two or more physically distinct and mechanically separable components, existing in two or more phases. Composites are heterogeneous systems prepared by mixing the components to give a controlled dispersion of one material into the other for the purpose of achieving desired or optimum properties. Mechanical properties of composites are superior to those of the individual components, and in some cases may be unique for specific properties. Composites may be divided into three broad categories.

1. Particulate filled materials in which fillers are dispersed into the matrix.
2. Fiber filled materials in which fibrous materials are embedded in a matrix to give fiber-reinforced composites. The fibers may be either continuous or short.
3. Interpenetrating network composites, which consist of two continuous phases.

Composite systems may be based on several kinds of materials such as metals, ceramics or polymers. There are numerous examples of naturally occurring composites such as wood, bamboo, bone, muscle, and various tissue combinations. Several tissues in the body which have high strength combined with flexibility consist of stiff fibers such as collagen embedded in a lower stiffness matrix. The fibers are aligned in a manner that provides maximum stiffness in the direction of high loads and are able to slide past each other to provide the flexibility required. Bamboo and wood show a fibrillar structure, which is the source of high strength and stiffness. Contemporary engineering materials are generally micro-composites such as metallic alloys, toughened plastics and fiber reinforced plastics. The combination of high strength materials into a matrix of tough ductile material is the basis of providing strength and toughness to metallic alloys and reinforced engineering plastics.

The importance of fiber reinforced composites arises from the fact that these systems have high specific strength i.e. they provide high stiffness and strength at relative low weight. The properties can be varied over a wide range by choosing the appropriate polymer-fiber ratio for a particular system. The field of fiber reinforced composites has grown very significantly over the years with the development of new fiber and matrix systems, advancement in processing methods and need for the replacement for the conventional materials by lighter weight substitutes.

The emphasis in this chapter is directed to fiber filled polymer composites. The properties of such materials are dependent on those of the components, shape and size of the filler fibers, the morphology produced, and the interface between the components. Thus a variety of property modifications can be achieved. Composites would in general provide one or more of the following:

1. Increased strength, modulus, and dimensional stability
2. Increased impact strength or toughness
3. Increased mechanical damping
4. Reduced cost
5. Modified electrical properties

22.2   FIBER FILLED COMPOSITES

22.2.1 Matrix Materials

Fiber reinforced composites may be made by using thermoplastics or thermosets as the matrix materials. Due to the requirements of processing, thermoplastic based composites use short fibers while thermosets matrix systems use long or continuous fibers. Polymer based composites may be sub-divided in the following categories.

22.2.1.1 Very Short Fiber Composites

For these composites, the length of fibers are generally 5 mm or less and are dispersed in a suitable thermoplastic and are processed by conventional processing methods, such as extrusion or injection molding. The fibers must be small enough so that they easily pass through the small gaps in an extruder and an injection mold. The fibers provide increased strength and modulus to the thermoplastic and produce materials that meet the more demanding requirements of engineering polymer applications. The most commonly used polymers are polyamides, polyesters, polycarbonate, acetal, and other engineering and high temperature thermoplastics. The increase in strength with about 30 wt% fibers is generally about 2 to 3 times that of the matrix.

22.2.1.2 Intermediate Length Fiber Composites

For longer fibers, thermosets become the natural choice for the matrix. The size of the fibers used is generally 10-100 mm. The fibers are referred to as chopped. The most common resin used is cross-linkable polyester. Other matrix resins include phenolics, vinyl esters, and epoxies. Batch processes are generally used for these composites.

22.2.1.3 Very Long and Continuous Fiber Composites

In some thermoset based composites the length of fiber is equivalent to that of the entire product. Epoxies are generally used with continuous fibers to produce high performance composites. The specific modulus and strength of these composites often surpasses that of metals. Such composites, with suitable reinforcing fibers, find application in the aerospace industry and other applications requiring high strength and modulus.

22.2.2 Reinforcing Fibers

The fibers that are used for reinforcement of polymers must be very strong, very stiff, and capable of being made in continuous length. There are three types of fibers that are generally used, namely, fiberglass, carbon or graphite, and organic fibers (the most common being aramides). The choice of the fiber depends upon the property enhancement required for the specific application and cost constraints. Glass fibers have average modulus and strength while carbon fibers are the stiffest (highest modulus). Aramid fibers are the toughest and possess the highest strength and elongation. Glass fibers are the least expensive while carbon fibers and aramid fibers are considerably more expensive. In some cases two types of fibers may be combined to give hybrid composites. Among these three types there are further classifications giving different properties for each specific type of fiber.

An important feature of reinforcing fibers is that they are coated for making them suitable for use in composites. The purpose of this coating is to provide a good bonding with the polymer matrix, to protect the fiber from damage and to hold the fibers in a bundle prior to composite formation.

22.3   MODULI OF FIBER REINFORCED COMPOSITES

There are considerable differences between the properties of reinforcing fibers and the matrix which results in complex distribution of stress and strain at the microscopic level upon application of load. Thus prediction of mechanical properties of composites is not straight forward. However, making some simple assumptions about the stress and strain distribution, it is possible to make reasonable prediction of properties of composites. The analysis is divided into continuous fiber composites and discontinuous fiber composites.

For the analyses of the mechanical properties of the composites, the following assumptions are made:

1. Fibers are aligned parallel to the direction of stress.
2. Fibers are surrounded by the matrix and fully bounded to the matrix.
3. The fibers are of equal shape and size, and do not overlap or touch each other.
4. Load is transferred from the matrix to the fibers by shear at the interface.

5. The composite elongates as a whole according to Hooke's law.

The rule of mixture applies to the composite systems if the above assumptions are applied. However, in practice deviations lead to less than theoretically possible mechanical properties. In this section the prediction of moduli for fiber reinforced composites and the mechanism of reinforcement is discussed briefly.

22.3.1 Continuous Fiber Composites

In the case of continuous fiber composites the fibers will be considered to be so long that effect of their ends may be neglected. Figure 22.1 represents a block of polymer reinforced with uniaxially aligned continuous fibers. The orthogonal axes are defined as 1, 2 and 3. Upon application of load it may be assumed that the stress and stain are uniform in each component, based on the consideration of mechanical coupling between the fibers and the matrix. The strain values for a stress in direction 1, σ_1, will be:

$$\varepsilon_1 = \varepsilon_f = \varepsilon_m \qquad (22.1)$$

where:

ε_f = strain in the fibers

ε_m = strain in the matrix

Since the fibers and matrix are of equal length, the total stress σ_1, would be equal to the weighted sum of stresses in the fi-

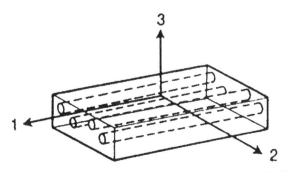

Figure 22.1 A Block of Polymer Reinforced with Uniaxially Continuous Fibers Along the Three Axes.

ber and matrix given as:

$$\sigma_1 = \phi_f \sigma_{f_1} + \left(1 - \phi_f\right)\sigma_{m_1} \qquad (22.2)$$

where:

ϕ_f	= volume fraction of fibers
σ_{f_1}	= stress in the fibers
σ_{f_m}	= stress in the matrix

Assuming both the fibers and the matrix to be perfectly elastic materials, the stresses may be expressed as:

$$\sigma_f = E_f \varepsilon_{f_1} \qquad (22.3)$$

$$\sigma_m = E_m \varepsilon_{m_1} \qquad (22.4)$$

$$\sigma_1 = E_1 \varepsilon_1 \qquad (22.5)$$

where:

E_f	= modulus of the fiber
E_m	= modulus of the matrix
E_1	= modulus of the composite

Substituting the stress values from Equations 22.3, 22.4, and 22.5 in Equation 22.2 and dividing by ε_1 gives the relation for the axial tensile modulus E_1 given as:

$$E_1 = \phi_f E_f + \left(1 - \phi_f\right)E_m \qquad (22.6)$$

If the modulus of the matrix is significantly less than of the fibers then,

$$E_1 \approx \phi_f E_f \qquad (22.7)$$

If the stress is applied in direction 2, Figure 22.1, the entire tensile force is assumed to be carried fully by both the fibers and the matrix and they will be equal. The overall stress would then become:

$$\sigma_2 = \sigma_{f_2} = \sigma_{m_2} \qquad (22.8)$$

Total strain would be additive and given as a weighted strain, given by:

$$\varepsilon_2 = \phi_f \varepsilon_{f_2} + \left(1 - \phi_f\right)\varepsilon_{m_2} \qquad (22.9)$$

Assuming that the fibers and matrix are purely elastic, the expression for stress may be written as:

$$\sigma_{f_2} = E_f \varepsilon_{f_2}, \sigma_{m_2} = E_m \varepsilon_{m_2}, \sigma_2 = E_2 \sigma_2 \qquad (22.10)$$

Substituting this in Equation 22.9, the modulus of the composite in the transverse direction is given as:

$$E_2 = \frac{E_f E_m}{\left(1 - \phi_f\right)E_f + \phi_f E_m} \qquad (22.11)$$

If E_f is much greater than E_m then E_2 becomes:

$$E_2 \approx \frac{E_m}{1 - \phi_f} \qquad (22.12)$$

Figure 22.2 shows the predicted values for E_1 and E_2 for a typical epoxy system reinforced with aligned glass fibers as a function of volume fraction of fibers. The value of E_1 increases linearly with ϕ reaching a value close to that of metals at very high glass fiber levels. E_2 is much lower than E_1 as is also observed in practice.

In a similar analysis it can be shown that the Poisson's ratio also obeys the rule of mixtures and for composites it may be expressed as:

$$\nu_{12} = \phi_1 \nu_1 + \left(1 - \phi_f\right)\nu_m \qquad (22.13)$$

where, ν_{12} is the Poisson ratio for contraction along direction 2, upon application of tensile stress in direction 1, ν_f and ν_m are the

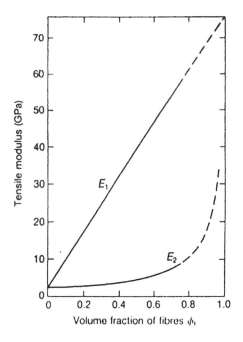

Figure 22.2 Predicted Tensile Moduli for Aligned Glass Fiber
Reinforced Epoxy [G1]

Poisson ratios for the fiber and matrix, respectively.

Now, consider the case when the fibers are not uniaxially aligned but are randomly placed. Let E_{max} be the maximum achievable modulus with uniaxial orientation of fibers. If the fibers are oriented randomly in two directions, i.e. 1-2 plane, then

$$E_1 = E_2 \approx \frac{3}{8} E_{max} \qquad (22.14)$$

If they are random in all three directions E is given by:

$$E_1 = E_2 = E_3 \approx \frac{1}{5} E_{max} \qquad (22.15)$$

22.3.2 Discontinuous Fiber Composites

When discontinuous or short fibers are used with thermo-

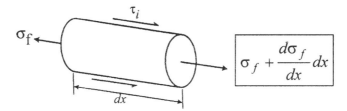

Figure 22.3 Stresses on a Short Fiber Embedded in Polymer Matrix

plastics, the ends must be taken into account. These ends are sites of high concentration and hence weak points. It is assumed that the surface area of end faces is small and hence the stress transfer from end faces to the fibers is negligible. Stress builds up along the length of each fiber from zero at the ends, to maximum in the center. Figure 22.3 gives a representation of stress on a short fiber embedded in a matrix.

It can be seen that uniaxial stresses of the polymer at its interface with the fiber would experience a shear stress of τ_i which transfers the tension to the fiber. Assuming equilibrium conditions, the forces acting on a fiber of diameter, D, and length, dx, may be written as:

$$\left(\sigma_f + \frac{d\sigma_f}{dx}dx\right)\frac{\pi D^2}{4} - \sigma_f\frac{\pi D^2}{4} + \tau_i\pi Ddx = 0 \qquad (22.16)$$

Rearranging Equation 22.16, we obtain:

$$\frac{d\sigma_f}{dx} = -\frac{4}{D}\tau_i \qquad (22.17)$$

Stress analysis of the fiber matrix composite under load is required to predict σ_f and τ_i.

Shear lag theory provides a simplified solution to help in analyzing short fiber composites. The theory assumes that there is perfect bonding between the fibers and the matrix. Let the fibers be of length l and diameter D such that their aspect ratio $a = l/D$. Analysis gives the prediction for σ_f as:

$$\sigma_f = E_1 \varepsilon_1 \left[1 - \frac{\cosh(2nax/l)}{\cosh(na)} \right] \qquad (22.18)$$

where, n represents a dimensionless group of constants and is given by:

$$n = \left[\frac{2G_m}{E_f \ln(2R/D)} \right]^{1/2} \qquad (22.19)$$

where, $2R$ is the distance from the fiber to its nearest neighbor. Differentiating Equation 22.18 and combining with 22.17, τ_i is obtained as:

$$\tau_i = \frac{n}{2} E_f \varepsilon_1 \frac{\sinh(2nax/l)}{\cosh(na)} \qquad (22.20)$$

It can be seen that na is the critical parameter which governs the stress build up along the fiber illustrated in Figure 22.4. Equation 22.18 clearly shows that for most efficient stress transfer to the fibers, na should be as high as possible. This in turn means that the aspect ratio a and the ratio G_m/E_f should be as high as possible. Typical values for actual composites systems, such as that of a polyamide with 30 volume % glass fiber are $a = 50$, $n = 0.25$, which gives $na = 12.5$.

The falling of fiber tensile stress to zero at the ends results in reduction of the axial tensile strength and modulus. The effect on modulus can be estimated by assuming that the fibers are parallel to the axial direction. Now if a line is drawn perpendicular to the fibers, it would intersect the fibers at a longitudinal position chosen at random. Thus the stress equation for the composite would become:

$$\sigma_1 = \phi_f \overline{\sigma}_f + \left(1 - \phi_f\right)\sigma_m \qquad (22.21)$$

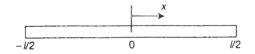

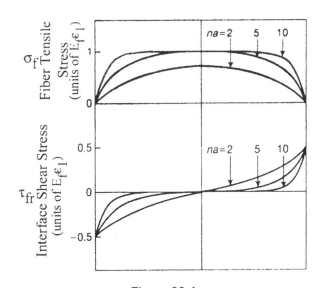

Figure 22.4
Distribution of Tensile Stress σ_f and Interfacial Shear Stress τ_i Along the Length of a Short Fiber Embedded in a Polymer Matrix

where, $\overline{\sigma}_f$ is the mean fiber stress intercepted, which is given by:

$$\overline{\sigma}_f = \frac{1}{l}\int_{-1/2}^{1/2}\sigma_f dx \tag{22.22}$$

Substituting σ_f from Equation 22.18 and integrating, we have:

$$\overline{\sigma}_f = E_f\varepsilon_1\left[1-\left(\tanh(na)/na\right)\right] \tag{22.23}$$

Dividing by ε_1 gives the tensile modulus of the composite:

$$E_1 = \eta_1\phi_f E_f + \left(1-\phi_f\right)E_m \tag{22.24}$$

where, η_1 is a correction factor for the short fibers and is given as:

$$\eta_1 = 1 - \left[\tanh(na)/(na)\right] \tag{22.25}$$

It can be seen that as na becomes large, η_1 approaches 1, as would be the case for continuous fibers. When na becomes less than 10, η_1 becomes significantly less than 1.

From Figure 22.4 it can be seen that the interfacial shear stress τ_i is concentrated at the fiber ends. As the strain increases, the interface first fails at these sites leading to debonding of fibers and matrix thus marking the beginning of shear failure of the matrix. This happens when τ_i reaches the interfacial shear strength. The factors that govern the interfacial shear strength are:

1. The strength of the chemical bond between the fibers and the matrix.
2. The friction between the fibers and the matrix due to the pressure exerted on the fibers by the matrix that arises from differential thermal contraction during cooling.
3. The shear strength of the matrix.

When the strain level increases further, the portion of the fiber that has debonded slips within its own hole and this is resisted by a constant friction stress τ_{fr}. The stress distribution now changes to the situation illustrated in Figure 22.5. As can be seen, the fiber ends have a constant interfacial shear stress, while the tensile stress builds up linearly from the ends up to the edge of debonded region, and is constant in the central region. The length, δ, of the debonded region can then be estimated by using Equation 22.16.

Rearranging, we have:

$$\delta = E_f \varepsilon_1 D / 4 \tau_{fr} \tag{22.26}$$

The debonded length, δ, will increase as the strain ε_1 increases. In

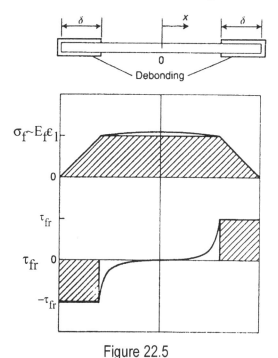

Figure 22.5
Distribution of Fiber Tensile Stress and Interfacial Shear Stress as a
Function of Debonding Length

the limit, the fiber would break when the strain is:

$$\varepsilon_1 = \varepsilon_f^* = \sigma_f^* / E_f \qquad (22.27)$$

which gives the debond length as:

$$\delta^* = \left(\sigma_f^* / 4\tau_{fr}\right)D \qquad (22.28)$$

However, this situation can occur only if the fibers are longer than
a critical length l_c which is given as:

$$l_c = 2\delta^* = \left(\sigma_f^* / \tau_{fr}\right)D \qquad (22.29)$$

This critical length is an important parameter in short fiber composites. If the fibers are shorter than the critical length, total debonding takes place before the failure of the composites and at failure the entire fiber simply slides out of the matrix. The fiber is not broken since its failure stress limit is not reached, thus it has not served its desired purpose. Therefore, it is important to ensure that the fibers are of a length greater than l_c. In general, reducing the fiber length reduces the strength. It is known that during extrusion or injection molding, fiber damage occurs and this should be minimized so as to have as many fibers as possible, longer than l_c.

The effect of the volume fraction of fiber in the matrix should be noted. At low ϕ_f, for fibers located with their ends within a distance, δ^*, of the failure plane, the proportion of fibers is l_c/l and they contribute a mean stress given by:

$$\overline{\sigma}_f = \sigma_f^* l_c / 2l \tag{22.30}$$

At low ϕ_f, from Equation 22.21, the strength becomes:

$$\sigma_1^* = \phi_f \, l_c / 2l \sigma_f^* + \left(1 - \phi_f\right)\sigma_m^* \tag{22.31}$$

At high ϕ_f, from Equation 22.21, the strength becomes:

$$\sigma_1^* = \phi_f \left(1 - l_c / 2l\right)\sigma_f^* + \left(1 - \phi_f\right)\sigma_m^* \tag{22.32}$$

For continuous or long fibers, $l_c/2l \rightarrow 0$, since $l \gg l_c$ and Equation 22.32 reduces to one analogous to Equation 22.2.

The discussion in this section centered on the prediction of modulus of fiber reinforced composites. The following section deals with other mechanical properties.

22.4 STRENGTH OF FIBER REINFORCED COMPOSITES

Understanding and prediction of the strength of fiber filled composites is not as well developed as in the case of moduli. The anisotropy and heterogeneity in these composite systems makes the fracture phenomena very complex. In addition, other factors that contribute to this complexity are interfacial adhesion, dewetting, fiber alignment, stress concentration at the fiber ends, overlap of adjacent fibers, and different modes of fracture of the fibers and the matrix. In systems where fibers are oriented in one direction (uniaxially), the strength of the composite in this longitudinal direction follows the simple additivity rule. Such composites exhibit three types of strength values due to three different modes of failure. These are the longitudinal tensile strength, the transverse tensile strength, and the shear strength. For most composite systems the longitudinal tensile strength of the composite is significantly higher than that of the matrix, the transverse tensile strength is less than that of the matrix, and the shear strength is comparable to that of the matrix. The importance of each of these in determining the failure of the composite depends upon the angle between the oriented fibers and the applied load. The failure mode at low angles is determined by the longitudinal strength, at intermediate angles by the shear strength, and at high angles by the transverse strength. In composites with a distribution of orientation of fibers the situation is quite complex and analysis is not straight forward.

Generally composites use brittle fibers but in some cases ductile fibers are also used. For ductile matrices the elongation at break is reduced by brittle fibers while ductile fibers affect this less and may even result in an increase in elongation. In composites with brittle matrices the ductile fibers do not result in a significant increase in elongation at break. However, this can be increased by reducing the adhesion between the fibers and matrix.

The tensile strength of composites is affected by other factors such as alignment of fibers, perfection of packing and the presence of voids. High strengths are obtained in composites containing long continuous fibers, such as those made by filament winding. Thermoplastic composites containing short fibers would have lower strength than continuous fiber composites due to a variety of reasons. Furthermore, the processing methods for these systems does not produce uniaxial orientation.

Inter-laminar shear strength is another important property used to determine the strength in composites, especially those using layers of fibers. This is measured using the flexural test method such that shearing force inside the composite results in longitudinal splitting of the sample resulting in failure. Inter-laminar shear strength increases with tensile and shear strength of the matrix and decreases with increased void content.

22.5   OTHER MECHANICAL PROPERTIES

Several other mechanical properties of composites that are of interest are: creep, fatigue, impact strength, heat distortion temperature, and coefficient of thermal expansion. These are briefly discussed in this section.

The level of creep of a matrix is very significantly reduced by the addition of fibers. An estimate of creep of a composite is that it is reduced by a factor equal to the ratio of tensile modulus of the matrix to the tensile modulus of the composite. However, this depends on other factors such as fiber orientation.

Fatigue of composites is much more complex than that of the matrix fatigue. In composites the fatigue failure would be associated with debonding of the fibers and/or generation of cracks in the matrix. Composites with ductile matrices generally give higher fatigue strength than those with brittle matrices. An increase in applied stress reduces the fatigue life. In uniaxial fiber composites the fatigue life is generally proportional to its tensile strength.

The impact strength of fiber reinforced polymer composites is more complex than that of matrix polymers due to the role of the fibers and the interface. During an impact the fibers may pull out of the matrix and dissipate energy by mechanical friction thus preventing localization of stresses along the fiber. Another possibility is the controlled dewetting of fibers which dissipates energy. Composites show a reduction in elongation at break and a reduction in the area under the stress-strain curve. This in turn reduces the impact strength or toughness of the composite. Stress concentrations around the fiber ends, areas of poor adhesion and overlap of fibers are some of the factors that contribute to impact strength. The phenomena of crack initiation and crack propagation in fiber filled composites is complex. The fibers tend to reduce resistance to crack initiation since they provide sites for stress concentration. On the other hand, the fibers reduce crack propagation by forcing the cracks to go around them or by bridging the cracks. In general composites with poorer adhesion of fibers to the matrix have higher impact strength than those with higher adhesion.

Heat distortion temperature (HDT) of a matrix increases significantly by the addition of fibers. This increase is greater in crystalline polymers than in amorphous polymers. In crystalline polymers the increase in HDT is due to increase in modulus in the presence of fibers and may reach close to the crystalline melting point. In amorphous polymers the HDT appears to be higher than the glass transition temperature. This apparent increase is due to a significant decrease in creep rate resulting from high modulus values and not due to an increase in the softening temperature of the matrix.

The coefficient of thermal expansion in composites depends upon the orientation of the fibers. In uniaxially oriented fiber composites the longitudinal coefficient of thermal expansion is low since the fibers have a lower value than the matrix thus preventing it from expanding. However, the transverse coefficient of expansion is considerably higher since this direction must accommodate the restricted expansion of the matrix in the longitudinal direction.

22.6   FORMING PROCESSES

Forming processes for reinforced plastics are critical since these are basically inhomogeneous systems. The forming process not only shapes the product being made but also positions the reinforcing fibers and determines their orientation. The properties of composites thus depend upon the direction of fiber orientation and resulting anisotropic behavior of the composite. Furthermore, the level and orientation of the fibers may vary from point to point and in turn the product may show variation in different locations. Manufacturing processes are designed to take full advantage of the reinforcing effects especially in load bearing engineering applications. In other words the mechanical properties are optimized by placement and orientation of the reinforcing fibers. In thermoset based composites, there are many more possibilities of arranging the fibers in desired directions. Typical processes include hand lay-up, spray-up, compression molding, vacuum bag/pressure bag autoclave, resin transfer molding pultrusion, filament winding, and reinforced reaction injection molding. In thermoplastics the placements of fibers are more random and governed by the flow behavior experienced in the processing operation. Extrusion and injection molding are the main processes for making articles from thermoplastic composites. Processes generally used for thermosets and thermoplastic based composites are briefly discussed in the following section .

22.6.1 Hand Lay-Up

Hand lay-up process is the simplest and the easiest of all processes used for thermoset based composites. In this, as the name implies, the reinforcing fibers are laid down manually by hand in the desired shape and arrangement. The resin is then applied by a brush over the fibers to impregnate them. The fibers may be in the form of a roving, a woven fabric or a chopped fiber mat. This is a slow labor intensive process but very versatile. It is used for several utility items and is convenient when small numbers of articles are to be made. For some complicated parts, the placement of fibers by hand lay-up may be the only possible method.

22.6.2 Spray-Up

A variation of the hand lay-up method is the hand spray-up in which the roving is fed to a spray gun which sprays chopped fibers along with the resin onto a prelaid surface. In this process a spray gun and fiber-chopping apparatus are used that lay down the fibers in a random fashion. The resin and initiator are mixed in the gun, sprayed from a nozzle, and the stream picks up the fibers that have been chopped so that they are carried onto a mold surface. A rough and uneven surface is obtained by the spray-up method. The gel coat is then applied on top to hide the fiber pattern underneath and give a smooth surface. This method is useful in making large products.

22.6.3 Compression Molding

Selected polymer resin and the glass fibers are premixed along with other additives. The premix is then formed into the desired shape by compression molding in a heated mold which also initiates the curing process for the resin. The premix can be made in a variety of ways giving different types of the alignment of the fibers. Typical premix systems include:

1. Dough Molding Compound - (DMC). This technique employs a mixture of resin with chopped fibers and has the appearance of dough. It is used for molding three dimensional articles in which the fibers are randomly placed throughout the article.
2. Sheet Molding Compound - (SMC). This technique utilizes a sheet made up of resin and chopped fibers, randomly placed on the plane of the sheet, to fabricate articles that are close to uniform thickness. This is analogous to the thermoforming process.
3. Tape - This technique utilizes a tape made of continuous fibers impregnated with the resins. The fibers may be placed uniaxially or in woven form.
4. Woven Fabric - This technique utilizes woven fabric impregnated with resin to provide strength in two directions.

22.6.4 Pultrusion

Pultrusion is a process of making continuous rovings of fibers that are impregnated with resin and is used to make long articles with a constant cross-section. The fiber roving is immersed in a bath of a pre-polymer, passed through a shaping die and then pulled continuously. Following this the pultruded product is cured to cross-link the resin. This is useful in making long-structural components.

22.6.5 Filament Winding

In filament winding, continuous rovings are pulled through a bath of resin and then wound on a rotating mandrel. The winding is done with the fiber moving alternately across the length of the mandrel thus resulting in a cross ply arrangement of the fibers. This results in cylindrical types of products like pipes and cylindrical vessels. Following the completion of the winding to a desired thickness the resulting product is cured to cross-link the resin. Innovative ways of winding can result in different types of fiber arrangements and different shapes.

22.6.6 Vacuum Bag / Pressure Bag Autoclave

In this method layers of the fibers or unidirectional sheets are pre-impregnated with partially cured resin to form a "pre-preg". The "pre-preg" sheets are stacked on the mold surface in the desired direction in predetermined orientations. These are then covered by flexible bags and cured using a vacuum or pressure bag technique in an autoclave at a preset temperature. This method is used mainly to manufacture high strength structural components and large hollow structures.

22.6.7 Reinforced Reaction Injection Molding

Reaction injection molding is gaining popularity and the polymer systems used can be mixed with short fibers along with one of the reactants. The flow mechanism generally randomizes fiber placement in the matrix.

22.6.8 Resin Transfer Molding

Resin Transfer Molding (RTM) is somewhat similar to traditional transfer molding and reaction injection molding (RIM). In RTM a fiber preform is made by laying the non-coated fibers into the mold of the product and then spraying with the binder to hold them in place. Chopped fibers, continuous fibers, or fiber woven mats may be used for this purpose. The mold is closed and the resin is then injected to fill the mold. Finally, the resin with the preform is cured to produce the final product. This process is a fast and versatile process.

22.6.9 Reinforced Thermoplastics

Extrusion and injection molding are the most common methods for making articles of thermoplastics reinforced with short glass fibers. In both cases the alignment of the fibers takes place due to the flow experienced by the composite system. The fibers are aligned to a maximum extent in regions experiencing high shear rates or experiencing elongation flow. In an injection mold the highest shear rate is along the mold surface, thus giving highest alignment in this region, while the center portion has the lowest shear rate. This provides a more or less random arrangement of fibers resulting in a combination of fiber orientations.

A recent innovative injection molding process uses two reciprocating screws, referred to as push-pull injection molding. One screw feeds the mold and also allows the material to flow out into a second screw which also feeds the mold and permits the material to flow into the first. This sequence may be repeated a number of times, each time forming a solidified layer with aligned fibers. The molded articles made by this process have high glass fiber orientation and thus have high modulus in the desired direction.

An important process feature of fiber reinforced thermoplastics is that the fibers undergo breakage when they experience high shear rates with high viscosity polymers. As has been shown earlier, if the size falls below the critical fiber length, the reinforcing is reduced. Therefore, the molds and dies used for these systems have to be carefully designed to minimize fiber damage during processing. The parameters chosen for processing play a significant part on fiber breakage, fiber alignment, and the resulting mechanical properties.

22.7 SUMMARY OF MAJOR PROPERTY INFLUENCES

As we have seen, fiber reinforced polymer composites are made using a number of different composite matrices. The resulting composites have a wide range of properties depending on the specific system. In composites having low fiber content and short length the properties are governed by the matrix phase, while in composites with long fibers and high glass fiber content the properties are governed by those of the fibers. The former are referred to as engineering composites and the latter as advanced composites.

Long and continuous fiber composites use thermoset matrices, examples being epoxies, phenolics, unsaturated polyesters and polyimides. Short fibers composites use thermoplastics matrices, examples of which are polypropylene, polyamides, polyesters, and polycarbonate. Fibers generally used are glass, carbon, or aramid fibers. The fibers are generally coated with a coupling agent to improve the bonding with matrices and a sizing material to prevent breakage during processing.

The fibers significantly influence the mechanical properties of the matrix, generally making them suitable for high strength and high modulus applications. The modulus and the strength increase with fiber addition and are highest in the fiber direction for composites with uniaxially oriented fibers. Other properties also vary and depend on a variety of factors including fiber length, fiber orientation and fiber bonding.

There are a number of processing methods for thermoplastic and thermoset systems. In all cases best results are obtained when the process does not damage the fiber and fibers are placed in the desired arrangement.

22.8   ABBREVIATIONS AND NOTATION

a aspect ratio
D fiber diameter
E modulus
E_f modulus of fiber
E_m modulus of matrix
E_1 modulus of composite
l fiber length

l_c	critical fiber length
τ_{fr}	friction stress
δ	debonded length
σ	tensile stress
$\overline{\sigma}_f$	mean fiber stress
ν_i	shear stress
ε	tensile strain
ε_f	strain in fibers
ε_m	strain in matrix
ϕ_f	volume fraction of fibers
ν_f	Poisson's ratio for fibers
ν_f	Poisson's ratio for matrix

22.9   GENERAL REFERENCES

1. McCrum, N.G., Buckley, C.P., and Bucknall, C.B., *Principles of Polymer Engineering*, Oxford University Press, Oxford, UK, 1997.
2. Strong, A. Brent, *Plastics Materials and Processing,* Prentice Hall, New York, N.Y., 1996.
3. Hull, D., *An Introduction to Composite Materials*, Cambridge University Press, Cambridge, UK, 1981.
4. Nielsen, L.E. and Landel, R.F., *Mechanical Properties of Polymers and Composites*, 2nd edition, Marcel Dekker Inc., New York, N.Y., 1994.
5. Agarwal, B.D., and Broutman, L.J., *Analysis and Performance of Fiber Composites*, John Wiley and Sons, New York, N.Y., 1980.
6. Asbee, K., *Fundamental Principles of Fiber Reinforced Composites*, Technomic Publishing Co., Inc, Lancaster, Pa., 1989.
7. Powell, P.C., *Engineering with Polymers*, Chapman and Hall, New York, N.Y., 1983.

INDEX

About the Authors:

Dr. D. J. David, Senior Monsanto Corporate Fellow and Manager of Exploratory Research, Saflex Technology, retired; CUMIRP** Research Professor, University of Massachusetts at Amherst, and Adjunct Professor, Chemical Engineering and Materials Science at the University of Minnesota.

Dr. Ashok Misra, Professor, Polymer Science; Dean of Alumni Affairs and International Relations; past Chairman of the Department of Materials Science; Indian Institute of Technology, New Delhi, India.

**Center for the University of Massachusetts and Industry Research Programs.